Reflections on the Astronomy of Glasgow

A Story of Some Five Hundred Years

David Clarke

EDINBURGH
University Press

Edinburgh University Press Ltd
22 George Square, Edinburgh EH8 9LF
www.euppublishing.com

Typeset in 10/12 Goudy Old Style by
Servis Filmsetting Ltd, Stockport, Cheshire, and
printed and bound in Great Britain by
CPI Group (UK) Ltd, Croydon CR0 4YY

A CIP record for this book is available from the British Library

ISBN 978 0 7486 7889 1 (hardback)
ISBN 978 0 7486 7890 7 (paperback)
ISBN 978 0 7486 7891 4 (webready PDF)
ISBN 978 0 7486 7892 1 (epub)

Cover note
A Glasgow reflection providing views of the past: Gainsborough House
(151 West George Street) – built in 2006 – provides a reflected sunlit image of the
old red sandstone edifice of the Connell Building (34 West George Street)

Contents

Preface

When I came to Glasgow University in August 1966 for interview, the Appointments Committee met in the Turnbull Room, on the south side of the Bute Hall. While the nerve-wracking procedure was underway, I realised that, overlooking the shoulders of the interviewing panel, I was faced by a beautiful longcase clock. My thoughts, seemingly working at a secondary level to the immediate task in hand, arrived at the conclusion that the clock must have had an astronomical connection. I said to myself, 'If I am awarded this lectureship in practical astronomy, I will make a point of looking further into this'. When the ordeal was over, I took the stairs down into the West Quadrangle to get some air, only to find another interesting clock on the face of the awe-inspiring building, tucked away in its south-east corner. In the middle of the same southerly face of the Quadrangle was a doorway with 'ASTRONOMY' painted over its arch in bold white letters. Inside was a lecture theatre with bench seats; it looked archaic, stark and forbidding. Would I soon be lecturing here?

As it turned out, I was lucky enough to have persuaded the Committee I was the right person for the post, and I duly received a letter of appointment. When I arrived to take up the lectureship, I quickly found out that the 'astronomy theatre' had been abandoned and that undergraduate lectures were given elsewhere on the campus. All thoughts of the clocks were put out of my mind. Some forty years later, away from the day-to-day pressures of teaching and research and with a little more time available, these images returned to my mind and I decided to undertake the long-forgotten intention. As soon as I had started, I realised the possibility of a grander project, one that explored the role of astronomy in Glasgow since the establishment of the University in 1451. Throughout my career, I had picked up snippets of historical anecdotes that seemed worthy of further investigation, and so the project of writing this book began.

The aim of the book is to relate the long history of the connections of astronomy with the University of Glasgow and with the city. Its purpose is to promote the unique story of the interplay of astronomy with a developing city over five centuries. Very broadly, it deals with the teaching of astronomy from the foundation of Glasgow University in 1451 to the present day, and it also includes stories of enterprises undertaken by local citizens who were inspired by astronomy to expand their education, of interactions with the Observatory and city authorities over the establishment of an accurate local time service and of business and trade along the Clyde associated with this public clock system. As well as having scientific content, it provides a social history relative to the development of the city and the University. Each of the twelve chapters contains references related to the source material for this. The flow is generally in chronological order, but with some minor deviations to link together special issues.

The words 'astronomy' and 'Glasgow' may seem an incongruous juxtaposition, and yet their link forms a fascinating 500-year-long thread involving education, at both institutional and public levels, and city commerce, with stories of personalities, some noteworthy, others notorious. It is remarkable to realise that the first astronomy teaching in the Glasgow College presented an Earth-centred Universe, prior to the Copernican revolution of the mid-sixteenth century. Some of the material in the book may be familiar, such as the telescope observations of sunspots made locally by Alexander Wilson in the 1760s. Less well known is the fact that ideas related to monochromaticity within light, to dew point and to hoar frost had Glasgow connections, and that Herschel's discovery of infra-red energy in solar radiation was made when using Glasgow-produced thermometers. The Glasgow 'Big Bang' of 1863, with a gun triggered from Edinburgh, provides a story of inter-city rivalry, with a newspaper article ridiculing and demeaning the title of 'Astronomer Royal for Scotland'. Different facets reflecting the 'Astronomy of Glasgow', and its accompanying social history, are presented.

Commencing with the textual materials of the thirteenth century contained in the Special Collections of Glasgow University Library, the story deals with the various important characters who promoted the local teaching of astronomy and its development through the centuries, including the likes of George Sinclair, the inventor of the diving bell, the Wilsons, observers of sunspots and researchers of the dew point, James Watt, John Pringle Nichol, a prolific writer, Robert Grant, producer of the Glasgow Star Catalogue, and Ludwig Becker, an inspirational teacher. Several of the portrayed characters were connected to the Royal Astronomical Society as Gold Medal winners, Presidents or Council Members.

The story is also told of the Clyde, the time-ball and the one o'clock gun, of Duncan McGregor, chronometer maker to the Admiralty, and the popular spread of astronomy education which led to the establishment of two civic observatories, something unique to any city. Since 1760, there have been eight local observatories of note that have made contributions to the development of astronomy, some at an international level, and these are all documented here. The second of these, at Garnethill, was visited by Sir William Herschel in order that he could set up telescopes that the local citizens had purchased from him. While undertaking the installations, he made observations of the Great Comet of 1811, visible in the Glasgow skies.

One of the striking outcomes of my research is the realisation of the many contributions made by Glasgow astronomers in terms of longer published texts. Several tens of treatises and books have appeared since the seventeenth century, the early ones by Sinclair remaining notorious to this day. Although Alexander Wilson produced only a short tract together with his published papers, his print type foundry within the University was the basis of the Foulis production of the Greek classics. In the nineteenth century Nichol's books were universally praised for their style and the interest they raised among scientists and the general reading public. In the twentieth century, the writing reins were taken up by Smart with the production of classic texts and other popular books. This has continued through the names of Ovenden, Roy, Clarke, Green, Brown and Woan. And now, finally, with my own modest contribution, all these works are put into context.

Several of the images in the book come from outside sources for which 'permission for use' charges apply. I am pleased to acknowledge the University of Glasgow for an award from the Wilson Fund, established by Professor Patrick Wilson on his retirement over 200 years ago from the Glasgow Regius Chair of Astronomy. Without this support, many of the images would have been culled, to the detriment of the book.

Many of the people I wish to acknowledge are anonymous, and I can only thank them under the umbrella of the organisation of their employ. These include staff members of the University Library, the Archive Services and the Hunterian Museum. If only they knew of the joy they have given by responding to my requests and presenting me with material in their archives. The same is true for the staff at the Mitchell Library in Glasgow who helped most generously in retrieving various articles from the *Herald*. I would like to thank Professor John C. Brown, Astronomer Royal for Scotland, for his help in clarifying and expanding some of the details of Chapter 12. A very special thank you goes to Margaret I. Morris, who very kindly allowed me to make copies of documents she holds in her library. Her enthusiasm for collecting astronomical memorabilia and her passion for the history of astronomy, particularly at the local level, infected me with an added purpose to undertake the writing of this book.

Finally, I would like to thank John Watson for guiding my proposal through the doors of Edinburgh University Press and overseeing the progress of the book through to its completion. I am greatly indebted to Anna Stevenson for her patience and skills in sorting out style and syntax problems and for occasional adjustments to the word order in my original typescript, so making clearer some of the ideas I wanted to convey. By highlighting phrases carrying esoteric scientific jargon, improvements have also been made to allow for a more general readership.

Legend has it that St Mungo, patron saint of Glasgow, whose original name was Kentigern, preached the sermon containing the words 'Lord, let Glasgow flourish by the preaching of the word'. In recent times this exultation has been truncated to the city's present-day motto 'Let Glasgow Flourish'. To this sentiment we might add 'May Glasgow's astronomy continue to flourish'!

Glasgow University Observatory, 1 January 2013
David Clarke

To those who have gone before,
To those who are here now,
And to those yet to come.

The Glasgow Ouroboros

Chapter 1

Glasgow Astronomy

1.1 INTRODUCTION

Before launching into the main themes of the story, it might be appropriate to give some general comments regarding the choice of title of this book.

The use of the word 'reflections' in the title is quite apt. As the cover of the book shows, reflections on the present allow us to see older things of the past. When astronomers apply reflecting telescopes with large mirrors to gain a focus, the recorded images of the heavens are formed from light that left its source many years ago. The reflections made in this book bring together a history running for the order of 500 years, a time interval required for the light to travel to us from many of the familiar bright stars in the night sky. It may be mentioned that the famous star cluster known as the Pleiades, or Seven Sisters, is referred to in the works of George Sinclair (see Chapter 2, p. 36. The light now providing our current sense of the presence of this beautiful stellar group began its cosmic journey some fifty years prior to the time when Sinclair was putting pen to paper in Glasgow in the late 1600s – a thought to reflect upon.

The combination of the words 'astronomy' and 'Glasgow' in the title of any book would seem a most improbable juxtaposition. For one thing, the climate of the west of Scotland is not conducive to the practice of optical observational astronomy. In addition, many outsiders see Glasgow as having a roughness to its population, perhaps associated with a lack of interest in things intellectual, one of the possible interpretations of its traditional description as 'No Mean City'. But this is a banner of the past, and undeserved by the greater majority of the citizens. Its hardworking and entrepreneurial people are generally interested in advancing their lot through education, a trait that has emerged periodically through involvement in astronomy. As well as the subject being associated with the excellent local universities, it has always had links to the educational aspirations of the general population, as this discourse will relate. The city of Glasgow is unique in this regard, with both professional and amateur astronomers making their mark. Education and research has been advanced at all levels with the building of observatories by the University and by public enterprise; the establishment and running of a local time service that supported trade and commerce in the city and along the Clyde demonstrated the practicality of astronomy to the local population.

These facts were acknowledged by the International Astronomical Union (IAU) through the assignation on 19 October 1994 of Minor Planet No. 5805 as 'Glasgow', in honour of the city and in celebration of Glasgow's illustrious Astronomical Society. This asteroid was discovered on 18 December 1985 by Dr Edward Bowell at the Anderson Mesa Station of the Lowell Observatory in Flagstaff, Arizona. The name was publicly announced on 23 November 1994, at a civic dinner hosted by Glasgow City Council that marked the centenary of the Society.

Asteroid 'Glasgow' orbits the Sun with a semi-major axis of 2.6AU, and has a modest eccentricity of 0.11, with the orbital plane at an inclination of 12 degrees to the ecliptic. Estimates of its size depend on its unknown surface reflectivity, or albedo; its diameter would be about 19km if it were a C-class asteroid (carbonaceous and dark), or 10km if it were a more reflective S-class type (stony/iron). No matter the detail, its surface area is just a little larger than the urban spread of the city whose name it now carries.

Why does astronomy have importance to Glasgow, and why has there been a celestial thread woven into the city's history? The purpose of this treatise is to set out the story of the interplay. Astronomy is not just a subject for intellectual exercise, simply for the sake of understanding the Universe and mankind's relationship to it. Yes, it is to be expected that astronomy would be part of the educational scene of any good seat of learning, and that it would have a presence within Glasgow's universities, particularly the most traditional one, with its long history. But, in addition to that, astronomy has connections with commerce, and these have been apparent within the nautical trade of the city. All these strands will be explored in the chapters here. Although the text is not designed to provide a trail of topographical astronomical memorabilia, there are several sites in Glasgow that are well worth investigating for their connections with the subject.

To many, the subject of astronomy is somewhat arcane and only pursued in the professional or amateur arenas by people with some peculiar bent. This attitude can sometimes be seen when momentous discoveries are made that revolutionise our thinking and understanding of the Universe. On these occasions, media announcements of such work are sometimes made with elements of quirkiness, the reports patching a vacant time slot, or space hole, and used to lighten the regular over-sensationalised, dark and depressing news. Astronomy is not alone in this regard, as colleagues in other scientific disciplines also complain of such trivialisation. Not all is bad, however, as some high-quality informative articles on space and celestial discoveries appear in newspapers and magazines and on television programmes, helped along under the umbrella of the 'public understanding of science'. As described in this monograph, Glasgow has been involved on many occasions in furthering the astronomical knowledge of its local population.

In the first instance, astronomy has always been applied to timekeeping and the running of the calendar. There is strong evidence that some of the standing stone monuments around the west coast of Scotland, as elsewhere, were erected for the purpose of providing a basic calendar, giving reference to key dates such as the summer and winter solstices. The origins of these structures are some five millennia old. What the understanding of the Universe was in those times is completely unknown, however, as there are no written records describing the activities of these megalithic societies. But we are left, nonetheless, with some magnificent stone circles and menhirs that stimulate our imagination about times past.

If there were stone circles, or monoliths, erected in what are now the precincts of the city, they have since disappeared under the land turnover of repeated urban development. Perhaps the nearest remaining astronomical monument to the city from this era is a small group of standing stones just to the north-west of Blanefield, north of Glasgow, and adjacent to the route of the West Highland Way. According to Professor Alexander Thom,[1] this simple stone arrangement was set to record the occasions of the summer solstice.

Surprisingly, though, there is an astronomical stone circle within the city itself, constructed in 1979 to mimic what might be considered a megalithic observatory. The stones stand on an old air vent for the Buchanan Street train line which ran underneath; the line is now disused. The circle is within Sighthill Park and was erected by the Glasgow Parks Department of Astronomy Project as part of the Jobs Creation/Special Temporary Employment Programme, 1978–9. It is readily accessible by walking westwards from Pinkston Road. A photograph of the monument is depicted in Plate 1, with the indications of what the stone astronomical alignments represent given in the plan of Fig. 1.1. A local organisation, Friends of Sighthill

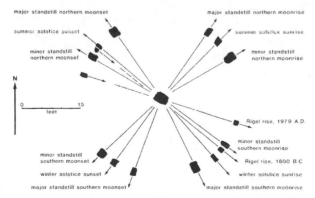

Figure 1.1 Relative to the circle centre at Sighthill, stones have been set pointing to the rising and setting directions of the Sun and Moon throughout the year, and also to the rising of the bright star, Rigel, as it would have been seen in 1800 bc, and also where it rises currently. The layout was designed by Duncan Lunan, and the plan above was drawn by Richard Robertson.

Stone Circle,[2] continues to promote the venture. Plans to revamp and extend the project have recently been discussed in an article in the *Herald*.[3]

1.2 THE INFLUENCE OF THE UNIVERSITY

We can only take up the documented story of astronomy and its association with Glasgow from the middle of the fifteenth century. At this time it became relevant to the general education of University graduates, many of whom took their learning abroad. The teaching of astronomy flourished within the University, and continues to do so in the present day. Some elements of astronomy are now taught within the more recently established local institutions of Strathclyde University and Glasgow Caledonian University. In addition, there have been a number of occasions when the clamour for knowledge about the heavens from people beyond the 'ivory tower' environment became so great that enlightened groups got together to encourage the promulgation of astronomical knowledge within the general local community, and this appears to be both remarkable and unique to Glasgow. Again, as astronomy played a special part in timekeeping in the eighteenth and nineteenth centuries, it has served the commercial activities of the Clyde and interplayed with the civic authorities. The associated buildings, specially designed for the purpose of astronomical observation, have also been important to the architectural heritage of the city.

Local astronomical history has had a long and continuous connection with the city's traditional seats of learning. There are several good accounts in the literature of the establishment of the College in Glasgow, now referred to as the University of Glasgow, its development and its eventual relocation to Gilmorehill in the city's West End nearly 150 years ago. A concise description of the College's foundation is given by Arthur Donald Boney[4] in his introduction to the story of the College's Gardens, which were an important feature in early educational activities relating to botany and medicine. As it turns out, the botanical activities within the College grounds and the running of the Observatory had many links, which Boney's text relates in detail.

A starting gun was fired through a public announcement on Trinity Sunday, 30 June 1451, at the Mercat Cross, situated at the bottom of the High Street, the main street running northwards to the cathedral. The assembled, and perhaps puzzled, Glasgow townspeople were told that Pope Nicholas V had issued a Bull of Foundation for a 'Stadium Generale' (the former customary name for a medieval university). The Bull, as depicted in Fig. 1.2, was issued on 7 January 1451, at the request of King James II. This was followed by the inauguration of William Turnbull, Bishop of Glasgow (from 1447 to 1454), who may be considered as the founder of the University. A room in the main University buildings at Gilmorehill is named after the Bishop, and is referenced again in Chapter 7 (see p. 197). Details of the original Latin text

Figure 1.2   A facsimile of the Papal Bull issued by Pope Nicholas V in 1451.

and a translation of the Bull, as well as the history of the loss of the original document in France, can be conveniently found on the internet.[5]

It may be mentioned here that the Julian calendar was in use at this time, following the reform of the Roman calendar by Julius Caesar in 46 BC. The year comprised 365 days, with an additional day every fourth year. In this era, New Year's Day was on 25 March, that is, 24 March 1450 would have been followed by 25 March 1451. It corresponded to the feast of the Blessed Virgin Mary, and occurred precisely nine months prior to Christmas Day. Changes to the calendar began in 1582 under Pope Gregory XIII. His reform corrected the slide of the calendar in respect of the timings of the equinoxes and solstices, significantly reducing the drift relative to the Julian system by not applying the extra date to those years divisible by 100 but not 400; the year 2000 was a leap year but 1900 was not, nor will 2100 be. The change-over to the Gregorian calendar also involved the loss of ten days, partly the accumulation of days added as a result of every fourth year being a leap year. It was adopted in Scotland and England in 1751, with the new year beginning on 1 January. In Scotland, 31 December 1599 was followed by 1 January 1600, but this change was not taken up by England until 1 January 1752. The University's Papal Bull dates back to 1450 according to the Julian calendar (old style), but to 1451 according to the more modern Gregorian one (new style). It may be noted that, even now, the current system is not perfect, as there is a marginal difference of 0.000125 days between the Gregorian calendar and the average year. To allow for this, a further correction, suggested by John Herschel, was that the year 4000 might be treated as a non-leap year. Any planned adjustment so far ahead, over and above the current leap year schemes, is at the mercy of the behaviour of the length of the vernal equinox year, which is unstable and unpredictable at these levels of precision.

The Papal Bull decreed that the new public place of study should be modelled on the University of Bologna, and envisaged the formation of a typical medieval university, having a basic Arts faculty through which all students were to progress before entering one of the other higher faculties (Divinity, Law or Medicine) for professional training. This concept for the institution is contained within the Bull in the following manner:

> but also of the indwellers and inhabitants of the whole kingdom of Scotland, and the regions lying round about, we, being moved with fatherly affection, and inclined by the supplications of the said king in that behalf, to the praise of God's name, and propagation of the orthodox faith, erect, by apostolical authority a University in the said City in all times to come for ever, as well in theology and cannon and civil law as in arts, and every other lawful faculty. And that the doctors, masters, readers, and students there may brook and enjoy all and sundry privileges, liberties, honours, exceptions, and immunities granted by the apostolic see, or otherwise in any manner of way to the masters, doctors and students in the University of our City of Bologna.

It is also interesting to note at the beginning of the Papal Bull that the possibility of understanding the workings of nature and the Universe and to have a more content life was open to all through study. This sentiment reads as follows:

> Amongst other blessings which mortal man is able in this transient life by the gift of God to obtain, it is to be reckoned not among the least, that by assiduous study may win the pearl of knowledge, which shows him the way to live well and happily, and by preciousness thereof makes the man of learning far to surpass the unlearned, and opens the door for him *clearly to understand the mysteries of the Universe* [author's italics], helps the ignorant, and raises to distinction those that were born in the lowest place.

The new institution did not reach its medieval ideal for some considerable time. The first meeting of the University, on an unknown date in 1451, was held in the Chapter House of the Black Friars in the High Street, when thirty-seven members of the General Chapter convened. The masters and students were to be governed by the Rector, who in turn was the representative of the Chancellor, the Bishop of Glasgow. The first Rector was David Cadzow, Precentor and Sub-dean of the Church of Glasgow; his successors over some thirty years were to be Canons of Glasgow. The term 'master' was to remain in general use until well into the eighteenth century, by which time the title 'professor' was often used alongside it. The system of 'Regenting' was introduced, by which one master took his students through the whole of the Arts course: the *trivium* of verbal arts (grammar, rhetoric and logic) and the *quadrivium* of numerate arts (mathematics, geometry, astronomy and music), this form being common throughout European establishments of learning. At that time, the scope of physics also included pure mathematics, astronomy and geography. For the first 100 years of its existence, the University remained so small that the Arts Faculty, with a Dean at its head, was the complete Glasgow College, and until the nineteenth century the 'College' (or 'University') and the 'Faculty' were one and the same.

Early teaching was undertaken in the crypt of the cathedral, in the Black Friars Chapter House and in the 'Auld Pedagogy', a small tenement building providing 'teaching, lodging and a common table'; this was situated in Rotten Row, a street lying south-west of the cathedral. In 1453, the schools moved to more spacious rented premises on the east side of the High Street. In 1460, the first Lord Hamilton gifted a tenement, also on the High Street, together with four acres of land to the rear, extending as far as the Molendinar Burn. This was added to by a gift from Thomas Arthurles in 1467. Later, Queen Mary gave some thirteen acres of land on the Dow Hill, east of the Molendinar Burn. This spot eventually provided the site of Glasgow's first observatory. All of this land allowed the physical expansion of the College's buildings and its Physick Garden.

The nature of the architecture and fabric of the College at the end of the

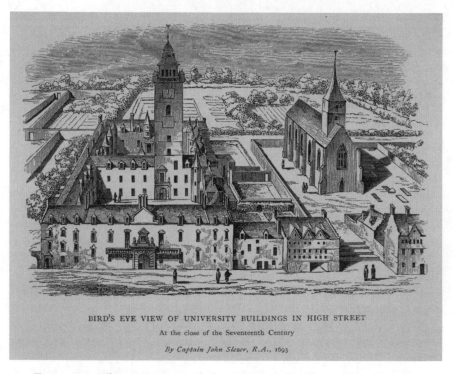

BIRD'S EYE VIEW OF UNIVERSITY BUILDINGS IN HIGH STREET
At the close of the Seventeenth Century
By Captain John Slezer, R.A., 1693

Figure 1.3 The engraving made by Captain John Slezer in his *Theatrum Scotiæ* (1693), giving a bird's-eye view of the University buildings at the end of the seventeenth century.

seventeenth century was documented in Slezer's *Theatrum Scotiæ*.[6] A copy of the image of the 'Colledge of Glasgow', as it was then known, is displayed in Fig. 1.3. It shows the extent of the 'campus' and garden as they were in 1693. Walled off from the University was Blackfriars Church. Used for worship by the University, it had previously been a Dominican church. At this time, there was no special building for astronomical observations, but an observatory was built in 1757, located on the upper right of the landscape, along the line of the main axis of the church. Blackfriars was demolished in the 1870s, together with the Old College.

According to Robert Sibbald's commentary in *Theatrum Scotiæ*:

> The Fabrick of the College is remarkable, consisting of divers Courts. The fore-part of it towards the City is of an excellent Structure being of hewen Stone. The Precincts of the College are enlarged by some Acres of Ground, purchased by some Money granted to it of late by the King and Estates of the Kingdom. It is separated from the rest of the Town by an exceeding high Wall.

The high wall was not indicative of the institution being an 'ivory tower', as there was considerable interplay with the city.

The establishment of the College and the demise of the buildings after some 400 years are neatly summarised by William Barr in his monograph entitled *Glaswegiana*:[7]

Slightly to the east flows Glasgow's sacred river, the Molendinar, where St Mungo baptised many of his converts in its pristine waters. He lived a simple life in a rough stone cell nearby. From the site rose Glasgow Cathedral, ever zealous of spreading learning and Christianity throughout the country and existing to this day as one of the four pre-Reformation Cathedrals in Scotland still in service as a place of worship. The source of the Molendinar is Hogganfield Loch. Now spurned and rejected, it has been covered up by a tarmac only appearing briefly at the city side of the Great Eastern Hotel. Acting as a sewer, it shortly plunges into a dark tunnel to be seen no more until joining the River Clyde.

High Street was opened in the year 1100 and so named because it led to the highest part of the town. It immediately gained importance when the University was built on its East side in 1460. In 1750 Glasgow boasted the possession of 13 streets but since then it has stretched out its ever-reaching tentacles and casually embraced towns, villages and spacious country regions into its commercial bosom and forming the network into a vast, vibrating vital city.

Paul Nicholas V, born in 1398 at Pisa, Italy, granted a papal edict for the erection of the first Glasgow University in 1451 which was duly built and stood for nine years at the north side of Rottenrow, before being transferred and rebuilt in High Street . . . we will notice a plaque[i] situated on the south end of the College Goods Station Railway Office building on which is written:– 'ON THIS SITE STOOD THE UNIVERSITY OF GLASGOW FROM 1460 TILL 1870. THE MAIN GATEWAY, NOW RE-ERECTED AT GILMOREHILL WAS OPPOSITE COLLEGE STREET.' The information is certainly correct for the place is marked by a large memorial stone set into the wall of the VISTA VENETIAN BLINDS office building (at one time the premises of Alston's Tea Rooms). The coat of arms of the University is colourfully displayed on the upper portion of the stone and was recorded by the Glasgow University in the year 1900.

It may stagger the imagination if one tries to visualise that some 150 years ago this area of the City was a scene of beauty and elegance. Within the University's boundaries, the stately buildings of the University, the splendid Hunterian Museum, the beautifully designed Blackfriars Church, the wealthily endowed Blackfriars Monastery, a conservatory[ii] and the historic Infantry Barracks.

The Infantry Barracks were situated between Hunter Street and Barrack Street on the site of the ancient Butts where the citizens practised archery, scoring marks at the popinjay.

. . . the old University stood on this site for 400 years until its removal to

---

[i] This embossed copper plate (see Fig. 1.6), marking the site of the Old College in the High Street, is now on display in a showcase at the entrance to the Bute Hall at Gilmorehill.
[ii] The word 'conservatory' above may be a typographical error, and perhaps should be replaced by 'observatory'.

Gilmorehill in 1870. Adjacent was the College Green, the recreational and fresh air lung of the University, and which stretched to the East as far as Hunter Street. It was tastefully studded with trees and the sacred Molendinar flowing through the green's spacious lawns, sweeping past the Observatory and its well tended gardens on his way to the River Clyde.

In 1846 the College entered into an agreement with a railway company to sell the site on the understanding that new grounds and a building would be provided at Woodlands, on the west side of the city. Adverse times for the development of the railways followed, and the contract fell through, with the College receiving substantial compensation. Eventually the College moved to its present site at Gilmorehill, and a goods station was built on the High Street in the mid-1870s by the City of Glasgow Union Railway Co. The Glasgow & South Western Railway (G&SWR) acquired the station in 1896 and this company was absorbed by the new London, Midland and Scottish (LMS) consortium in 1923 (see Fig. 1.4).

After nationalisation in 1948, the station was operated by British Rail until its closure twenty years later, after which the area became a car park. The wheels of progress continued to turn and, at the beginning of the twenty-first century, the site was set aside for an extensive building enterprise called Collegelands. In the summer of 2007, the old College grounds, close to where the Observatory once stood, were covered by a gigantic pile of rubble. The entrance had a large hoarding advertising the forthcoming development (see Fig. 1.5).

Major proposals for the site were announced on 10 December 2009 in the *Herald*,[8] followed by a plea from a Mr Robert D. Campbell[9] suggesting that the Collegelands developers might reinstate the plaque referred to above, as it was part of an enterprise in 1921 of the Pen and Pencil Club to mark historical sites within the city. Its present location is at the entrance to the Bute Hall on the University campus at Gilmorehill; the details of the inscription can be clearly seen in Fig. 1.6.

1.3 THE EARLY SYLLABUS

Returning to the first 150 years following the establishment of the College, some form of astronomy would have been taught under the umbrella of the 'numerate Arts'. It staggers the imagination to note that this Glasgow seat of learning is so old that the early teaching was based on pre-Copernican concepts, with an Earth-centred Universe. The astronomy of the day would have embraced the long-established ideas of Aristotle and Ptolemy. According to Durkan and Kirk,[10] with the completion of the study of logic, the student was believed to be grounded in a common method applicable in any scientific enquiry.

Again following Durkan and Kirk,[11] the next stage was the intermediate

Figure 1.4 The upper photograph shows the offices of the London, Midland and Scottish Railway company's College goods station in the High Street, c. 1950. The Old College was just north of this block. Note that Alston's Tea Rooms, mentioned in Barr's *Glaswegiana*, can be seen just to the left of the policeman. The lower image reveals the current scene some sixty years later and one hundred and forty years after the removal of the College; the block of the goods station has been developed and the Glasgow trams no longer run.

Figure 1.5   The entrance to the Old College in the summer of 2007, advertising the latest development for the site.

Figure 1.6   The Pen and Pencil Club's plaque, dated 1921, commemorating the move of the Old College in the High Street to Gilmorehill in 1870.

or mathematical sciences which did not demand experience nor require an effort to deal with purely intellectual exercises. They were intermediate in the sense that they called for the application to the material of physics in the form of music theory, astronomy and optics. Arithmetic and geometry were not applied mathematics in the same sense. There was, however, no medieval divorce between philosophy and science, as in modern mathematical physics; the one looked for its verification in the other. In the Glasgow scheme, the quantitative branches of knowledge were relegated to extraordinary lectures and were left incomplete. If they were read, the themes would have been 'perspective, *algorismus* and the principles of geometry'. The term 'perspective' here means 'optics'; *algorismus* involved arithmetic and fractions; and the geometry would have been the Euclid.

An indispensable book, though read extraordinarily in the special sense as above, and common perhaps to all the establishments of learning in Europe covering the *quadrivium*, would have been that by Johannes de Sacrobosco, namely *The Sphere of Sacrobosco* or the *Sphaera mundi* (c. 1230). It was mentioned in the curriculum of around 1575 of James Melville, as commented on below. The *Sphaera mundi*, with various appendages, was widely used in manuscript form in Europe and later in printed versions. Within the Special Collections of Glasgow University Library, there are several editions of this famous work,[12] the oldest dating to 1478; a full listing of the Sacrobosco collection is provided under Reference 12. No doubt some of these treasured copies would have been in circulation within the Glasgow College soon after their printing, and they show the pens of annotation and graingerisation, as was the common practice of the time.

As an example of the work, Fig. 1.7 provides copies of two pages from an edition published in Venice in 1543, and Fig. 1.8 provides further examples of the figures contained in an edition of 1500, the top image carrying a substantial annotation.

The works of Johannes de Sacrobosco were key to the curriculum in Glasgow for the 150 years following the establishment of the College, and it is important to make some comment on them and on their author. It is common practice that any person of merit is claimed as a son of the country of his birth. So it is with Johannes de Sacrobosco. However, although Pedersen has presented some evidence[13] to suggest that he was 'British', the more detailed claims that he was English, Irish or Scottish have little or no foundation. The 'Sacrobosco' part of his name is thought to be a translation of 'holy wood'. His birthplace has been claimed as Halifax in England, Artane (now a suburb of Dublin) in Ireland, Holywood in Northern Ireland[14] and Holywood in Dumfriesshire, Scotland; it has also been related that he became a Canon of the Order of St Augustine at the Monastery of Holywood in Nithsdale. Several commentators suggest he was educated at the University of Oxford but, again, there is no direct evidence to substantiate this. In fact, most material on his origins and life derives

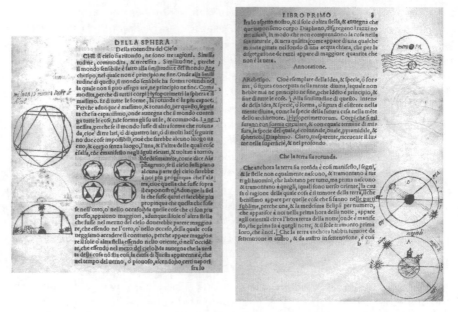

Figure 1.7 Two pages from an Italian translation of the first book of *Sphaera mundi* by Johannes de Sacrobosco from an impression made in Venice in 1543 by Francesco Brucioli, & i Frategli.

from the speculations and inventions of sixteenth- and seventeenth-century antiquarians. That he taught in Paris is perhaps more definite, as following his death, in around 1256, a memorial was constructed in the Monastery of St Mathurin, closely associated with the University of Paris. This was decorated with an astronomical instrument, perhaps an astrolabe, and a few lines of Latin verse which referred to Sacrobosco's calendrical work; the monument has now been lost through the ravages of time. What does seem a little more certain is the genuine authorship of four works as outlined by Pedersen.[15] These are formally entitled *Algorismus*, *Tractatus de sphaera*, *Compotus* and *Tractatus de quadrante*.

The first manuscript is a brief treatise on elementary arithmetic using Arabic numerals. In the first section, a description and explanation is given of the ten numerals (0 to 9), with zero referred to as *theta*, *circulus*, *cifra* or *figura nihili*, as it signifies nothing in itself, but gives significance to other numerals according to its relative position. Sacrobosco stressed the fact that any number, no matter how large, could be written using only these ten digits. This characterisation provided a much simpler way of undertaking any kind of arithmetic relative to the awkward Roman system. It might be said that his proposals set the scheme for the universally accepted denary scale. Other chapters of his work dealt with addition, subtraction, multiplication, division, square roots and cube roots.

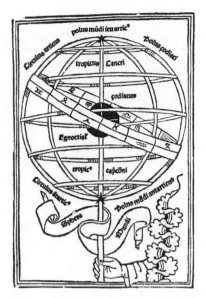

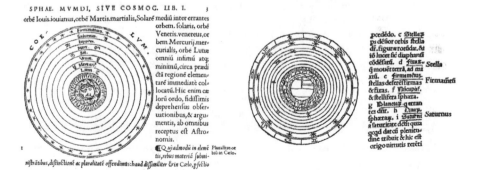

Figure 1.8 The top image is a frontispiece, common to many editions of the *Sphaera mundi*, copied from one printed in 1518, within the collection of many similar tomes archived in the Glasgow University Library. Many of these books are heavily annotated or graingerised, the script below the image being typical of such added comments by a reader. The lower left-hand image displays the Aristotelian/Ptolemaic Earth-centred model of the Universe, as presented in the same tome; the right-hand image of the then currently believed structure of the Universe is from an edition printed in 1500.

Sacrobosco's *Sphaera mundi* offered a basic account of spherical geometry, underpinning the mathematical astronomy of Ptolemy and his Arabic commentators. It was divided into four books which (1) described the shape and place of the Earth within a spherical Universe; (2) charted the various circles on the celestial sphere; (3) described the phenomena caused by the diurnal rotation of the sky, such as the rising and setting of heavenly bodies, as seen from different geographical locations; and (4) gave a brief introduction to Ptolemy's theory of the planetary motions and eclipses. The whole work comprised a comprehensive account of the Earth as a sphere at the centre of the Universe. It rapidly achieved popularity, being, in 1472, the first-ever astronomy manuscript to be printed. It was reproduced in abundance in various forms in many countries, and in translation. Successive impressions appeared, with more than 200 editions, not all of them yet fully catalogued. It might be considered the most popular astronomical book of all time, including current times, remaining a standard teaching text until the seventeenth century.

The *Compotus*, sometimes referred to as *Compotus Ecclesiasticus*, *De anni ratione* or *Compotus Philosphicus*, was a treatise on timekeeping, the calendar and calendrical computation. This book deals with many aspects of time, studying the day, week, month and year as well as the phases of the Moon and the ecclesiastical calendar. Sacrobosco maintained that the Julian calendar was ten days in error and should be corrected. Regarding what it contained in terms of scholarship, understanding and notions for rationalising and reforming the calendar, it was perhaps his most important book. Again, treasured copies can be found in the Special Collections of Glasgow University Library. The fourth text, *Tractatus de quadrante*, was relatively short, describing the construction and use of the so-called *quadrans vetus*, or old quadrant.

As is the case today, the structure of the University, the detail of the curriculum and the methods of teaching were constantly under review. Independent and radical thinking has always been a trait associated with Glasgow and, on being appointed as Principal soon after his return to Scotland in 1574, James Melville, with his dynamic personality and academic brilliance, set himself the task of establishing a good educational system. Prior to his arrival, the College had remained small and impecunious. In his *nova erectio* he expanded the curriculum, establishing Chairs in languages, science, philosophy and divinity, which were confirmed by charter in 1577. According to Durkan and Kirk,[16] Melville commented that:

> it is not our will that these three Regents change every year into new courses, as is the custom in other colleges of our kingdom, whereby it comes to pass that while they profess many branches of learning they are found skilled in few; but they shall exercise themselves in the same course that young men who ascend step by step may find their precentor worthy of their studies and gifts.

Accordingly, the first Regent was to provide instruction in rhetoric and Greek so that students might become 'more fit to receive the principles of philosophy'. The second Regent was, therefore, assigned to the teaching of dialectic and logic, and moral philosophy from Cicero, Plato, Aristotle and similar 'best authors'. He was also required to teach arithmetic and geometry 'which are of no small importance for the acquisition of learning and sharpening the intellect'. Natural philosophy was to be taught by the third Regent, who was also 'to profess geography and *astronomy* [author's italics] and likewise general chronology, or history, and computation of time from the creation of the world'. In the previous curriculum up to 1573, emphasis had been firmly placed on the philosophies to the exclusion of arithmetic, geometry and astronomy, geography and history, and Greek and Hebrew. The *nova erectio* redressed this imbalance.

Again from Durkan and Kirk,[17] the fourth master, in accordance with the statutes, was to lecture on the physics of Aristotle, which came in eight tomes, on the book known as the *Sphaera mundi* and also on cosmography, the science of the geographical world. The science of the Universe which Melville taught at Glasgow, and which then consisted of a mixture of astronomy and astrology, was, however, silently dropped from the curriculum. As a testimony to the man, there is a meeting room called the Melville Room within the main University building at Gilmorehill.

1.4 GLASGOW UNIVERSITY LIBRARY

Reference has been made to the large collection of works by Sacrobosco kept in the Special Collections department of the Glasgow University Library [GUL]. In addition, there are many other astronomical treasures shelved there, some tinged with astrology, the books and manuscripts being far too numerous to be listed individually here. A recent enterprise has been to digitise some of the pages of these valuable tomes and present them via the University's website;[18] several of these images are also available on the Flickr photo-sharing site.

Some of the items would no doubt have been obtained at the time of publication, but others would have accrued from personal library collections. Of particular importance to the development of astronomy is the famous treatise by Copernicus entitled *De Revolutionibus Orbium Caelestium*, published in 1543. The Library has three copies of this work, and the Special Collections catalogue[19] provides details of their annotations and provenance. Pages from the copy originally in the Hunterian Collection have been digitised and can be readily viewed via the University's website.[20]

Also immediately available for access are a few pages of *Tabulae Astronomicae*,[21] published in 1492 for Alfonso X, King of Castile and Leon. The GUL catalogue provides an extensive provenance for this tome, with details suggesting that it passed through the hands of several students,

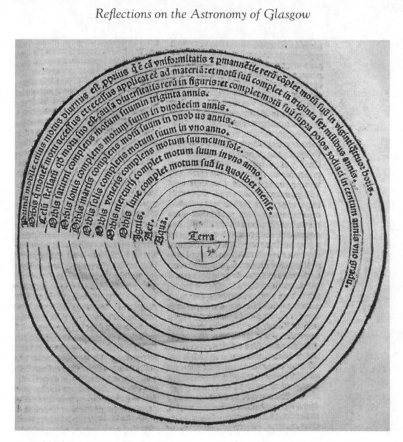

Figure 1.9   A woodcut of 1494 depicts the structure of the
Universe according to Aristotle.

including one of the James Moores, matriculating in the late sixteenth
century. Two copies of the *Rudolphine Tables* of 1627 by Johannes Kepler[22]
are housed within the Special Collections, and, again, some pages are immediately available to view on the website.[23] This work by Kepler includes data
from Tycho Brahe's observations.

At the time of the establishment of the University, before the thesis
of Copernicus, the Earth was thought to be at the centre of the Universe,
with a structure following the teaching of Aristotle. The University Library
has several treasured volumes, other than those of Sacrobosco, illustrating
this old system. Fig. 1.9, a print[24] of a 1494 woodcut, provides a beautiful
example.

Another gem is the *Kalendarium* of 1492 by Joannes Regiomontanus.[25]
Within this tome there is a 3D pop-up page, providing an instrument with
volvelles, or paper wheels, and a brass pointer which can be manipulated to
show the motion of the Moon. An image of this exquisite piece is presented
in Fig. 1.10.

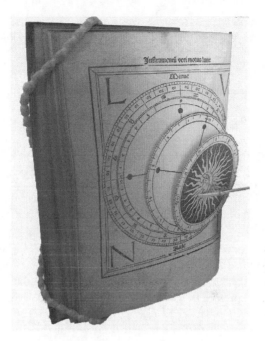

Figure 1.10 Folio 28v from the *Kalendarium* of 1492, by Joannes Regiomontanus, displays a device with volvelles, or paper wheels, which can be manipulated to show the motion of the Moon.

1.5 GLASGOW UNIVERSITY HUNTERIAN MUSEUM

The Hunterian Museum was established in 1807, built on Dr William Hunter's founding bequest. Originally it was sited within the Old College grounds, as can be seen in Fig. 3.11 and Fig. 3.12. The Museum is now located at the campus on Gilmorehill. The Hunterian collections include scientific instruments used by people who established Glasgow's heritage, such as James Watt, Joseph Lister and Lord Kelvin.

Many of the artefacts have astronomical connections; these include old telescopes, sextants, quintants, celestial globes and so on. Some of the pieces are not fully catalogued, and it is not always possible to match those that are to archived inventories associated with astronomy teaching. For example, the University Observatory inventories of the nineteenth century refer to the possession of celestial globes. As it turns out, the Hunterian depository contains three such globes, two of them by Carey, but it is impossible to relate them to the old records.

Items have accrued from both the old Natural Philosophy Department and the Observatories of the Department of Astronomy. One piece of interest is a sextant that formerly belonged to the father of Lord Kelvin.

For the record, relevant documented items appearing in the catalogue available via the Hunterian Museum website are listed in the Appendix to this chapter.

1.6 A CLANNISH PEOPLE

In dealing with the general history of the eighteenth-century College in Glasgow, John Duncan Mackie[26] has commented on the influence of kinship and 'kindness' among the Scots. In providing a multitude of examples, he says:

> The very name of Wodrow stands for probity; but James Wodrow, Professor of Divinity, was succeeded by his son Alexander in 1705, while the more famous son, Robert, acted as University Librarian between 1697 and 1701 before he began his long ministry at Eastwood (1703–34).

As will be commented on in the next chapter, it was the Revd Robert Wodrow who exposed the plagiarism of Professor George Sinclair's claim to authorship of a religious work written by another person.

In the realm of people associated with astronomy, it may be noted that the first Professor of Natural Philosophy, Robert Dick, was succeeded in 1751 by his son, also named Robert. Alexander Wilson, the first Professor of Astronomy (1760–84), managed to promote his son Patrick as his successor to the same chair, with the support of the Faculty, and originally against the wishes of the Crown. The renowned engineer James Watt owed his introduction to the University to his kinship with Robert Muirhead (Professor of Humanity 1754–73). Several of the books written by John Pringle Nichol, Professor of Astronomy (1836–59), contained plates prepared by his younger brother, William, who was a famous lithographer. Professor Nichol's son became Professor of English Literature at the University. As we shall see later, Professor Robert Grant, who succeeded Nichol, only achieved his position thanks to the outside intervention of 'connections', after a local candidate had been put forward.

But we are beginning to present the history glibly and too quickly, so at this point, this introduction must draw to a close in preparation of a more systematic presentation.

Appendix

Hunterian Museum Artefacts (GLAHM)
Note that the abbreviation OA refers to the overall dimension of the noted piece.
Cat. No: 105684. Reflecting Telescope. Manufacturer: Short, James, 1743. Surry-Street, The Strand, London. 29/372 = 12. The first number is the model number, the second the serial number and the last the focal length in inches. Materials: brass; steel; glass; ivory. Size: OA when the tube is

horizontal 530mm × 265mm × 440mm. This telescope was thought by Bryden[27] to have originated from the Macfarlane bequest, but there is no direct evidence for this. See Chapter 3.

Cat. No: 113782. Reflecting Telescope. Manufacturer: Short, James, 1754. Surry-Street, The Strand, London. 222/899 = 9.4. Size: OA boxed 420mm × 75mm × 103mm. According to Bryden,[28] this instrument was listed in Professor Dick's inventory of equipment of 1756 as item 86B and used before the establishment of the Macfarlane Observatory – *A Gregorian Telescope in a small mahogany chest in the Upper Optical Division.*

Cat. No: 105681. Reflecting Telescope. Manufacturer: Short, James, c. 1759. Surry-Street, The Strand, London. 250/1103 = 9.6 – Alt-Az mounting on a pillar and claw stand. Materials: metal; glass; ivory. Size: OA when the tube is horizontal 390mm × 275mm × 315mm. This telescope[29] is thought to have been bought by Professor Alexander Wilson.

Cat. No: 105670. Refracting Telescope. Manufacturer: Ayscough, James, c. 1750. Ludgate Street, London. Materials: box-wood; brass; glass; ivory; paper; cardboard. Size: OA 816mm × 50mm diameter.

Cat. No: 105136. Reflecting Telescope. Manufacturer: Miller [John, Jnr], between 1769 and 1803. Edinburgh. Materials: brass; speculum-metal (copper-tin alloy); paper. Size: Objective 97mm; OA 774mm × 112mm × 230mm.

Cat. No: 113568. Refracting Telescope. Manufacturer: T. Harris & Son, early 19th century. London. Materials: brass; glass. Size: OA 960mm × 320mm × 455mm.

Cat. No: 113533. Refracting Telescope. Manufacturer: James White, c. 1895. 16, 18 and 20, Cambridge Street, Glasgow. Materials: brass; bronze; glass; alcohol. Size: OA 390mm × 115mm × 180mm.

Cat. No: 113913. Refracting Telescope. Manufacturer: Unknown. Materials: glass; brass; steel; mahogany. Size: OA 345mm × 62mm × 28mm.

Cat. No: 129197. Refracting Telescope. Manufacturer: Unknown. Materials: brass; cast iron; glass. Size: OA 272mm × 225mm × 225mm.

Cat. No: 105683. Sextant. Manufacturer: Dudley Adams, c. 1800. 22, Fleet Street (near Charing Cross), London. Materials: ebony; ivory; brass; glass. Size: OA 410mm × 410mm × 110mm.

Cat. No: 113373. Quintant. Manufacturer: Troughton, c. 1800. The Sign of the Orrery, 136 Fleet Street, London. Materials: ebony; brass; silver; glass; velvet; mahogany. Size: OA in box 340mm × 290mm × 130mm.

Cat. No: 113911. Quintant. Manufacturer: Henry Hughes & Son Ltd., c. 1800. The Sign of the Orrery, 136 Fleet Street, London. Materials: ebony; brass; silver; glass; beize; mahogany; card. Size: OA in box 288mm × 282mm × 136mm.

Cat. No: 105679. Hadley's Octant. Manufacturer: Gardner & Neil – early nineteenth century. Belfast. Materials: brass; glass; mahogany; deal; ivory. Size: OA boxed 320mm × 300mm × 105mm.

Cat. No: 113912. Artificial Horizon. Manufacturer: Cary, c. 1890. 181 The Strand, London, England. Materials: glass; brass; aluminium; mahogany; beige; paper. Size: OA 204mm × 164mm × 124mm.

Cat. No: 113923. Celestial Globe. Manufacturer: Cary, c. 1890. 181 The Strand, London, Materials: mahogany; brass; pasteboard; paper; plaster of Paris. Size: OA 700mm diameter × 1175mm. The globe is 550mm in diameter. A damaged second Cary Globe is also in the archive.

Cat. No: 105661. Specific-Gravity Beads. Manufacturer: Wilson, Alexander, December 1754. Glasgow. Materials: glass; wood; velvet. Size: OA 90mm diameter × 40mm approx.

References

1. Thom, A. (1967), *Megalithic Sites in Britain*, Oxford: Oxford University Press, pp. 97–101, at p. 98 [Glasgow University Library (hereafter GUL) Archaeology D80.M3 THO – two copies].
2. See www.sighthillstonecircle.net
3. The *Herald*, 2 June 2010, 'Good Heavens . . . astronomer bids to rejuvenate stone circle'.
4. Boney, A. D. (1988), *The Lost Gardens of Glasgow University*, Bromley: Christopher Helm [GUL Botany A75.C7 – two copies].
5. See www.universitystory.gla.ac.uk/papal-bull
6. Slezer, J. (Captain) (1693), *Theatrum Scotiæ: containing the prospects of His Majesty's Castles and Palaces, together with those of the most considerable Towns and Colleges; the ruins of many ancient Abbeys, Churches, Monasteries and Convents within the said Kingdom*, London: n.p. [GUL Sp Coll Hunterian Ce.2.12; GUL Sp Coll EA5-x.9; GUL Sp Coll Bi8-a.1].
7. Barr, W. W. (1988), *Glaswegiana*, Glasgow: Richard Drew Publishing (reprinted 1990) [GUL History DX200 BAR4].
8. The *Herald*, 10 December 2009, 'New £200m project to put derelict city site back on map'.
9. The *Herald*, 16 December 2009, 'Commemorative plaques would be a nice touch by Collegelands project developers'.
10. Durkan, J. and Kirk, J. (1977), *University of Glasgow 1451–1577*, Glasgow: University of Glasgow Press, p. 89 [GUL Education S271 1977-D – five copies].
11. Ibid., p. 89.
12. Glasgow University has an excellent and extensive collection of editions of *Tractatus de Sphaera* or *Sphaera mundi*, the *De Algorismo* and *Tractatus de anni ratione*, appearing as separate books, or collected together and loosely bound. In some cases, the citations below are in reduced form; more extensive descriptions are available on the GUL website and can be accessed using the individual call numbers.

    (1478), Iohannis de sacrobusto anglici uiri clarissimi Spera mundi feliciter incipit Sacro Bosco, Joannes de, fl. 1230. Sphaera mundi. Impressa Venetijs: per Franciscu[m] Renner de Hailbrun [GUL Sp Coll Bk5-g.22].

(1482), Ioannis de SacroBusto sphericum oposculum ... Other Authors – Peurbach, Georg von (1423–1461), Theoricae novae planetarum. Regiomontanus, Joannes (1436–1476). Disputationes contra Cremonensia, Venetiis: Erhardi Ratdolt Augustensis [GUL Sp Coll BD7-f.13].

(1485), Noviciis adolfscentibvs [sic]: ad astronomicam rempu[blicam] capessenda[m] aditu[m] i[m]petra[n]tib[us] ... Sphaera mundi. Sacro Bosco, Joannes de, fl. 1230, Venice: Impressum est hoc opusculum mira arte & diligentia Erhardi Ratdolt Augustensis. Anno salutifer[a]e incarnationis [GUL Sp Coll Hunterian Bx.3.41].

(1495), Eximij atq[ue] excellentissimi physicoru[m] motuum cursusq[ue] syderei indagatoris Michaelis Scoti sup[er] auctor[em] Sperae ... Expositio super auctorem Spherae, Bentiuolo: per Iustinianum de Ruberia [Two copies: GUL Sp Coll Ferguson Al-a.6; GUL Sp Coll Ferguson Al-c.9].

(1498), Uberrimum sphere mundi commentum intersertis etiam questionibus Domini Petri de Aliaco. [Una cum ... uberrimo commentario Petri Cirvelli ...] Sacro Bosco, Joannes de, fl. 1230, Parisiis: Guy Marchant, for Jean Petit (1498) [GUL Sp Coll Bh8-e.1].

(1500), Sphaera mundi – Figura sphere: cum glosis Georgii de Monteferrato ... Sacro Bosco, Joannes de, fl. 1230, Venetiis: impensis Georgii de Monteferrato [GUL Sp Coll Hunterian R.6.16].

(1501), Algorismus domini Joa[n]nis de Sacro Busco noviter impressus. Cum gratia et priulegio Sacro Bosco, Joannes de, fl. 1230, Venetijs: per Bernardinum Venetum de Vitalibus [GUL Sp Coll Hunterian R.6.16].

(1511), Textus de Sphera Johannis de Sacroboseo cum additione ... adjecta: novo commentario ... illustratus cum compositione Anuli astronomici Boni Latensis et Geometria Euclidis Megarensis Sacro Bosco, Joannes de, fl. 1230, Parisiis [GUL Sp Coll Veitch Eg8-a.17].

(1518), Sphera cum commentis in hoc volumine contentis ... Ptolomeus de speculis Sacro Bosco, Joannes de, fl. 1230, Venetiis: Impensa heredum quondam domini Octaviani Scoti Modoetiensis ac sociorum [GUL Sp Coll Mu54-a.4].

(1518), Sphera cum commentis ... videlicet ... Sacro Bosco, Joannes de, fl. 1230. Venetiis [GUL Sp Coll Veitch Eg7-a.6].

(1518), Sphera cum commentis in hoc volumine contentis ... Sacro Bosco, Joannes de, fl. 1230, Venetijs: impensa heredum Octaviani Scoti (1518) [GUL Sp Coll Ferguson Ai-x.9].

(1531), Spherae tractatus Ioannis de Sacro Busto / Gerardi Cremonensis ... [et aliorum] Sacro Bosco, Joannes de, fl. 1230. Other Author – Gerardus Cremonensis (1114? –1187), Venetiis: In aedibus Lucantonii Iunte Florentini [GUL Sp Coll Mu41-a.4].

(1534), Liber de sphaera Sacro Bosco, Joannes de, fl. 1230, Venetijs: apud Ioan. Ant. de Nicolinis de Sabio [GUL Sp Coll Mu56-f.20].

(1543), Ioannis de Sacrobusto Libellus de Sphaera ... Sacro Bosco, Joannes de, fl. 1230. Other Author – Melanchthon, Philipp (1497–1560), Antuerpiae: Excudebat Ioannes Richard [GUL Sp Coll Mu56-f.21].

(1546), La sphère de Iehan de Sacrobosco, tradvicte de Latin en langue fran-
coyse . . . augmentée de nouuelles figures auec vne préface contenant arguments éuidents, par lesquels est prouuée l'utilite d'astrologie Sacro Bosco,
Joannes de, fl. 1230, Paris: Iehan Loys [GUL Sp Coll Hunterian Au.1.19].

(1550), Annotationi sopra la lettione della Spera del Sacro Bosco . . . Authore
M. Mauro Fiorentino Mauro, fra, ca. 1490–1556 [GUL Sp Coll Hunterian
R.5.22].

(1566), Sphaera emendata / cum additionibus . . . et . . . scholijs . . . Sacro
Bosco, Joannes de, fl. 1230. Other Authors – Giuntini, Francesco (1523?–
1590), Vinet, Élie (1509-1587), Nunes, Pedro (1502-1578), Antverpiae: Apud
Ioannem Richardum [GUL Sp Coll Mu54-h.24].

(1571), La sfera / di Messer Giovanni Sacrobosco . . . Sacro Bosco, Joannes de,
fl. 1230. Sphaera mundi. Other Authors – Dante de' Rinaldi, Pietro Vincenzo.
Danti, Egnatio. Italian, Fiorenza: nella stamperia de Giunti [GUL Sp Coll
Mu55-f.20].

(1572), Sphaera Sacro Bosco, Joannes de, fl. 1230, Venetiis: apud hæredes
Melchioris Sessæ (1572) [GUL Sp Coll Ferguson Af-f.57].

(1591), In sphaeram Ioannis de Sacro Bosco commentarius Clavius, Christoph,
1538–1612 [GUL Sp Coll Ferguson Al-b.57].

(1607), In Sphaeram Joannis de Sacro Bosco commentarius. Nunc quinto
. . . ab ipso . . . in locis locupletatus. Accessit geometrica, atque vberrima de
crepusculis tractatio Clavius, Christoph, 1538–1612, Lyon [GUL Sp Coll
Bl5-f.14].

(1608), In Sphaeram Ioannis de Sacro Bosco, commentarius . . . Clavius,
Christoph, 1538–1612, St Gervais [GUL Sp Coll Bi5-h.1].

[Tractatus de anni ratione] / [Sacro Bosco] Sacro Bosco, Joannes de, fl. 1230
[GUL Sp Coll Bl4-k.7].

(1838), Johannis de Sacro-Bosco Anglici de arte numerandi tractatus. Nunc
primum edidit ex antiquo manuscripto J. O. Halliwell / [Sacro Bosco] Sacro
Bosco, Joannes de, fl. 1230, Cambridge [GUL Sp Coll BD8-e.37].

(1839), Rara mathematica; or, A collection of treatises on the mathematics and
subjects connected with them, from ancient inedited manuscripts / edited by
James Orchard Halliwell. London: John William Parker (1839) [Two copies:
GUL Sp Coll Mu19-y.18; GUL Sp Coll Farmer q170].

(1912), Konrads von Megenberg Deutsche Sphaera . . . / herausgegeben von
Otto Matthaei Sacro Bosco, Joannes de, fl. 1230. Konrad, von Megenberg,
1309–1374. Matthaei, Otto, Berlin: Weidmann [GUL German J5 KON Vol
23].

(1914), Regimento do estrolabio e do quadrante; tractado da spera do
mundo / réproduction fac-similé du seul exemplaire connu appartenant à la
Bibliothèque de l'état de Munich, Munich: Carl Kuhn (1914). [GUL Research
Annexe I6-x].

(1915), Tratado da sphera com a Theorica do sol e da lua e ho primeiro livro
da Geographia de Claudio Ptolomeo . . .; tratado que ho doutor Pero Nunez
fez sobre certas duvidas da navegação Nunes, Pedro, 1502–1578, Munich:

J. B. Obernetter [GUL Research Annexe I6-x].(1980), Die deutsche Sphaera/ Konrad von Megenberg; hrsg. von Francis B. Brévart Konrad, von Megenberg, 1309–1374, Tübingen: M. Niemeyer [GUL German JK130.S6 1980-B].

13. Pedersen, O. (1985), 'In quest of Sacrobosco', *Journal of the History of Astronomy*, 16, 175–220.

14. Drennan, A. S. (1991), 'Sacrobosco in Ulster', *Irish Astronomical Journal*, 20, 78–83.

15. Pedersen, 'In quest of Sacrobosco', 175–220.

16. Durkan and Kirk, *University of Glasgow 1451–1577*, p. 284.

17. Ibid., p. 315.

18. Digitised material from the GUL Special Collections can be accessed at www. lib.gla.ac.uk, www.gla.ac.uk/services/specialcollections/virtualexhibitions/revo-lutionsinastronomy/ or www.flickr.com

19. Copernicus, N. (1543), *De Revolutionibus Orbium Caelestium*, Nuremberg: Petreius, Johann [GUL Sp Coll Hunterian Cz.1.13; GUL Sp Coll Ea6-a.3; GUL Sp Coll Bk2-e.5].

20. See Reference 18.

21. Alfonso X, King of Castile and Leon (1492), *Tabulae Astronomicae*, Venice: Hamann, Johann [GUL Sp Coll Bl9-g.25].

22. Kepler, Johannes (1627), *Tabulae Rudolphinae*, Ülm: Jonas Saur [GUL Sp Coll Ea6-x.15; GUL Sp Coll f155].

23. See Reference 18.

24. Orbellis, Nicolaus de (1494), *Cursus librorum philosophiae naturalis [Aristotelis] secundum viam Scoti*, Basel: Michael Furter [GUL Sp Coll Veitch Eg6-d.8].

25. Regiomontanus, Joannes (1482), *Kalendarium*, Venice: Erhardus Ratdolt [GUL Sp Coll BD7-f.13].

26. Mackie, J. D. (1954), *The University of Glasgow, 1451–1951: A Short History*, Glasgow: Jackson, p. 186 [GUL Education S271 1954-M – six copies].

27. Bryden, D. J. (1968), *James Short and His Telescopes*, Handbook and Catalogue of the Commemorative Bicentenary Exhibition, Royal Scottish Museum Edinburgh.

28. Ibid.

29. Ibid.

Chapter 2

Some Early Astronomy

2.1 A MATHEMATICAL PROFESSOR

The pursuit of astronomy, being part of the studies of an educated person, was certainly followed in Glasgow before the turn of the seventeenth century, as related in the previous chapter. Originally, within the University, the teaching of the subject was allotted to a professor with a background in natural philosophy. The first person of note who was involved in astronomy and its teaching, and who was recognised beyond his University post, was Professor George Sinclair. He was born around 1630, probably in the East Lothian area, and was educated at the University of Edinburgh, taking his Master's degree in July 1649. Following an appointment as Professor of Philosophy at the University of St Andrews, he moved to Glasgow and was admitted as Professor of Philosophy on 8 April 1654. He was ejected in 1662, however, when University professors were required to submit to the Episcopal form of Church government. He became a mineral surveyor and engineer, chiefly in the employ of Sir James Hope, and then a schoolteacher in Edinburgh. In 1670, he was employed by the magistrates of Edinburgh to supervise a project involving piping fresh drinking water from Cormiston into the city. In 1688, Sinclair returned to Glasgow University as a Regent, having declared his readiness to swear the oath of allegiance to King William III. He was Professor of Mathematics and Experimental Philosophy from 1691 until his death (c. 1696). He gifted and lent money to the fund for the building of the Old College. A Chair of Mathematics called the George Sinclair Chair was established by the University in 1984.

In *A History of the University of Glasgow*, Coutts[1] noted that around 1655, and into the time when Sinclair was in office, the College scene could be described as follows:

> Masters and students were enjoined to wear their gowns in the College, and students to do so in the streets as well. The common table was to be maintained as hitherto, and it was recommended that as many of the students as possible should sleep and diet in the College. Students were directed to speak Latin, and a fine was denounced against those who spoke Scots. Students in arts were recommended to practise the good and useful exercise of oratory and rhetoric, and were to be taught

as formerly logic, metaphysics, physics, and ethics, the text of Aristotle being diligently and succinctly gone through. They were also to be instructed in some good abridgement of the various parts of mathematics, at least arithmetic, geometry, geography, and astronomy. The principal and the professors of theology – there was only one such professor at the time, but the Commissioners recommended the addition of another – were to see that their students went through the whole body of theology and the Holy Scriptures, and to exercise them in homilies; as well as to see that they had some knowledge of antiquity, the fathers, councils, ecclesiastical history, and chronology; and that they were proficient in Hebrew and Greek, and had some touch of Chaldee and Syriac.

In recording the contributions made by Professor Sinclair, Coutts[2] says:

> Sinclair was a man of notable industry and perseverance, and one of the earliest Glasgow professors who showed a decided turn for physical and applied science, as is shown both by his professional work and his writings. He was the author of a number of treatises on Mathematics, Hydrostatics, Coal, Astronomy and Navigation, and other subjects, and of *Satan's Invisible Works Discovered*, a book on witchcraft and ghosts. Even his books on science sometimes contain a mixture of superstition, but he was not the only author of the age liable to this charge.

Three of his publications caused controversy over aspects of plagiarism, accuracy and fantasy. His work entitled *Ars Nova et Magna Gravitatis et Levitatis; sive . . . indicbus*,[3] published in Rotterdam in 1669, was noted in the *Philosophical Transactions* with a suggestion that an earlier work by Robert Boyle, famous for his gas laws, had been plagiarised. An anonymous[4] vindication of this charge appeared in around 1669; the source of this was difficult to ascertain. In 1672, Sinclair[5] published his treatise in English under the new title of *The HYDROSTATICKS, or, The weight, force, and pressure of FLUID BODIES, Made evident by Physical, and Sensible Experiments*, and included discussions on coal mining. As will be related below, the cudgel of criticism of this work by Professor Gregory of St Andrews was very heavy and blunt. Later, as will be described in the last section of this chapter, there was a scurrilous episode, with Sinclair passing off a work as his own that he had blatantly taken from another author.

In his first publication, the *Ars Nova*, some astronomy was presented in the form of dialogues, with four voices taking part. In *Libri IV, Dialogues*,[6] comets were discussed, and the work included an engraving depicting a cometary tail within a star field.

2.2 HYDROSTATICS AND THE DIVING BELL

According to Laing's[7] study of the history of diving bells, Professor Sinclair was involved in attempts to raise 'treasure' off the Isle of Mull from one of the wrecks of the Spanish Armada of 1588, under the auspices of the Duke of Argyle [*sic*], and he was partially successful. In his *Ars Nova* on subjects including the weight and the pressure of air, in the fifth dialogue of book 2d, Sinclair[8] described a newly invented machine; it was in the form of a bell, and as it contained a diver or, to use an old-fashioned term, a 'urinator', it was called a *Campana Urinatoria*. Whether he used this device is unknown, but three pieces of ordinance were raised, two of them cannons 11 feet in length and 8 fingers' breadth in diameter. One of the chief findings was that the ball projectiles were made of stone rather than iron.

In *The HYDROSTATICKS*, Sinclair[9] describes his own invention as:

> A Wooden Ark, constructed with a leaden foot-stool, on which the diver's feet might stand, and cause it to go under water. — This invention, then, is for *diving*, a most excellent art, for lifting guns, ships, and any other things that are drowned below the water. And it is in imitation of the *diving bell*, already found out and made use of the success.

In constructing this ark, he proposed to insert a window or two, fitted with a glass, to enable the diver to look around him. 'A little shelf might hold a compass, with a magnetic needle; and, in one of the corners may hing a little bottle with some excellent spirits, for refreshing the stomach under water'. In summarising this discourse, he says:

> As the *Ark* is a most useful device for profit, so 'tis excellent for pleasure and recreation, if a man were disposed to see the ground and channels of deep waters, or were inclined to find out hydrostatical conclusions, a knowledge very profitable, and what few have attained to. Though it seems somewhat difficult to enter the Ark, and go down below the water, yet a little use will expel all fear.

Laing[10] comments that the learned professor nowhere states that he himself had been disposed either to take such 'pleasant recreation' under water, or to perfect his own attainments in hydrostatics.

At the time of the publication of Sinclair's 1672 treatise, Professor James Gregory of St Andrews University attacked him derisively, particularly in relation to his proposed and pretentious theorems on hydrostatics. The criticisms appeared in a pamphlet[11] of 1672 entitled, in the form of a parody, *The Great and New Art of Weighing Vanity: or a Discovery of the Ignorance and Arrogance of the great new Artist in his Pseudo-Philosophical Writings*, but carefully abstained from questioning an account of the witches of Glenluce which Sinclair had very incongruously brought into his own work. According to Laing,[12] in an article in the *Encylopaedia Britannica*, Sir John Leslie commented that:

> It is painful to remark that James Gregory, the inventor of the reflecting telescope, who, although endowed with the talents of highest order, appears to have had a keen temper, and to have imbibed a hereditary attachment to Royalty and Episcopacy, should have stooped to attack an unoffending and less fortunate

rival. It is a piece of low scurrility, and memorable only for a very short Latin paper appended to it, containing the series first given to represent the motion of a pendulum in a circular arc.

As well as deriding Sinclair's whole presentation of hydrostatics and ridiculing the imaginative design of his 'Diving Ark', Gregory was very dismissive of his writing on astronomy, optics and mathematics.

For example, in 'OBSERVATION III',[13] Sinclair describes the equipment he constructed to measure the difference between the rates of the secular easterly movements of the Moon and those of the Sun. It was done using an optical device that projected large images of the objects on a wall in his room. To demonstrate the efficacy of the arrangement, he calculated how the differences of image drift with time would be represented in terms of movement expressed in 'inches', without reference to how this related to angular measure on the sky. This, and other reports of astronomical observations, were pounced on by Gregory. Setting out his criticisms, with additional accusations that Sinclair practised plagiarism, Gregory wrote:[14]

> If our *Author* think that he was well exercised, when he was making his observations of the *Comet*, he should judge a part of his time well spent, in letting the world know for what they served: but he seems to intend no more; then to make men believe, that he is not ignorant of a degree or a minut, altho he reckons the Suns motion by inches.
>
> I question not, that a *Coal-hewer* is more useful to the Countrey than he and I both: and therefore he is obliged to me, for giving him a more useful trade, then he now driveth. Nor can I deny, but he justly deserved it; for a *Coal-hewer* is one who maketh gain by digging in another mans mine; and so hath he done; for that History of *Coal* which he hath printed, is none of his, altho he hath made advantage by publishing the sale thereof. But this is no great wonder, since the most part of the truths contained in his writings, are digged out of other mens works. And that the Author of his History may not escape the fate of others with whom he maketh with his doctrine, some mistakes of his own, and particularly that erroneous application of *Euclid* above mentioned in *page* 4 of this Book.

Referring again to Sinclair's observational work on the solar and lunar easterly drift, Gregory[15] continues:

> Seventhly, when he hath observed his inches, he reduceth them not to degrees, minuts, or seconds; &c. for the Suns motion is not reckoned in inches.
>
> Lastly, suppose he had done all these things aright; this method hath been ordinarily practised above these thirty years: Let him look *Hevelij Selenogriphia*, *Scheineri Rosa Ursina* and *Doctor Wallace* in the end of his *Arithmetica infinitorum*.

It is evident from Gregory's[16] castigation that he had been in correspondence with Sinclair. In the preface to the Reader, he presents a letter he had dispatched earlier, which notes:

but it is not like that you are fit for that purpose, who so surely believe the Miracles of the West, as to put them in print; and record the simple meridian altitudes of Comets, and that only to halfs of degrees, or little more as worth noticing. However, if ye do this last part concerning Coal-sinks well, and all the rest be but an Ars magna & nova, ye may come to have repute of being more fit to be a Collier than a Scholar.

Sinclair prepared a very substantial answer to Gregory's attack in the same coarse style, but it was not published. The handwritten manuscript of this reply, bearing the quaint title of 'Cacus dragged from his Den by the heels',[17] is preserved in the Glasgow University Library. The signature of 'Georgius Sinclarus' appears both on the fly-leaf and twice on the inside of the back page. Its cover carries a handwritten note saying that it was given to the Library in 1692 by 'Geo. Sinclarus'.

At the beginning of the preface to the Reader, Sinclair says:

> Do not marvel, that after so much delay I have to turned an answer, to that odious pamphlet, published by some of the masters of the Colledge of St Andrews, under the Bedellus name as Author. I acknowledge there are some reasons why I should not . . .

That Sinclair's reply was not published may perhaps have been related to Gregory's early death in October 1675, at the age of thirty-seven. On page 2 of his preface, Sinclair notes that 'Gregor is gone. But surely, his diable pamphlet is his living . . .' At the end of the preface, the bottom of the page has been pasted over, and he says: 'If Gregory had been alive . . .' Sinclair addressed with vigour various criticisms and points in his response, also deriding Gregory and charging him with a total want of skill in the use of astronomical instruments. He commented that 'though, by help of subscriptions, he had erected a sort of Observatory at St Andrews'.

In a summary of Sinclair's character, Coutts[18] also notes:

> It was less to Sinclair's credit that in 1684 he published under the title of *Truth's Victory over Error*,[19] a translation of David Dickson's *Prealectiones in Confessionem Fidei*, apparently with a design that it should pass for his own. On ceasing to be a Regent in 1666, he set up as an engineer and surveyor, and in this capacity he was extensively employed in connection with mines in the south-west of Scotland, and is said to have first suggested effective means of draining them. Round about 1670, he was employed by the magistrates of Edinburgh to superintend the execution of the works for the first water supply brought into the city; and if not the first, he was among the first who endeavoured to measure the height of mountains by means of variations of atmospheric pressure as indicated by mercury. In this way he measured, in 1668 and 1670, the heights of Arthur's Seat, Leadhills, and Tinto.

More will be said about Sinclair's plagiarism of David Dickson's work later in this chapter.

2.3 SINCLAIR THE PRACTICAL NATURAL PHILOSOPHER

Sinclair lived at a time when the understanding of science was developing from its medieval base of simply following the Greeks, Aristotle and Ptolemy, to making new investigations with ingenious instruments. Thus, he was of the new generation of scientists, and was a natural philosopher in a most traditional sense, conducting experiments, making observations of natural phenomena and putting his knowledge to practical ventures. In his treatise entitled *Natural Philosophy Improven by New Experiments*,[20] his writings show a remarkable spread of subject matter, but with most of the material taken from his earlier works. In the field of astronomy, he describes the juxtaposition of the planet Jupiter to the bright stars in Gemini and the oval shape of the setting Sun, commenting on the effects of atmospheric refraction. He developed several devices to determine the north–south meridian, and constructed an optical device using two lenses and a long tube to investigate the retrograde motion of the planets and of the Moon, outside solar eclipses. He measured the magnetic deviation of a compass needle relative to true north. He also discussed the behaviour of dead bodies in water – how they originally sink but come to the surface after a few days!

2.3.1 Determining Latitude

One of the techniques for determining latitude is to observe the altitude of a circumpolar star at lower culmination, A_l, and upper culmination, A_u, at its meridian passages. Such an exercise was described by Sinclair under 'OBSERVATION XV' in *Natural Philosophy Improven by New Experiments*;[21] the details of his measurements made during the winter of 1669–70 are presented in Fig. 2.1. It is instructive to investigate and to interpret these records.

Sinclair refers to observations of the foremost 'guard-star'. Although it is not immediately obvious which star was the target, based on the indicated approximate local sidereal time, a likely candidate is Kochab, or β UMi. Kochab and Pherkad, or γ UMi, are sometimes referred to in combination as the 'Guardians of the Pole'. The measurements cannot refer to Arcturus, or α Boö, the 'Guardian of the Bear', as this star is not circumpolar at the local latitude.

Although Sinclair was out of post in 1669–70, a question might be asked as to whether he undertook the observations at the Old College. A quadrant device was probably available there, as, according to Gavine,[22] it appears in an inventory of 1727, although some of these listed instruments might originally have been Sinclair's and later donated to the College by him.

The latitude of his observing site was calculated using the standard formula appropriate to the pair of culminations occurring north of the zenith, that is taking the mean of the two culmination altitudes: $(A_u + A_l)/2$.

OBSERVATION XV.

DEcember. 17. 1669, I obſerved with a large *Qua-drant*, half 9 a clock at night, the formoſt *Guard-ſtar*, when it was in the *Meridian*, and loweſt, to have 41 degrees

---

𝕸iſcellany 𝕺bſervations.   229

---

grees 22 minuts of altitude. And on *January* 7. 1670 at
7 a clock in the morning, I found it, when it was in the *Me-ridian*, and higheſt, to have 70 degrees, 27 minuts Hence
I conclude the *elevation of the Pole* here to be 55 degrees,
54 minuts, 30 ſeconds: and conſequently as much at
Edinburgh; becauſe both the places are upon one and the
ſame Parallel.

Figure 2.1 Altitude measurements of a star at upper and lower culmination made by Professor George Sinclair in the winter of 1669–70. Their interpretation provided a determination of the local latitude.

This provides a value of 55° 54′ 30″, with the modern Ordnance Survey (OS) map giving a value of 55° 51′ 51″ for the location of the Old College. The difference between these noted parallels of latitude corresponds to the observing site being north of the College by about 5km. Accurate timings for the culminations are not available, and so an estimate of the longitude cannot be made.

As there is no reference to refraction corrections, it is unlikely that these had been applied. A correction can be readily made to first order using the refraction formula

$$z = \zeta + R,$$

where z is the true zenith distance, ζ the observed zenith distance and $R = k \tan\zeta$, with the value of $k \approx 58$ arcsec. After converting the observed altitudes to zenith distances, applying the correction formula and reconverting to altitudes, the revised values for the culminations are:

$$A_u = 70° \, 26′.7 \; ; \; A_l = 41° \, 20′.9 \; .$$

The corrected latitude determination is 55° 53′ 48″, bringing the site to about 3.5km north of the College, still at a significant distance in relation to the measurement uncertainties. It therefore seems unlikely that Sinclair's observations were conducted at the College. He comments that his determined latitude is 'as much at Edinburgh'. In his various discourses on weather events, he frequently referred to Midlothian and East Lothian, and this seems to be where he was based.

The formula used to determine the local latitude implies that the observed star had a high declination, and that both upper and lower culmination occurred on the meridian at points north of the zenith. From the two measurements, the declination, δ, of the star may be calculated from:

$$\delta = 90° - \frac{A_u - A_i}{2}.$$

Using the culmination altitude values, corrected for refraction as above, the measured star's declination is calculated as 75° 27′.1. According to the 2000.0 epoch, β UMi has coordinates: $\alpha = 14^h\ 50^m\ 42^s.3$, $\delta = 74°\ 09′\ 20″$, revealing a substantial difference in declination of 1° 17′.8, the equivalent of more than two apparent lunar diameters. What is the reason for this?

The coordinates of all stars change progressively as a result of the Earth wobbling on its axis, such that the projection of its pole (one of the origins of the stellar coordinate system) executes a circle with a period ~26,000 years. The process is referred to as the 'precession of the equinoxes'. The apparent drift in the declination of a star caused by this may be expressed approximately by:

$$\Delta\delta = n \cos(\alpha),$$

where $n = (20″.0431 - 0″.000085\ t) \times t$, with t expressed in years. In order to calculate the change in declination from one epoch to another, the procedure involves application of the formula in steps of short intervals, and updating the progressive changes that also affect the value α. This iteration procedure is tedious, and for the purpose of the exercise here, a simplistic approach is applied by simply setting $t = -330$ yr, to convert the declination for 2000 to that of 1670. With α converted to degrees (222°.68), the value of Δδ is 1° 21′, giving $\delta_{1670} = 75°\ 30′.3$, this being sufficiently close to Sinclair's measured value (75° 27′.1), so confirming the target star as β UMi.

2.3.2 The Great Comet of 1680

In his preamble of *Natural Philosophy Improven by New Experiments*,[23] Sinclair describes a range of observations of comets, eclipses, the position of Jupiter, and so on, and gives details of the behaviour of the Great Comet of 1680. At the beginning of this description, he refers to his observing location as having a latitude of 55° 54′ 30″, this being the value he had determined some ten years previously. Prior to returning to Glasgow as Professor of Mathematics, it appears that his experimental work, at least that concerned with astronomy, was conducted from a regular location, nearer to Edinburgh.

Sinclair's sketches (see Fig. 2.2) and his description of the comet's movement are difficult to interpret with any useful accuracy, which is concordant

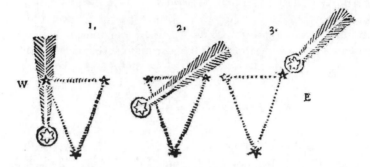

Figure 2.2   Drawings made by Sinclair of the Great Comet appearing
in the winter of 1680–1. Sketch No. 1 is dated 18 January 1681, No. 2 is
dated 20 January 1681 and No. 3 is dated 27 January 1681.

with Gregory's criticisms. From the location of the letters 'W' and 'E', it
might be concluded that the comet's position was recorded on a celestial
globe, rather than on a star atlas as it directly appeared against the stellar
background. The drawing has the perspective of someone looking from the
outside of the celestial sphere, rather than that of an observer looking up
at the sky from the ground. He also notes that the comet was seen on 22
November 1680 at Plimmouth [*sic*] in England.

In fact, this comet (C/1680 VI) is famous for two reasons. Firstly, it was
seen by Gottfried Kirch in Germany on 14 November 1680, and was the first
to have a telescopic discovery. Secondly, according to Brandt and Chapman,[24]
the study of its orbit was key to establishing the universality that all orbiting
objects follow paths in the form of conic sections. From observations made
by Flamsteed, Halley and himself, Sir Isaac Newton determined its orbit using
a graphical method. Halley later performed numerical calculations resulting
in orbital details at much greater accuracy. Encke refined the calculations
in 1818, showing that the comet came within about 0.006AU of the Sun,
making it a sun-grazing object. Newton's work was key to demonstrating that
cometary orbits are simply under the constraint of solar gravity and follow a
path of a conic section with the common elliptical form, even to the extreme
of being parabolic. In the later edition of his *Principia*, Newton wrote:

> the orbit is determined . . . by the computation of Dr Halley, in an ellipse. And
> it is shown that . . . the comet took its course through the nine signs of the
> heavens, with as much accuracy as the planets move in the elliptical orbits given
> in astronomy.

Thus, Newton concluded, 'comets are a sort of planet revolved in very
eccentric orbits about the Sun'. For the above reasons, this Great Comet is
sometimes referred to as Newton's Comet, but following the convention of
naming it after the discoverer, it is also referred to as Kirch's Comet.

2.3.3 *Astronomy and Navigation*

In addition to his observational work, Professor Sinclair also provided descriptions of the Celestial Sphere. In 1688, he produced *The Principles of Astronomy and Navigation, – OR, A Clear Short, yet Full Explanation, of all Circles of the Celestial, and Terrestrial Globes, and their Uses, being the whole Doctrine of the Sphere, and Hypotheses to the Phenomena of the Primium Mobile.*[25] Its title page gives the author as 'George Sinclair, sometime Professor of Philosophy in the Colledge of GLASGOW' (see Fig. 2.3). Appended to the tome are articles entitled 'Mercurial Weather-Glass', discussed later, and 'Buoying up a Ship of any Burden from the Bottom of the Sea', already referred to above.

Within the text, it is interesting to note that the equinoctial points, or equinoxes, are said to fall on 10/11 March and 13/14 September, these dates relating to the old-style calendar prior to its revision in 1752 (see Chapter

THE
PRINCIPLES
OF
Aſtronomy and Navigation:

O R,

A Clear, Short, yet Full Explanation, of all Circles of the Celeſtial, and Ter-reſtrial Globes, and of their Uſes, being the whole *Doctrine of the Sphere*, and Hypotheſes to the Phenomena of the *Primum Mobile.*

To which is Added

A Diſcovery of the Secrets of Nature, which are found in the *Mercurial-Weather-Glaſs*, &c.

A S A L S O

A New Propoſal for Buoying up a Ship of any Burden from the Bottom of the Sea.

By *GEORGE SINCLAR*, ſometime Profeſſor of Philoſophy in the Colledge of *GLASGOW.*

Edinburgh, Printed by the Heir of *Andrew Anderſon,* ⸻ ʼɔ Hls moſt Sacred *Majſty,* Anno *Dom.*1688.

(I)

THE
PRINCIPLES
OF
Aſtronomy and Navigation.

CHAP. I.

Of the Circles of the Sphere in general.

THE material Sphere, is an Inſtrument, where-by all the Phenomena of the *Primum Mobile,* are moſt eaſily known. It is compoſed of Ten Circles chiefly, which are, the Equinoctial, the Zodiack, the two Colures, the Horizon, the Me-
B   ridian,

Figure 2.3 The title page of the treatise published in 1688 by Professor George Sinclair on the Celestial Sphere, with the first page of the first main section.

1). Sinclair's discourse on the celestial sphere is descriptive but definitive, and is based on the use of a model orb that might be found in a study. The differences between Great Circles and Small Circles are clearly defined, with those of particular importance described, including the Great Circles of the Equator and Meridian, and the Small Circles related to the Tropic(k)s and the Artic(k)s, the latter now referred to as the Arctics.

Some of the defined terms are quaint and are now rarely used. These include Anceti – 'those who dwell under opposite Parallels, equally distant from the Equator, towards both Poles, under the same Meridian, but not in Points *diametraliter* opposed, but in a Semi-circle of the Meridian intercepted between Poles'.

The seasonal times of the rising and setting of the stars are referred to in terms of Cosmice, Acronyce and Heliace, with descriptions given to cement the definitions. Examples for each of the terms are given with reference to the classical poets. Sinclair's writings are not overly clear on these matters, and, even today, there is debate on how to interpret what the classical poets are saying when they describe particular apparitions of a star or constellation relative to the position of the Sun. For some geometries, two of the terms seem equally appropriate to classify a situation. Sinclair suggests that the rising of a star Cosmice relates to the time when a star appears above the horizon at the same moment as the Sun rises, and the setting of a star Cosmice corresponds to the time when the star is setting at sunrise. Acronyce refers to a star rising just as the Sun sets, or a star setting at the same time as the Sun. The term Heliace is perhaps more familiar to the modern era as it is used with reference to Egyptology. Modern astronomical dictionaries refer to a 'heliacal' rising of a star or planet as the occasion when the object rises shortly before the Sun. The heliacal rising of the star Sirius played a significant role in ancient Egypt by indicating the approach of the annual flooding of the Nile. According to Sinclair's work, such a position of the Sun would also make Sirius a star rising Cosmice at this time.

To portray the condition of a star Cosmice, Sinclair refers to Virgil's fifth book of the Georgicks in Latin, but provides a translation:

> When *Pleiades*, the Maiden seven,
> Are set into the Western-Heaven,
> When *Ariadnes* Crown full bright
> Before the Sun comes in our sight,
> Then sowe your Ground with *Wheat* and *Rye*
> 'Tis of *October* eighteen day.

Paraphrasing this, Sinclair notes:

> Before you sowe your Ground (says he) let the *Pleiades*, or Seven-stars be set, which about October 18, go down in the West, in the same Moment of Time, while the Sun is Rising in the East, which is called *Occalus Cosmicus*.

2.3.4 The Barometer

Sinclair's passion for the use of the mercury barometer is very apparent. In July 1670, he measured a 2-inch difference in barometric pressure between the top and bottom of the Cheviot. He comments:

> From this Experiment we cannot learn the determinat hight of the Air, because the definit hight of the Mountain is not know. I know there are some, who think that the Air is indefinitely extended, as if forsooth, the Firmament of fixed Stars were the limits of it, but I suppose it is hard to make it out.

As referred to earlier, Sinclair was one of the first people in Britain to measure the height of mountains using barometric principles. He gave the name 'baroscope' to his tube of mercury which he took to the top of Arthur's Seat, to Leadhills and to Tinto Hill; later, he adopted the name 'barometer'.

The understanding of the barometer's behaviour in relation to the weather and its ability to make forecasts fascinated Sinclair. In the subsection 'Proteus bound with Chains' of *The Principles of Astronomy and Navigation*,[26] he commented on the usefulness of the barometer as a means of forecasting the weather for the man (and woman) in the street and for the farmer. One of the paragraphs declares:

> 'Tis said, that the Ladies, and Gentlewomen at *London* do Apparel themselves in the Morning by the Weather-Glass. Whatever be in this, 'tis certain, that when a man riseth in the Morning, he may know infallibly, what sort of Weather will fall out ere Night.

He puzzled to understand the physics behind the behaviour of the barometer, and on its reliability to forecast the weather. In the section entitled 'Mercurial Weather-Glass', he comments:

> This Weather-Glass was esteemed infallible; but now it is known to be most fallacious, and uncertain. The reason for this, because it is acted and moved, not only with Heat and Cold, but with greater and lesser pressure of the Air. If it be demanded, how shall I know, whether it be the coldness of the Air, or the greater pressure of the Air, which causes the Water to ascend: and whether it be, the warmness of the air, or the lesser pressure of the Air which causes the Water to descend? I answer, It is difficult to know; for both do sometime concur, that is, the weight and greater pressure of the Air, and the coldness too. And sometimes the lesser pressure of the Air with warmness, make the Water fall down. And sometimes the greater pressure of the Air raiseth the Water, without addition of Cold: And sometimes the addition of Cold, without any alteration in the Air, as to more weight, will raise it. And tho by this means it be fallacious, yet many notable Phenomena do appear from it. By the help of this, and the Mercurial-Weather-Glass, and the Sealed one, and by the help of the Hygroscope, which marvellously shews the least alteration in the Air, as to Moisture and Dryness, and by contemplating the Heavens, the Sky and the Clouds, and considering the Winds,

how they blow, the various Aspects of the Planets one to another, (as Star-gazers do affirm) men might come to foretell the Weather particularly'.

Our understanding of meteorology has certainly developed in the intervening 330 years, but weather forecasting is still not 100 per cent certain because of the complexities of the system, as Sinclair noted. One component does not appear in modern computer programmes, though; no provision is made in the software for planetary configurations, as this would allude to astrology.

2.3.5 Other Texts

Mention was made earlier of Professor Sinclair's misdemeanour in passing off as his own a translation of an original work in Latin by Professor David Dickson. The Edinburgh edition of this work, *Truths Victory over Error*, appeared in 1684.[27] The left-hand section of Fig. 2.4 provides an image of the title page.

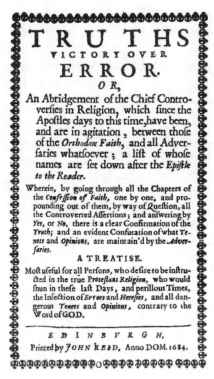

Figure 2.4 The title page of the 1684 edition of *Truths Victory over Error*, with Professor George Sinclair claiming authorship, is shown in the left-hand column. In the 1725 edition, the Preface has been revised by Robert Wodrow, who exposed Sinclair's plagiary, as shown in the extract on the right-hand side of the figure.

It is interesting to note that in the 'Epistle Dedicatory' of this edition Sinclair boastfully says:

> I have made some small attempts, during the twelve years I taught *Peripatetick* and *Experimental Philosophy*, and since, for the Announcement of Learning among others, which have not wanted success, whereby the Author hath been encouraged; especially by the kind acceptance, his *Writings* have met with the greatest *Philosophers* and *Mathematicians* in this Age, in *England, Holland, Germany* and *France*. It is yet recent in the minds of many Noble and worthy Persons what esteem His *Royal Highness* had of my Observations, of the great *Blazing Star*, which appeared in *December* 1680, which since have been published. I do not mention this for applause, or out of vanity, but for some peculiar reasons hinted at below. But these studies being only *Hand-maids*, and *Subservient* to *Design Knowledge*, and not so generally useful, and I have now given them a Manumission, unless I be animated by the benign, and favourable aspect of those, who may and can. I move in a distinct Sphere from Masters of Universities. They teach in *Philosophy, the Causes and Reasons of Things*. What I write is but *Practical* and *Mechanical*, for the promoting of natural Knowledge and Learning, as do the *Virtuosi*. But in stead of such I present your *Honours* with a small bundle of *Orthodox Truths* confirmed by plain Scripture Testimonies, wherewith the true Christian Church hath in all Ages scattered the swarms of dark Errors, the damnable Heresies, Locusts from a bottomless pit.

His comments on 'the great Blazing Star' relate to the appearance of the Great Comet of 1680. His recordings of this object were presented earlier (see p. 33), where it was revealed that they had only meagre usefulness.

In the London edition of this work, printed in 1688,[28] the deception of authorship remained, as did the pomposity or vanity. The preface begins with:

> The Author of the following *Treatise* is well enough known by his *Writings*. After he had studied his Philosophy, he was invited in the year 1654, to be one of the Masters in the *Colledge of Glasgow*, where he Taught twelve years during which time, he Published his *Tyrocinia Mathematica*. After this he resolved to study *Natural and Experimental Philosophy* and he Wrote his *Ars Nova & Magna*; a Book in much esteem abroad, both for the *Matter* and Form.
>
> In the year 1662, hearing of the Fame, and Learning of the *Royal-Society*, he went to *London*, and was kindly received at *Gressom Colledge*. Here he was advised to Translate his *New Experiments* into *Latin*, which first he had written in English. While he prospered in his Studies at *Glasgow*, and his *Schollars* much advanced in solid and true learning far beyond many of their Neighbours; some began to look awray upon him: wherefore those of the *University* (all which are now at rest, except one) did with him . . .

In later editions, the authorship of the text was corrected and the book thus bore the name of the true writer, David Dickson, Professor of Divinity at the University of Edinburgh. As mentioned in Chapter 1, this particular plagiarism appeared to be exposed by the Revd Robert Wodrow, who had been Glasgow University Librarian between 1697 and 1701. That Sinclair had the effrontery to use David Dickson's notes as his own, particularly as their Latinised version had been available to his students in Edinburgh, defeats the imagination.

In the edition printed in 1725[29] and collated by Robert Wodrow, the text was prefixed by an account of the true author's life. Its preface also notes:

> This Book was first published in the Year 1684, by G.S. that is, Mr *George Sinclare*, well enough known by his several Books published both in *Latin* and *English*. What led him to translate another Man's Book, and send it abroad under the initial Letters of his own Name, I shall not determine. I am willing to leave Ashes of the Dead in quiet, especially those of a Person, who, in this Time, was taken Notice of with some Applause by Learned Men Abroad, as well as at Home; and wrote several Things in Philosophy, Mathematicks and History, in his own Way, not without their Use in the Time when they were published.
>
> I knew him in his old Age and declining Years, when much decayed, yet still retaining the serious and religious Dispositions I hope he had thro' his whole Life; and tho' I cannot intirely vindicate this low and mean Piracy, in publishing the Work of another, in such a Manner; yet I hope it was his Regard to the great Truths in our excellent Confession of Faith, and his Desire that common People, in a Time when we were in imminent Hazard of Popery, a bigotted Papist being on the Throne, should be guarded against the Errors then breaking is like a Flood upon us; that put this good Man to take the Pains to translate the Dictates of the Learned Professor *Dickson*, from *Latin* to *English*.
>
> If he had the poor View of a little Glory to himself, by publishing these in his own Name, it happened to him as generally it does to self-seeking and private-spirited Persons, even in this present State, their Naughtiness is discovered, and they miss their Mark: But we shall charitably suppose he had higher and better Aims.
>
> There were several Copies of what the Revd. Mr *Dickson* had dictated to his Scholars, in Latin, upon the Confession of Faith, in the Hands of Ministers and others; and it came to be known, that *Truth's Victory over Error* was only a Translation and the venerable Author's Name suppressed. This made one to dash down the following Lines upon the Running Title before the Title Page in the former Edition, in pleasantry:
>
> <div align="center">'<i>Truth's Victory over Error</i>
No errors in this Book I see,
But G. S. where D. D. should be'.</div>

Another of Professor Sinclair's important treatises was his *Tyrocinia mathematica*,[30] written in Latin and presenting the basic rules of arithmetic. No doubt it would have been used as a text in the College at Glasgow. The rudiments of addition and subtraction were then an important part of education, and the text covers these. Any courses containing the principles of differential calculus were yet to come, as this subject was only just being formulated by scholars elsewhere at the time. Thus, with the production of his book on astronomy, which was a discourse on the use of the celestial sphere, and his treatise on arithmetic, he might be considered as the local Sacrobosco of his time.

Perhaps his most cited work, as the results of any modern search of his achievements tend to emphasise, is *Satan's Invisible World Discovered*,[31] first published in 1685 and reprinted several times. It records various local episodes related to witches, ghosts and diablerie, and forms a curious record of the then current notions on those subjects. It had wide circulation in Scotland. The fact that this book is still referred to today, in preference to any other of his treatises, probably reflects the general continuing fascination in the paranormal more than any ongoing interest in his science, which now is only relevant to historians.

2.4 THE TURN OF THE CENTURY

Towards the end of the sixteenth century, astronomy continued to be taught within the University under the umbrella of physics. A collection of course notes[32] by Thomas Clarke [Clericus] for the 1687–8 session shows that astronomy was well covered by the Regent, John Tran. The book of course notes (see Fig. 2.5 for an image of its front cover) gives the names of the students in the class and an index to the discussed topics. Some of the pages of the index relating to the early part of the course appear to be missing. A pair of pages demonstrating the required geometry for lunar and solar eclipses are beautifully caricatured; the drawings are shown in Fig. 2.6. The section for 18 January 1688 begins with *Mundo & Caelo*; for 7 February the title is *Astronomia*. The entry for 10 February provides an Earth-centred map of the Universe with copious notes on the *Hypothosi Ptollimaira*. What is surprising at this late date of the seventeenth century is that there is a lack of detail given with respect to *De Systemto Copernicans*; although the latter is mentioned, and a plan drawn, as seen in Fig. 2.5, there is a paucity of material, with no strong emphasis on it being the correct system, nor any apparent mention of Kepler whose laws had then been established for eighty years.

According to David Murray,[33] John Law was the last of the Regents admitted to the post following the procedure of 'disputation' of his presented *Theses Philosophicae*[34] at the graduation ceremony in 1698. The short document relates largely to physics and astronomy, with the last section referring to a new form of sundial he devised. It is interesting to note that even then he wrote about the motions of the planets in terms of Ptolemy and epicycles. The following year, he published *Calendarium Lunæ Perpetuum*,[35] a work related to a new method of calculating the times of lunar and solar eclipses from the birth of Christ to the current time (1690), as well as to future years. An image

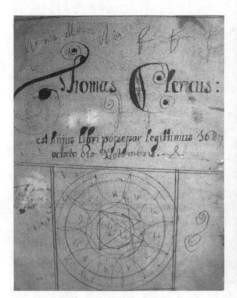

Figure 2.5 The front cover of Thomas Clarke's notes of material presented by the Regent, John Tran, in the session of 1687–8. The diagram from p. 79 traces out the Copernican system for the Solar System, with the orbits of the Galilean satellites of Jupiter clearly shown.

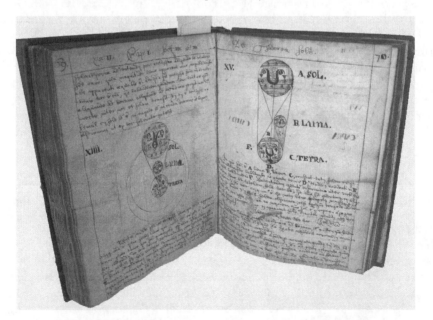

Figure 2.6 Pages 69 and 70 of Thomas Clarke's notes show the geometries required for lunar and solar eclipses.

Figure 2.7 The cover page of *Calendarium Lunae Perpetuum* is presented together with the beginnings of a table carrying the predictions of eclipses visible in Scotland from 1699 to 1740.

of the cover page is presented in Fig. 2.7, together with the start of a table showing the predictions of eclipses visible in Scotland from 1699 to 1740.

Murray[36] notes that judging from the dictata of the Regents and the graduation theses, astronomy occupied a considerable place in the course of study in the Magistrand class, that is the fourth or last year before graduation. There may have been some practical work in that class, as the University already had a telescope in 1693, and in that same year acquired another, 8 feet in length. Again, Murray[37] comments that Stirling's Library in Glasgow once held a copy of Mr Law's *Demonstrationes Logicae* transcribed by a William Stirling, who was a first-year student (Bajan class) in the year 1699–1700. A manuscript of this work,[38] together with other items, is available in the Glasgow University Library. John Law later wrote a number of works relating to financial systems and trade.

The teaching of astronomy, with practical elements, continued into the eighteenth century and extended when the Magistrand class became the class of natural philosophy in 1727 with the appointment of Professor Robert Dick. These developments will be dealt with in the following chapter.

2.5 EARLY TIMEKEEPING

Although the great era for promoting the display of time within the city was the second half of the nineteenth century, as described in Chapter 8, an

Figure 2.8   The sundial (c. 1670) on the southerly side of the Provand's Lordship is set at a small angle relative to the main wall. The asymmetry, either side of noon, of the numbering of the hours and the lines running to where the missing gnomon would have been, suggests that the face of the stone plate on which the shadow fell does not exactly face south.

early sundial remains attached to the south-western gable of the Provand's Lordship, the oldest standing house in Glasgow, situated at the top of Castle Street, nearly opposite the cathedral. According to Gemmell's *The Oldest House in Glasgow*,[39] published just over one hundred years ago, a broken sundial could still be seen with the initials 'W. B.', its inclined face receiving the light from the western Sun. The initials refer to a William Brysone Taylor, who took possession of the house in 1642 and in 1670 extended it with wings, one of which held the sundial. In a sketch of the house made by 'Crimea' Simpson in 1843, the sundial was intact with its gnomon. At the time of its installation, the University Tower (see Fig. 1.3) appears to have supported an external clock face, and the sundial would not have been the main public display of local time. A current image of its repaired plate, set into the house's southern wall at a slight angle, is provided in Fig. 2.8. According to Cleland,[40] the clock in the steeple of the Merchant's Hall was made in 1659.

References

1. Coutts, J. (1909), *A History of the University of Glasgow (From its Foundation in 1451 to 1909)*, Glasgow: James MacLehose & Sons, p. 149 [GUL Education S271 1909-C – two copies; GUL Sp Coll Mu21-a.2; GUL Sp Coll Bh12-e.14; GUL Sp Coll MacLehose 824].

2. Ibid., p. 171.

3. Sinclair, G. (1669), *Ars Nova et Magna Gravitatis et Levitatis; sive dialogorum libri sex de aeris vera ac reali gravitate, etc.; quibus accessere De instrumentis hydragogicis libri duo: de hygroscopio, t chronoscpio seu pendulo liber unus nec non Palladis gymnasium . . . in fine additis indicibus*, Rotterdam [GUL Sp Coll Mu5-g.4; GUL Sp Coll Mu55-g.19; GUL Sp Coll Bn7-h.6].

4. Anonymous (c. 1669), 'A vindication of the preface of the book intituled, *Georgii Sinclari*, quondam in Universitate *Glasguensi* Philosophi Professoris, *Ars Nova & Magna Gravitatis & Levitatis*, from the challenges and reflections of the Publisher of the Philosophical Transactions, as they are to be found, *Numb. 50*. date, *August 16. 1669*', [GUL E-Book].

5. Sinclair, G. (1672), *The HYDROSTATICKS, or, The weight, force, and pressure of FLUID BODIES, Made evident by Physical, and Sensible Experiments. By G.S.* Edinburgh: George Swintoun, James Glen, and Thomas Brown [GUL Sp Coll Ferguson Af-a.38; GUL Sp Coll Ferguson Al-a.28; GUL Sp Coll Ea6-f.13; GUL Sp Coll Bi8-h.25; GUL E-Book].

6. Sinclair, *Ars Nova*, p. 360.

7. Laing, D. (1857), 'XXXVIII Notice of a Scheme, with the Warrant of King James VII., and the Lords of Privy Council, for a Patent to be granted to certain Merchants in London, for Weighing up and Recovering Ships in the Scottish Seas. 26th May 1686', *Archaeological Scotia*, 4, 429.

8. Sinclair, *Ars Nova*.

9. Sinclair, *The HYDROSTATICKS*.

10. Laing, 'Notice of a Scheme', 430.

11. M. Patrick Mathers, Arch-Bedal to the University of St Andrews (1672), *The Great and New Art of Weighing Vanity: or a Discovery of the Ignorance and Arrogance of the great new Artist in his Pseudo-Philosophical Writings; to which are annexed some Tentamina de Motu Penduli et Projectorum*. Glasgow: Robert Sanders [GUL Sp Coll Mu35-h.13; GUL Sp Coll Bi3-l.31; GUL Sp Coll Ferguson Ak-e.34; GUL Sp Coll Ea8-g.24; GUL Sp Coll Mu54-h.23; GUL Sp Coll BG56-e.25; GUL E-Book].

12. Laing, 'Notice of a Scheme', 431.

13. Sinclair, *The HYDROSTATICKS*, p. 201.

14. Mathers, *Great and New Art of Weighing Vanity*, p. 14.

15. Ibid., p. 46.

16. Ibid., p. 46.

17. Sinclair, G. (1672), 'Cacus dragged from his Den by the heels' [GUL Sp Coll MS Gen 351].

18. Coutts, *A History of the University of Glasgow*, p. 171.

19. Sinclair, G. (1684), *Truths [sic] victory over error, or, An abridgement of the chief*

controversies in religion which since the apostles days to this time . . . the infection of errors and heresies, and all dangerous tenets and opinions, contrary to the word of God, Edinburgh edition, Edinburgh: John Reid [GUL E-Book].

20. Sinclair, George (1683), *Natural Philosophy Improven by New Experiments touching the mercurial weather-glass, the hygroscope, eclipsis, conjunctions of Saturn and Jupiter, by new experiments, touching the pressure of fluids, the diving-bell, and all the curiosities thereof: to which is added some new observations, and experiments, lately made of several kinds: together with a true relation of an evil spirit, which troubled a mans family for many days: lastly, there is a large discourse anent coal, coal-sinks, dipps, risings, and streeks of coal, levels running of mines, gaes, dykes, damps, and wild-fire* [GUL Sp Coll 615; GUL Sp Coll Ferguson Af-c.12; GUL Sp Coll Mu54-f.31; GUL Sp Coll Mu54-f.30; GUL Sp Coll Mu54-Bl.k1; GUL E-Book].

21. Ibid., pp. 229–30.

22. Gavine, David M. (1981), 'Astronomy in Scotland 1745–1900', unpublished PhD thesis, Open University, 55.

23. Sinclair, *Natural Philosophy Improven by New Experiments*.

24. Brandt, J. C. and Chapman, R. D. (2004), *Comets*, Cambridge: Cambridge University Press.

25. George Sinclair (1688), *The Principles of Astronomy and Navigation, or, A clear, short, yet full explanation of all circles of the celestial and terrestrial globes and of their uses: being the whole doctrine of the sphere and hypotheses to the phenomena of the primum mobile : to which is added a discovery of the secrets of nature which are found in the mercurial-weather-glass &c.: as also a new proposal for buoying of a ship of any burden from the bottom of the sea by George Sinclair*, Edinburgh: n.p. [GUL Sp Coll Ferguson Af-c.40; GUL Sp Coll Mu54-h.8; GUL Sp Coll Bi3-l.1; GUL Sp Coll Ea8-g.12; GUL E-Book].

26. Ibid.

27. Sinclair (1684), *Truths victory over error*, Edinburgh edition [GUL E-Book].

28. Sinclair, G. (1688), *Truths victory over error*, London edition [GUL Sp Coll Bh5-l.9; GUL E-Book].

29. Sinclair, G. (1725), *Truths victory over error*, Printed by William Duncan [GUL Sp Coll T.C.L. 417]. Later editions of *Truths victory over error* with corrected authorship include
    (1764) [GUL Sp Coll Mu38-i.11; GUL E-Book]
    (1772) [GUL Sp Coll Robertson Bf64-k.26; GUL Sp Coll BDA1-b.4; GUL E-Book]
    (1787) [GUL Sp Coll Robertson Bf62-m.31; GUL Sp Coll T.C.L. 1024; GUL E-Book].

30. Sinclair, G. (1661), *Tyrocinia mathematica, sive, Juniorum ad matheses addicendas introductio in quatuor tractatus videlicet arithmeticum, sphaericum, geographicum & echometricum divisa*, Glasguae: ex bibliopolatum typography [GUL Sp Coll Mu31-i.34; GUL Sp Coll Ea-g.31; GUL Sp Coll Bh9-l.6; GUL Sp Coll Ferguson Af-a.40; GUL E-Book].

31. Sinclar, Georg (1685), *Satan's Invisible World Discovered, or, A choice collection of modern relations proving evidently against the saducees and atheists of this present age*

... *to all which is added, that marvellous history of Major Weir, and his sister: with two relations of apparitions at Edinburgh*, Edinburgh: John Reid [GUL Sp Coll Ferguson Af-d.14; GUL Sp Coll Ferguson Al-b.92; GUL E-Book]. Other available editions of the above title are as follows:
(1769) [GUL Sp Coll Ferguson Ak-e.34; GUL Sp Coll Ferguson Al-b.13; GUL Sp Coll Ferguson Al-b.65; GUL E-Book].
(1779) [GUL Sp Coll Ferguson Al-b.5].
(1780) [GUL E-Book].
(1789) [GUL Sp Coll Ferguson Ak-e.34; GUL Sp Coll Ferguson Al-b.6; GUL Sp Coll Ferguson Al-b.55; GUL E-Book].
(1808) [GUL Sp Coll T.L.C. 3155].
(1814) [GUL Sp Coll Ferguson Al-b.91].
(1815) [GUL Sp Coll Ferguson Al-c.46].
(1831) [GUL Sp Coll BD13-e.25].
(1871) [GUL Sp Coll Ak-d.68; GUL Lib Res Store 23206]
32. Thomas Clark [Clericus] (1687/80), Lectures given by John Tran, session 1687/8, on Natural Philosophy, Glasgow: 1687–8. The student scribe was Thomas Clark [Clericus] [GUL Sp Coll MS Gen 765].
33. Murray, David (1927), *Memories of the Old College of Glasgow – Some Chapters in the History of the University*, Glasgow: Jackson, Wylie and Co., p. 260.
34. Joannes Law Praeses (1698), *Theses Philosophicae*, Glasguae: Excudebat Robertus Sanders [GUL Sp Coll Mu3-c.3]. The same document is also catalogued as *Illustrissimo Et Potentissimo Domino, Alexandro D. Montgomerio, Antiquissim, Dignissim, & Nobili* – Law, John, 1671–1729. Glasgu: Excudebat Robertus Sanders, unus regis, & solus urbis, nec non academi typographus, Anno Dom. M. DC. XCVIII [1698] [GUL E-Book].
35. Law, Js. P. P. (1699), *Calendarium Lunæ Perpetuum*, Glasguæ: Excudebat Robertus Sanders, unus è regiis typographis, & prostant apud Bibliopolas ibidem [GUL Sp Coll Mu11-d.30; GUL Sp Coll MS Gen 38; GUL Sp Coll RB 1200; GUL E-Book].
36. Murray, *Memories of the Old College*, p. 260.
37. Ibid., p. 260.
38. John Law (1704–5), *Demonstrationes logicae*, May 1704; *Disputationes metaphysicae praemium*, 19 Feb. 1705; *Tractatus theologiae*; *Theses metaphysicae* . . . 1705; *Theses logicae*, and *An explication of* . . . *geometry*, in two other hands at end. Inscribed inside front board: John Law, 1714. Author: John Law fl.1691–1705; Regent, University of Glasgow, 1691–1714 [GUL Sp Coll MS Gen 873].
39. Gemmell, W. (1910), *The Oldest House in Glasgow*, Glasgow: Hay Nisbet & Co. Ltd [GUL History DX 205 GEM].
40. Cleland, James (1840), *The Rise and Progress of the City of Glasgow: comprising an account of its ancient and modern history, its trade, manufactures, commerce, and other concerns*, Glasgow: John Smith, p. 256 [GUL Sp Coll Mu22-b.9].

Chapter 3

Establishing Astronomy

3.1 EXPERIMENTAL PHILOSOPHY

Prior to the turn of the eighteenth century, and into its first fifty years, the teaching of astronomy and its associated research in the University was under the care of the professors of natural philosophy. According to James Coutts,[1] the beginning of science in the College can be discerned by the employment of apparatus and instruments for teaching purposes. In the statement of the needs of the University, which Principal Dunlop was commissioned to lay before the King in 1691, apparatus for experimental philosophy was included. In 1693, a telescope 8 feet long, a prism and tubes for weather glasses were set aside in the library, with a number of books donated about this time by George Sinclair, Professor of Mathematics, who had decided scientific leanings related to physics, astronomy and allied subjects (see Chapter 2).

The first person of note after the turn of the eighteenth century was Robert Dick (Snr). He had previously acted as a tutor to the Master of Belhaven, having been elected a Regent in philosophy, and was admitted to office on 22 October 1714. When Regenting was abolished in 1727, he became Professor of Natural Philosophy, probably because, in the reorganisation, the other two Regents, Gershom Carmichael and John Loudon, felt more at home in philosophical subjects than in physics. Though it lacked a firm liaison with mathematics, which is now regarded as an essential for the study of physical sciences, the subject of natural philosophy still retained much of the precedence it had enjoyed when it was the senior class under the old Regenting system.

According to John Duncan Mackie,[2] even before 1727 the Faculty were acquiring 'instruments'; this has been confirmed by Gavine,[3] who has consulted a 1727 inventory of equipment belonging to the natural philosophy faction that lists 'a pair of ~16-inch globes with hour circle and quadrant of altitude, a small armillary sphere, a brass quadrant with moving index, plummet and four lights [this possibly refers to shades] and small unspecified telescopes'. The Library also listed some astronomical works donated by Professor George Sinclair. Purchase accounts amounting to over £300 can be traced for a period of about thirty years. Arrangements were put in place for the equipment's careful handling and safe custody. The pieces were used for demonstrations; the 'boyes' did not perform experiments themselves. Coutts[4] notes that, as early as 1730, there was a class in experimental philosophy distinct from the regular course. The appointment in that year of Henry Drew, a Hammerman (a member of the incorporation that included engineers and mechanics), to look after the instruments and help with experiments, marks the first appearance of a laboratory assistant, or demonstrator, at the University. Professor Dick, who was described by 'Jupiter' Carlyle as a 'very worthy man, and of agreeable temper', made considerable use of the telescope, and in 1749, the University purchased for him apparatus for 'electrical experiments'. A recorded anecdote[5] of Carlyle relates that when Dick brought out his telescope in order that all interested might view the Great Comet of 1744, he invited Mr James Purdie, Rector of the High School, then located in Greyfriar's Wynd. The good grammarian took a stand at the instrument and, turning to the Professor, said, 'Mr Robert, I believe it is *hic* or *haec cometa*, a comet.' The appearance of this comet provided an additional stimulus for the study of astronomy within the University.

3.2 EARLY ASTRONOMICAL EQUIPMENT

Professor Robert Dick was succeeded in 1751 by his son of the same name; Robert Dick Jnr held the Chair until 1757, enjoying a well-deserved reputation as a scholar in his own right. The regular annual budget for experimental philosophy was £12, but Dick (Jnr) was awarded a grant of £150 from the College to enhance the instrument collection. In the summer of 1754, he went to England with a view to purchasing instruments for use in the natural philosophy class, in which physical astronomy was taught, and notes related to the journey have been summarised by Gavine.[6] Dick commented that it was difficult to 'pin down the people of science', but he ordered a 30-inch mural quadrant from Bird, a Short micrometer, a Gregorian reflector (possibly also made by Short), a transit instrument and a Shelton clock. His funds did not stretch to a 65-guinea equatorial, which he had considered. At Greenwich, he inspected the quadrants by Graham and Bird, and then travelled to Oxford with James Ferguson and his pupil, a Mr Gordon. In a letter dated 14 September 1754, addressed to Robert Simson, the Clerk of Senate, he described the Physick Garden there and listed the exotic plants under cultivation, from which Boney infers[7] there was criticism of the poor state of the Glasgow garden. In another letter, he mentioned Mr Campbell's private observatory at Greenwich, built for only £10, but questioned whether the College would be able to build one for the same budget. He continued the drive for opening an observatory and, on 24 June 1754, the University approved a plan for raising public subscriptions for a building in which the instruments should be placed for use. Before preparing a wooden model of the intended observatory for Glasgow,

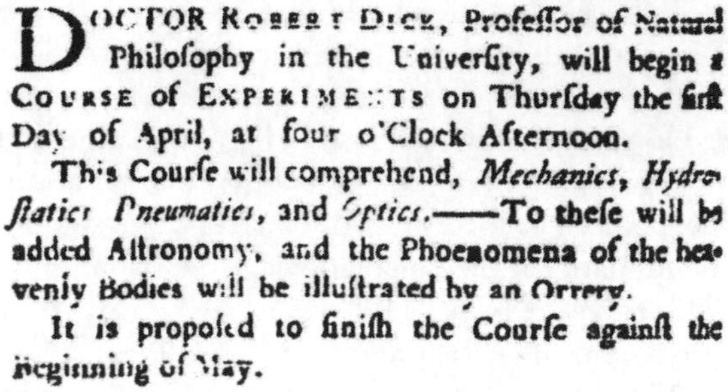

DOCTOR Roᴇᴇᴛ Dɪᴄᴋ, Profeſſor of Natural
Philoſophy in the Univerſity, will begin a
Coᴜʀsᴇ of Exᴘᴇʀɪᴍᴇɴᴛs on Thurſday the firſt
Day of April, at four o'Clock Afternoon.
Thɪs Courſe will comprehend, *Mechanics, Hydro-
ſtatics Pneumatics*, and *Optics*.——To theſe will be
added Aſtronomy, and the Phoenomena of the hea-
venly Bodies will be illuſtrated by an Orrery.
It is propoſed to finiſh the Courſe againſt the
beginning of May.

Figure 3.1 The *Glasgow Courant* of 22 March 1756
advertised Dr Robert Dick's 'Experimental Course' which
included some astronomy, with an orrery used to illustrate the
phenomena of the heavenly bodies.

Dick proposed to consult James Bradley, the Astronomer Royal, and Lord
Macclesfield at Shirburn Castle.

In 1755, Professor Dick advertised an evening course of 'Philosophical
Lectures and Experiments' commencing on 23 December at 7 pm. It was
announced that 'The Course will comprehend Mechanics, Hydrostatics,
Pneumatics, and Optics as usual. To these will be added Astronomy, and
the Phenomena of the Celestial Bodies will be illustrated by an Orrery.'
This series of presentations probably marks the first occasion of extra-mural
lectures on astronomy sponsored by the University. A similar lecture series
took place in the 1756–7 session; the *Glasgow Journal* of 29 March 1756
noted: 'Professor Dick had in addition an experimental course from 1st
April till the beginning of May comprehending Mechanics, Hydrostatics,
Pneumatics, Optics and Astronomy' (see Fig. 3.1, which shows the original
advertisement for the course). Partly as a result of fortuitous events, Dick
(Jnr) was able to act as patron and helper of the young James Watt, who later
achieved worldwide fame as an engineer and inventor.

3.3 THE MACFARLANE BEQUEST

Nothing further appears to have been recorded with respect to the subscription fund for an observatory, but the matter came to a head with the
announcement on 20 January 1756 of the death of Alexander Macfarlane, a
merchant in Jamaica, and the bequest of his astronomical equipment, with
his whole collection of instruments left to Glasgow University. Macfarlane's
taste for astronomy had led him to establish an observatory in Jamaica fitted
with the best instruments the times could afford. Formerly educated at
Glasgow, he wished to make a gift of these instruments to his alma mater.
On receiving this information, the University marked its gratitude by conferring upon the donor's brother, Walter, Laird of Macfarlane, the degree of

LLD, attested by a handsome diploma enclosed in a silver casket bearing the University arms. The diploma and casket were presented to the University in 1931 by George McFarlane, an indirect descendant of Walter Macfarlane, Alexander's brother, presumably because the direct male line had terminated with the death of the twenty-fifth laird in 1866. The fragile pieces are now in the safekeeping of the University Archives.

Alexander Macfarlane graduated from Glasgow University in 1728 and emigrated to Jamaica. By November 1735, he had become a successful merchant, landowner and assistant judge, and he was appointed the island's first Postmaster-General. His house in Port Royal included a specially built and well-equipped observatory. Apparently Macfarlane had purchased his astronomical apparatus from another Jamaican merchant, Colin Campbell, FRS, who had an observatory built there in 1731 to complete the catalogue of southern stars. The latter was a friend of Sir Edmond Halley and a scientific collaborator of James Bradley, the Astronomer Royal. According to Bryden,[8] Campbell was born in Jamaica, the eldest son of John Campbell of Inveraray, Argyll, a survivor of the Darien expedition. He was dispatched to Europe for his education and attended Glasgow University; he would have been taught the elements of astronomy by Professor Robert Dick (Snr). In 1731, he returned home in the company of Joseph Harris who, according to Bradley, went to assist Campbell in designing and erecting an observatory at Black River for the improvement of astronomy and the promotion of other studies of natural knowledge of Jamaica. Harris had real experience as a practical astronomer and had a book published on the theory and practice of navigation and nautical astronomy. Campbell's scientific activities are known only through the work of Harris who, after only a year, was forced by ill health to return to England, bringing with him a journal of observations made during his stay. The instruments contained in Campbell's observatory had some notoriety as mention was made of them by Robert Smith, Professor of Astronomy at Cambridge, in his text *A Compleat System of Opticks*,[9] published in 1738, but making reference to the standing of the subject around 1730–1. It is noted that there was a mural arch 4 feet in radius, a smaller version of one used at Greenwich; with the addition of a small extension of the circular segment, its operation would have been able to provide accurate stellar positional measurements over the whole sky at Jamaica's more southerly latitude. Smith's description of the instrument and details of how he expected Campbell to succeed with his astronomical venture, being a graduate of Glasgow University, are presented in Fig. 3.2.

Certainly by 1743, the instruments were in the possession of Alexander Macfarlane and were housed in his observatory. In November of that year, Macfarlane sent a letter to James Short, the famed Scottish telescope maker, with a report of some observations. This was later read by Short to the Royal Society, and the members heard that Macfarlane had 'erected

870. As to the Catalogue of the southern constellations, partly invisible in our latitude, no doubt it will soon receive an accurate correction and increase, by an excellent Mural Arch of 4 foot radius, made by Mr. Sissons in the *Strand*, exactly after the model of this at *Greenwich*; excepting that the limb of it is continued about 12 degrees towards the south of the plumb-line, for observing the places of the moon and planets in passing towards the north from the zenith of *Jamaica*; where this quadrant is to be fixt, for the use of *Colin Campbell* Esquire; a young gentleman of whom the learned in the Sciences have great expectations: not only from his own genius and application, but also from the advantages of his situation, his fortune and education; in which last he has had the happiness to be directed by no less a judge than the Right Honourable the Earl of *Ilay*; to whom he has also the honour of being related.

Mr. *Campbell*'s Mural Arch at *Jamaica*.

Figure 3.2 Robert Smith's entry in his book *A Compleat System of Opticks* describes the mural arch used in Jamaica and how he expected Glasgow graduate Colin Campbell to map the skies at the southerly latitude of Jamaica.

an observatory . . . and furnished it with instruments carried over thither by Mr Campbell, which he has purchased'. The location of the house and observatory (see Fig. 3.3) can be found on the *Plan of Kingston*, published 1738–47, and with the title deed to His Excellency Edward Trelawny Esquire.

The building was an imposing colonial residence, similar in design to the then current style except that, in place of a gently pitched roof, there was a flat platform surrounded by a hand rail. Here, the 4-foot mural arch was erected on a wall; a 5-foot transit telescope and Campbell's one-month regulator clock were similarly built in. A 5-foot portable zenith sector was found difficult to use, and Macfarlane designed a horizontal reflecting zenith sector which was built for him in Jamaica in 1755 by Pierre Martel, a Swiss–Jamaican instrument maker and surveyor who cared for Macfarlane's equipment. According to Swinbank,[10] after Macfarlane's demise, the Jamaican observatory was turned into the county jail.

As for Alexander Macfarlane himself, he was the youngest of four sons. During his education at Glasgow University, he would also have attended lectures given by Professor Robert Dick (Snr). As a result of his work in Jamaica, although fairly minimal in modern terms, he was proposed as Fellow of the Royal Society on 20 November 1746, the certificate of recommendation reading as follows:

Mr. Alexander Macfarlane of Jamaica, A Gentleman well versed in Philosophical & Mathematical Learning, who has the Valuable Collection of Astronomical Instruments which was carried over there by Mr. Colin Campbell a member of this Society, in order to make Observations to complete the Catalogue of the Southern Constellations, We whose names are underwritten being either personally acquainted with him, or believing from his Character that he will be a useful

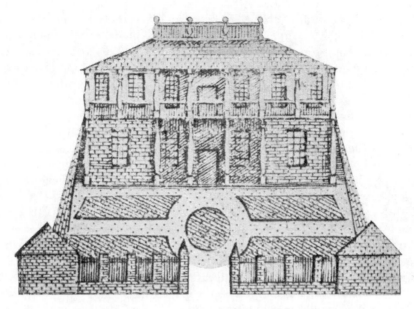

Figure 3.3   The north-facing front of Alexander Macfarlane's house and observatory in Kingston, Jamaica. Although there is no obvious 'dome' for a telescope, it is likely that the wall structure surrounded by the hand rails would have supported the attachment of a quadrant.

and valuable Correspondent do recommend him as every way qualified to be a member of this Society.

London, Novr. 20, 1746.

James Short   Alexr. Ouchterlony   Richd. Graham
Peter Davall   James Burrow

The Gentleman above mentioned having been besides what I have understood from others, strongly recommended by the Honble. Mr. Legge, I hereby not only Set my own Name but also testify his desire to have Subscribed had he been present.

Martin Folkes [President of the Society]

It may be noted that James Short, mentioned earlier, was the first proposer. After the letter of proposal had been duly displayed at the requisite number of meetings of the Royal Society, Alexander Macfarlane was elected a Fellow of the Society on 19 February 1746 (old style). He had already communicated astronomical observations made in Jamaica in 1743–4 to the Royal Society, and he continued to transmit reports of his observations until his death in October 1755. He left much of his considerable estate to his two surviving brothers, but bequeathed his astronomical apparatus to the University of Glasgow. The motto of the Macfarlane family was '*Astra castra, Numen lumen*', or 'The stars my camp, the Lord my light'.

It is of interest to note that Macfarlane's observations of a lunar eclipse and a Transit of Mercury, dispatched to Short, provided a value for the longitude of Kingston. This had great relevance to the 1761–2 voyage to Jamaica, arranged by the Board of Longitude, as a trial of John Harrison's Marine Chronometer, H-4 (see Sobel[11] for a beautifully written account of the whole story). None of the vociferous participants in the subsequent 'Longitude Affair' cast doubts on the validity of Macfarlane's observations, though the supporters of the Board of Longitude pointed out that longitudes based on observations of lunar eclipses and Transits of Mercury could not be expected *per se* to be highly accurate. Harrison's supporters attempted to dispute this, although they somewhat grudgingly accepted that the longitude of Jamaica was not sufficiently known.

While neither Campbell's nor Macfarlane's Jamaican observatory can be said to have added much to the general level of astronomical knowledge and understanding, Macfarlane, in his bequest, ensured that the instruments would have a useful life, continuing into the nineteenth century. The bequest was gratefully accepted, and the University acted with commendable speed, with the collection being shipped on the *Caesar* to Glasgow at a cost of £8. The ship's master, Captain Wylie, was provided with a gratuity of £2 for taking care of the consignment.

The original manuscript inventory remains in the safekeeping of the Glasgow University Archives.[12] It was written in French by Pierre Martel, already mentioned as having cared for Macfarlane's instruments in Jamaica. Fourteen cases, marked A to O, were dispatched, as shown on the first page of the shipment manifest in Fig. 3.4. A sample listing for one of the boxes is also displayed in the same figure. A translated short summary of the contents of the shipment is contained in an appendix to Gavine's thesis.[13] Case A is noted as carrying a 5-foot quadrant which may have been the mural arch mentioned by Smith.[14] As well as telescopes, other instruments in the bequest included a miscellaneous collection of lenses and micrometers, variation compasses, an astrolabe, an electrostatic machine and a camera obscura.

According to Bryden,[15] one of the James Short telescopes,[16] currently in the possession of Glasgow University, is the only instrument remaining from the Macfarlane collection, identified as 'un telescope reflechissant avec toutes les pieces', in the contents list of case L. In his comparison of the quality and quantity of instruments in Jamaica with respect to the Royal Greenwich Observatory, Bryden[17] also lists this telescope as being at Campbell's original Black River Observatory. There is argument that these statements of association are incorrect. Firstly, the contents list for case L also includes a reference to an iron bracket for fixing the telescope to a wall; the Short telescope in the University's collection is free standing on foldable legs, and it is unlikely that it would be used by attaching it to a wall. Secondly, the instrument's date is 1743, and thus it could not be part of the original inventory of Campbell's expedition. There is no evidence of it being purchased later and separately

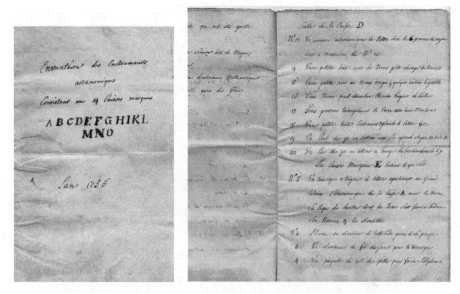

Figure 3.4 The left-hand page provides the list of crates containing the Macfarlane bequest of astronomical instruments. The middle part of the right-hand image shows the last items for Case D and the beginning of the list for Case E, the first item being a brass telescope with lenses.

by Macfarlane after its date of manufacture. Thirdly, although not a concrete argument, James Watt had connections with James Short when he was in London seeking to learn the trade of an instrument maker; on the inspection and cleaning of the consignment, it might be expected he would have made special reference to any Short instrument, but there is no mention of it in his invoice itemising his restoration work.

Some of the Macfarlane instruments had corroded during the sea voyage, and the young James Watt was given the task of cleaning them and putting them in order. On 25 October 1756, James Watt wrote a letter[18] from Glasgow to his father in Greenock. In it he said: 'I would have come down [to Greenock] to-day, but there are some instruments that are come from Jamaica that Dr. Dick desired that I would help to unpack, which are expected to-day'.

A minute of the University Senate of the following day (26 October) reads:

> Several of the instruments from Jamaica having suffered by the sea-air, especially those made of iron, Mr. Watt, who is skilled in what relates to the cleaning and preserving of them, being accidentally in town, Mr. Moor and Dr. Dick are appointed to desire him to stay some time in the town to clean them, and put them in the best order for preserving them from being spoiled.

University records for 2 December 1756 show that 'a precept was signed to pay James Watt five pounds sterling for cleaning and refitting the instruments lately come from Jamaica'. A copy of Watt's invoice duly receipted is shown in Fig. 3.5. This fee is very likely the first monies earned by him on his own account since the termination of his brief apprenticeship in London. As noted by Mackie,[19] the task, which earned him £5 sterling, was soon at an end, but it proved to be the beginning of a happy alliance with the University.

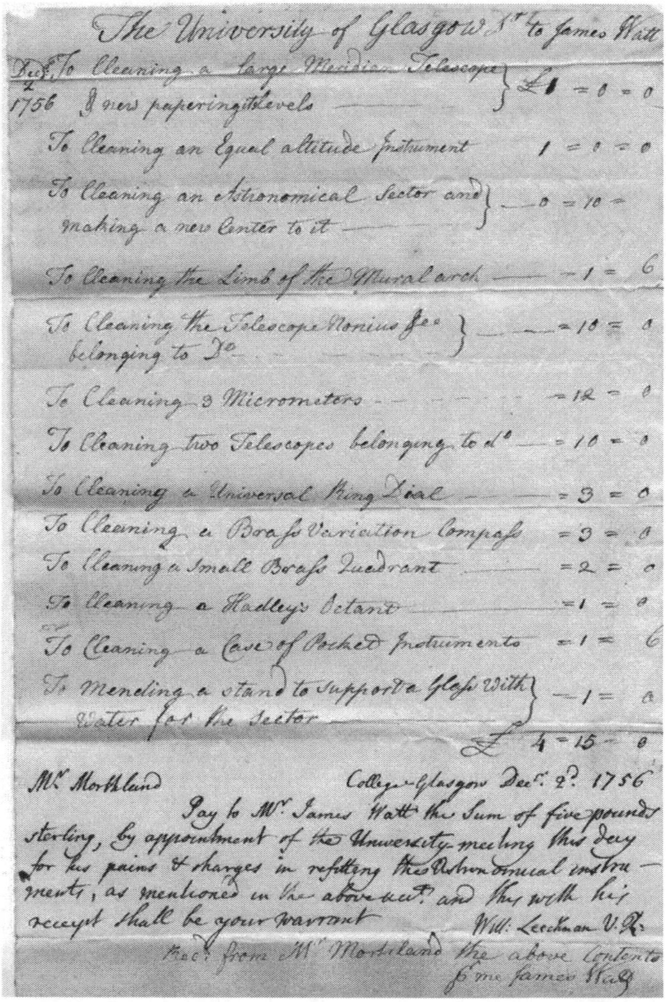
The University of Glasgow Dr to James Watt
Dec 2 1756 To Cleaning a large Meridian Telescope & new papering the levels — £1 = 0 = 0
To Cleaning an Equal altitude Instrument — 1 = 0 = 0
To Cleaning an Astronomical Sector and making a new Center to it — 0 = 10 =
To Cleaning the Limb of the Mural arch — 1 = 6
To Cleaning the Telescope Nonius &c belonging to D° — 10 = 0
To Cleaning 3 Micrometers — 12 = 0
To Cleaning two Telescopes belonging to d° — 10 = 0
To Cleaning a Universal Ring Dial — 3 = 0
To Cleaning a Brass Variation Compass — 3 = 0
To Cleaning a Small Brass Quadrant — 2 = 0
To Cleaning a Hadley's Octant — 1 = 0
To Cleaning a Case of Pocket Instruments — 1 = 6
To Mending a stand to support a Glass with Water for the Sector — 1 = 0
£ 4 = 15 = 0

Mr Morthland College of Glasgow Decr 2d 1756
Pay to Mr James Watt the Sum of five pounds sterling, by appointment of the University meeting this day for his pains & charges in refitting the Astronomical instruments, as mentioned in the above acct. and this with his receipt shall be your warrant
Will: Leechman V.R.
Recd from Mr Morthland the above Contents
p me James Watt

Figure 3.5 The first ever invoice (£4 15s) prepared by James Watt, for the work related to the cleaning of the Macfarlane instruments recently received from Jamaica, with the mandate for payment of £5 attached and receipt added.

3.4 JAMES WATT

Although Watt's first commission had only a minor connection to astronomy in Glasgow, the story of how he developed with the succour of the University deserves a little expansion. Details of his life and achievements can be found in his biography by James Patrick Muirhead.[20] A concise summary relating his marriages and movements within the city has been written by George MacGregor.[21] As a youth, he was interested in astronomy,[22] particularly the instruments associated with it, and in navigation. He subsequently had some minor associations with the subject through his design of eyepiece micrometers for surveying telescopes,[23] with one of his devices specially made for use in the Macfarlane Observatory. He retained connections with Glasgow throughout his life. A particular friendship with the second Professor of Astronomy, Patrick Wilson, is well documented, and references to this are made in the following chapter.

According to Mackie,[24] Watt was a kinsman of Professor George Muirhead, and he came from good stock. His grandfather had taught mathematics in Greenock, and his uncle had coupled the same occupation with working as a surveyor in Glasgow. His father was a merchant and played a considerable part in local politics. At the time of his youth, however, things were not going well with his family. His uncle died in 1737, his elder brother was drowned at sea and his father's business was in an uncertain state. As a youth of eighteen, he went to Glasgow in 1754 to spend some time with his maternal relatives, and he remained there until May 1755. He was glad to use the good offices of his mother's relative, Professor Muirhead, with a view to obtaining employment as a mathematical instrument maker.

Professor Dick was kind to him, and, although Watt had not served a regular apprenticeship and did not intend to do so, Dick managed to 'place' him (not without cost to his father) with a firm in London, where he learned his business for a year. According to Turner,[25] Watt arrived in London with a letter of introduction to the Scottish telescope maker, James Short. He was not employed by Short directly on his arrival from Glasgow; Short probably did not make the mounts for his telescopes and would have had no need to employ a craftsman for that purpose. With Short's help, however, Watt eventually found a teacher in John Morgan, an optician in Finch Lane, Cornhill. In a letter to his father dated 2 September 1755, Watt wrote:

> If it had not been for Mr Short, I could not have got a man in London that would have undertaken to teach me, as I now find there are not above five or six that could have taught me all I wanted.

His stay in the capital was very much centred on learning and on developing skills for making mathematical and nautical instruments. He related his fear of moving around the streets in case of being scooped up by the press gangs for service in the navy. He left London in August 1756 and two months later was in Glasgow and available for the cleaning of the Macfarlane consignment.

Watt desired to set up business in Glasgow, but, though it was abundantly clear that he had 'real' qualifications, he lacked the conventional credentials. Neither being the son of a burgess nor having, as yet, married the daughter of one, nor having served a regular apprenticeship to a craft to become a Hammerman, Watt could not, without great expense, pursue his vocation in the city of Glasgow, which remained under medieval trading restrictions until 1846. He was prohibited from opening a shop for advertising his craft within the burgh. It was the friendly help of the University, with its privileges, that set him free from the control of the city. By midsummer 1757, he had received permission to occupy an apartment and to open a shop within the precincts of the College, in the north-west corner of the Inner Quadrangle, and to use the designation of 'Mathematical-instruments-maker to the University'. Although it does not appear that any contemporaneous record has been preserved in the University Archives of a specific date when the workshop was assigned to him, that of 27 November 1759 shows that directions were given for having 'the room above Mr Watt's workshop' repaired. An advertisement dated 22 October 1759 for an engraved map of the river Clyde carries the phrase 'as sold by James Watt, at his shop in the College of Glasgow', showing that he was established within the University precincts.

The determination of Watt to succeed in business was very apparent, as can be seen in his letters to his father. In one dated 15 September 1758 (see Muirhead[26]) he discussed the prospect of the profits that could be made from the production of Hadley quadrants with the possibility of referring his sales to Liverpool or London. He continued to operate from within the College for some seven years in partnership with a John Craig. He resided there until 1763, when he left to get married, and he maintained his workshop in the precinct until 1773.

The younger Professor Dick died on 22 May 1757. By this time Watt had commended himself not only to Robert Simson, Professor of Mathematics and Clerk to the Senate, but also to Joseph Black, newly appointed Lecturer in chemistry, and to John Robison, a student who later was to succeed Black. Coutts[27] has noted that not long after becoming Professor of Natural Philosophy (see below), John Anderson employed Watt to carry out repairs on his instruments, for which work a sum of £5 5s was ordered to be paid on 26 June 1760. The day before, Anderson had been authorised to spend £2 to recover the model of the steam engine from a Mr Sisson, a highly skilled maker of astronomical instruments in London. Somehow the model engine did not satisfactorily fulfil its purpose.

Watt's attention had been directed to steam engines in 1759 by John Robison, who suggested the idea of using them to propel wheeled carriages. Soon after, Robison left the University, and the pursuit of the idea was abandoned. Watt conducted further experiments with steam in 1761 and 1762, and, in the following year, Anderson asked him to repair the model of a Newcomen engine, paying him £5 15s in June 1766. He worked and experimented on this engine for some time, receiving occasional advice from Dr Black. As a result, in 1765 he discovered the great defect in its principle of operation, and how it might be remedied. Condensation had been effected in the cylinder by injecting cold water, which cooled not only the steam but the whole metallic chamber, involving a great waste of power and enormous consumption of fuel. To remedy this, he contrived a separate vessel for condensing the steam, with a valve to the cylinder that opened each time the steam was condensed in the cycle, and thus the cylinder was always kept in a suitable condition for working hot and dry. For years after, Watt continued to experiment and contrive further improvements, but the separate condenser was his most vital contribution. Ten years later, Watt's engine was first used for pumping, and in 1780, steam was first employed to drive a loom. It is somewhat ironic that the laboratory where Watt conducted his experiments and business was demolished when the University vacated its site in the High Street, with the real estate purchased by the City of Glasgow Union Railway Company for a goods station serviced by steam engines. At the time of the closing of the historic buildings, the sentiment was not missed by the *Glasgow Herald*,[28] which commented:

> Deep-brooding Watt, sitting in his academic shop, studying great physical powers, evoked from his brain the very spirit . . . which is about to lay the walls of his student's cell in ruins. It is to the railway that the University is about to yield up its ancient dwelling-place, and, in a few months, there will sweep over the spot where the great philosopher sat the very spirit which he was then chaining to the car of civilisation.

His attention was taken up not only by the development of steam power. He also invented a machine for drawing in perspective. In 1762, soon after his arrival at the University, he made a small organ[29] for Black; this instrument remains in working order at the People's Palace in Glasgow. Though he had no musical gifts himself, in his College workshop he went on to make flutes, guitars, harps, violins and many of the instruments that go to form an orchestra. In 1769, he prepared for the Town Council a survey of the history of the Clyde, with a view to opening the river for navigation, and in the same year he submitted estimates for a canal 9 miles long to bring coal cheaply from Monklands to Glasgow. The University contributed £200 to the execution of this design. Watt himself supervised the construction work between 1770 and 1772.

As for his foray into making telescope micrometers used in his surveying work, the details of his designs are given in Muirhead.[30] In addition to simple ones with cross-hairs, in the late 1770s he invented one containing a very small angled prism. Drawings related to the descriptions of his micrometers are provided in Fig. 3.6. He described the system involving a thin prism in the focal plane of the telescope as follows:

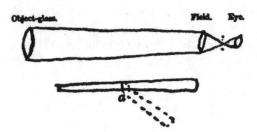

Figure 3.6 The upper section is Watt's drawing of a telescope and micrometer with cross-hairs. Below is his invention using a thin prism which has been cut, allowing one part to be set at an angle relative to the other.

It consisted of a thin prism almost parallel, (say of one degree or two). This prism was cut by a diamond into two parts, which, when they were fixed in the same plane, refracted all the rays which passed through them equally: but one of them remained fixed, and the other moving on a centre at *a*, (according to the pricked line), would refract that portion of the rays which passed through it more than those which passed through the fixed part; and being fixed in the focus of the object-glass of a telescope, two images were formed of each object, by which its diameter could be measured. An index and divided sector of a circle served to measure the comparative refractions. This instrument I made with the sector and radius of wood, and gave it to Professor Anderson, of Glasgow College, and I suppose it is still among his apparatus, which he left to a public institution.

Watt also noted that a few years after designing this kind of optical system, 'the Abbé Rochon afterwards (1783) published a description of some micrometers with prisms, but I think they were upon somewhat different principles in their construction'. Searches for Watt's device within the archived collections of Strathclyde University, once known as Anderson's Institution or College and becoming Anderson's University in 1828, would suggest that the micrometer is no longer extant.

A memorial statue of James Watt can be seen in the south-west corner of George Square, Glasgow (see Plate 3). It is a bronze replica of the one presented by his son to the University and located in the Hunterian Museum at Gilmorehill. The dedication reads:

This Statue of
James Watt,
Fellow of the Royal Societies
Of London and Edinburgh,
And Member of the Institute of France,
Is presented by his Son,
To the University of Glasgow,
In Gratitude for the Encouragement
Afforded by the Professors
To the Scientific Pursuits
Of his Father's Early Life.

3.5 THE REGIUS CHAIR OF PRACTICAL ASTRONOMY

The death of Robert Dick (Jnr) in May 1757 created various financial and political difficulties in finding a successor for the Chair of Natural Philosophy. A fuller account of these problems and the machinations related to the new appointment can be found in Emerson.[31] Robert Dick (Snr) was still alive, and technically was still the holder of the Chair that had been granted to him and his son, with the right of survivancy to the one who lived longer. This put constraints on the monies that could be made available to attract a suitable candidate. His Grace Archibald, the Duke of Argyll, being a keen amateur astronomer himself, desired a person who could profitably use to advantage the equipment recently received by the Macfarlane bequest. One person put forward was a John Bevis, but he was not favoured by the Faculty. He also seems to have refused to take the required religious oath, and dropped out of consideration.

By the end of June, the Duke was considering the candidacy of Alexander Wilson, very well known to him for his instrumentation skills related to the manufacture of thermometers, barometers and specific gravity beads. He had continued his patronage of Wilson since they had first met in London (see Chapter 4), and especially as he had brought the art of letter-founding to Scotland. It is thought that Wilson was thinking of returning to London about this time with his type-founding business.

William Rouet, who at the time was Professor of Ecclesiastical History, reported on 30 June 1757 that he had been in contact with the Duke concerning the possible appointment of a James Buchanan to the Chair of Natural Philosophy. He noted that:

> As to our vacancy, I find by all accounts that Mr Buchanan is the most universally agreeable to my colleagues . . . that though he might be a good Mathematician, yet perhaps others understand better the use of Astronomical instruments. I told his Grace that Mr Wilson was a very good coadjutor [assistant] . . . and if we could afford it, deserved highly of the University to have some small allowance for his assistance in the matter.

The Duke appeared

> well disposed to befriend Mr Wilson and family . . . he offered to provide his Son in a Commission in ye Army, & twice putt it to me what way he could serve ye University in giving Mr Wilson some inducement to stay at Glasgow & give his assistance in ye using the Astronomical Instruments.

The Duke wrote to Robert Simson, Clerk of Senate, telling him that he should 'acquaint Mr Wilson that Mr Ruat [Rouet] & I have had some conversation about him & I am forming a Scheme to mend his situation'. These ploys were intended to discourage Wilson from leaving Glasgow. At this stage, though, it was considered that Wilson might be appointed only as an assistant or an astronomer; no one seems to have thought of him becoming the Professor of Natural Philosophy.

Emerson[32] relates the full protracted saga of the candidacy considerations for the Chair of Natural Philosophy in respect of James Buchanan, a James Williamson and eventually, by the summer of 1757, John Anderson, mentioned earlier in this chapter. The machinations dragged on into the autumn, with any decision postponed while the Duke continued a head-hunting exercise looking for a suitable person and sought a way to provide for Wilson. The Faculty came under pressure to settle the issue for the teaching year, so as not to lose the fee-paying students. A lobby was formed to encourage people to vote for Anderson, but this engendered the complication of replacing his current post of Professor of Oriental Languages. During this immediate period of indecision, another candidate emerged in the person of the Revd William Wilkie, who was seen as a suitable candidate either for the Chair of Natural Philosophy or for that of Oriental Languages. By this stage, most of the Faculty assumed that the Chair of Oriental Languages would go to Buchanan, and that Anderson would be appointed Professor of Natural Philosophy.

The debate continued, with heated meetings, involving procedural matters such as Anderson's right to vote for himself, and postponements; the meeting scheduled for 30 September was cancelled as there was no quorum. The delays allowed the Duke more time to try to find funds for a stipend to pay Wilson. The majority of professors signed a letter to the Duke in which they remonstrated over his delays and emphasised that the College would have to choose a professor soon as the course needed to be taught. The wrangling continued, elegantly summarised by Emerson.[33] Finally the issue was resolved on 21 October, with a slim victory for the Anderson–Buchanan camp over Wilkie, upheld by the casting vote of the Rector and that of Anderson himself. For the record, Anderson became known as 'Jolly Jack Phosphorus' and was very much involved with the politics of the College and renowned for being fiery and argumentive.[34] Further details of his life and scientific exploits are provided by Coutts;[35] there is also a concise biography, written by MacGregor.[36] Anderson died on 13 January 1796, bequeathing most of his estate to establish a rival institution in the city to that of the older University. Anderson's University has since evolved to become Strathclyde University. With the Chair of Natural Philosophy settled, the finagling for an appointment for Wilson continued. During this summer of discontent, the foundations of the Observatory had been laid, and the Duke of Argyll had spoken to the Duke of Newcastle about a fund to keep Wilson at the University. Nothing immediately came of this, but the idea was not dropped. In 1758, a means of doing this arose in that the Earl of Hopetoun made a proposal to Professor Rouet to go abroad as a travelling tutor to the Earl's two sons. John Anderson, who had been talking to the Duke of Argyll in London, wrote from there, explaining the proposal in the following way:

[Hopetoun's] scheme is this, If the College can be prevailed upon to give Leave of Absence to Mr Ruat, the Duke of New[castl]e has agreed to accept of £400 & found a new Professorship in our University the Salary of which is to be £50 per An, that is to say the King will grant £30 a Year out of the Exchequer which joined to Lord H—n's £400 will make a Salary of £50 for a new Professor who is named by the Crown . . . & Mr Wilson is to be the Man so that Wilson will have a Seat among us.

It is to the credit of the majority that they declined to accept the offer in full. Again there were bitter arguments, as described by Emerson.[37] By 1759, the deal outlined above was accepted in principle, although most professors did not vote for giving Rouet his extended leave. Rouet took leave without permission, and was then deposed from his position in the College. The issue of the sacking then went to the courts, with litigation continuing until 1767.

The Duke of Argyll, having recently succeeded to this title, used his influence with government and procured from His Majesty George II Wilson's nomination and appointment as Regius Professor of Practical Astronomy and Observer in the College, with an annual salary of £50 payable out of the Exchequer. In this way, the Chair established a means of providing a custodian of the College's astronomical equipment and Observatory, and, at the same time, allowing Alexander Wilson to remain in Glasgow. It was the only new professorship within the University during the thirty-four-year period from 1727 to 1761, as commented on by Coutts.[38] The teaching of the aspects of practical astronomy were still considered to be in the charge of the Professor of Natural Philosophy. Astronomy, then, definitely took its place in the Faculty, and thanks, no doubt, to the good equipment of the Observatory, the subject was taught, at least creditably, if not with outstanding distinction, in an age when astronomy was little studied elsewhere. This prestigious position had its title changed to the simpler one of 'Regius Chair of Astronomy' in 1893 by Ordinance 31 of the Universities Commission (1889), the office remaining to this day in the University.

3.6 THE MACFARLANE OBSERVATORY

Determined that the instruments from Jamaica should be used to the best advantage, both Coutts[39] and Mackie[40] note that in May 1757 the College resolved to build an observatory at a cost of £400 on the Dow Hill (Dovehill), beyond the south-eastern corner of the existing College's grounds across the Molendinar Burn. To preserve the site's amenity, and to keep the view from being obstructed, two neighbouring plots of ground at Dow Hill and the Butts were purchased for about £500. The Town Council, having been asked to aid an institution so useful in promoting navigation and commerce, made the gift of a third plot. The land thus obtained carried the University

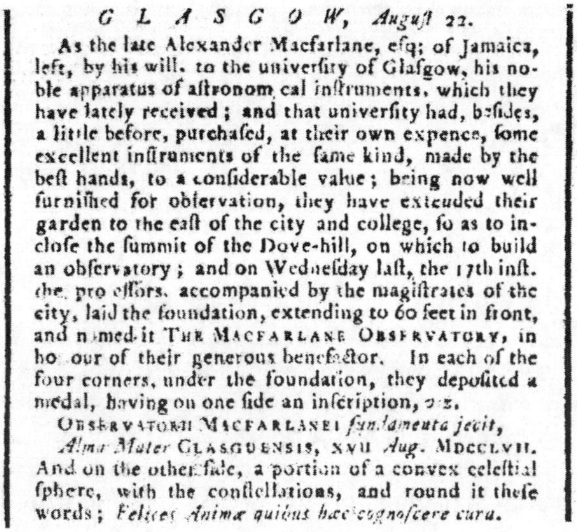

> *G L A S G O W, August 22.*
>
> As the late Alexander Macfarlane, efq; of Jamaica, left, by his will. to the univerfity of Glafgow, his noble apparatus of aftronomical inftruments, which they have lately received; and that univerfity had, befides, a little before, purchafed, at their own expence, fome excellent inftruments of the fame kind, made by the beft hands, to a confiderable value; being now well furnifhed for obfervation, they have extended their garden to the eaft of the city and college, fo as to inclofe the fummit of the Dove-hill, on which to build an obfervatory; and on Wednefday laft, the 17th inft. the profeffors. accompanied by the magiftrates of the city, laid the foundation, extending to 60 feet in front, and named it The Macfarlane Obfervatory, in honour of their generous benefactor. In each of the four corners, under the foundation, they depofited a medal, having on one fide an infcription, viz.
>
> Obfervatorii Macfarlanei fundamenta jecit, Alma Mater Glasguensis, xvii Aug. Mdcclvii.
> And on the other fide, a portion of a convex celeftial fphere, with the conftellations, and round it thefe words; Felices Animæ quibus hæc cognofcere cura.

Figure 3.7   The *Glasgow Courant* for the week beginning Monday, 15 August 1757 reports on the opening ceremony for the Macfarlane Observatory.

precinct south to the Gallowgate, and so appreciably increased the value of the whole property.

The foundation stone was laid on 17 August 1757 by the professors, accompanied by the Magistrates, and the building was named the Macfarlane Observatory. A copy of the report of the ceremony as recorded in the *Glasgow Courant* of that week is given in Fig. 3.7. In each of the four corners, under the foundation, a medal was deposited that had the following inscription on one side:

OBSERVATORII MACFARLANEI FUNDAMENTA JECIT
ALMA MATER GLASGUENSIS XVII. AUG. MDCCLVII

On the other side there was a portion of a convex celestial sphere, with the constellations, around which were the words:

FELICES ANIM QUIBUS HAEC COGNOSCERE CURA

None of these items appears to have been rescued[41] when the building was demolished in 1856, the equipment having been moved out some ten years earlier.

After the building had been completed, the order was given that the instruments should be installed in it during June 1760. Coutts[42] has noted that the building turned out to be rather damp, being situated close to the Molendinar Burn, and efforts were made to overcome this by means including building ditches around the basement and painting the walls with linseed oil. Ironically, the Acre Road Observatory, built nearly 200 years later in the 1960s, had a basement that was permanently flooded until it was lined with waterproof cement. Unfortunately no plans or drawings of the old building can be found. Murray[43] notes that the Observatory had a frontage measuring 60 feet long. According to King,[44] the Observatory was 'a quaint and picturesque building of moderate size'. Reference to it is also made in Wade,[45] who in 1821 described it as follows:

> This simple little structure, which for an observatory, appears both to stand and to be rather lower than could be wished, consists of a quadrangular centre, with a projection, also quadrangular, but lower than the centre and surrounded by a balustrade, on the east and west. The gentle rise on which this edifice stands, and various other spots in these agreeable pleasure-grounds, afford interesting views of the College Edifice, the Cathedral Spire, College Church, and the enriched tower belonging to the new and neighbouring Church of Saint John.

At the time of its completion, it appeared to be isolated relative to the city, as indicated in Fig. 3.8 which shows W. Douglas's 1777 plan of the College grounds. Its strong rectangular footprint is clearly shown on a map contained in Gibson's book,[46] also published in 1777, and it is also seen in the left-hand image of Fig. 3.9.[47] Around this time, the city was expanding quickly to the west of the High Street, as can be seen from the two maps in this latter figure. The building may not have been strictly rectangular; instead, it might have had turreted structures on its north face corners, as the overlay map in Fig. 3.13 suggests.

The accounts of John Bryce, once the Bedellus but later Master of Works, provide details of the expenditure necessary for the preparation of the observatory building in 1757. Sample entries include:

> 1. 1757 1 July – Pd John Muirhead for 100 dozen of stobs and 3½ dozen of rafters £4 7s 3d
> 2. 2 July – to four men six days each carrying the stone dyke away from the Observatory 14s 0d. To Do. Men drink money to get done this night 6d
> 3. 6 July 1757 – to sundry masons and wrights & other men for carrying a stone & a piece of big timber for the meridian line taking etc. 1s 3d
> 4. 9 July Pd. John Chalmers for part of the day attending on Mr Moor at meridian line taking in at Observatory. 6d

Obtaining the true north–south direction of the meridian line was very important. Much of astronomy in this age was related to timekeeping, using instruments such as transit telescopes. Again from the accounts of John Bryce:

> 5. 17 Aug 1757 – To Mr Moor – drink money to the masons at laying the foundations of the Observatory – 10 6d.

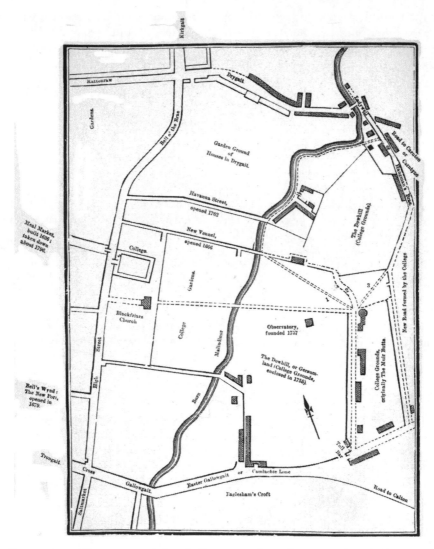

Figure 3.8 The extent of the College grounds is shown in the 1777 survey by
W. Douglas; the plan clearly indicates the position of the Observatory built some twenty
years previously.

From the accounts it appears that it was common to provide the labourers
with 'drink money'. A check of the modern University finance system fails
to reveal a code for such expenditures, suggesting that this kind of payment
no longer operates. If these monies were spent immediately, then it is likely
that the local Old College Bar (see Fig. 3.10) would have been the eventual
beneficiary.

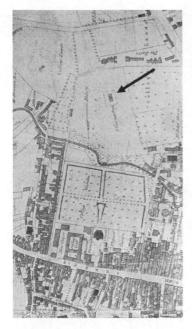

Figure 3.9 Plans of Glasgow c. 1770 show the location of the College in the High Street with the Observatory located eastwards across the Molendinar Burn. The picture on the left is part of the 1778 Plan of the City of Glasgow, Gorbals and Calton by John McArthur; the picture above is from a directory of 1791.

Figure 3.10 Opposite the old entrance to the College, a public house dating from c. 1531 still stands, bearing testament to the fact that seats of universal learning last longer than ivory towers. This might well have been the location where the masons' drink money was spent.

The University was thus supplied with admirable apparatus at very little expense to itself, and instruments previously used in the teaching of natural philosophy were added to those of the Macfarlane bequest. Overall, the Observatory boasted[48] a 10-inch Newtonian reflector, one of Herschel's design and 20 feet long, a 3-inch refractor by Dollond, various altitude and azimuth circles and a mural quadrant by Bird. The Transit Room in the West Wing contained two transit instruments and two clocks.

Once the Observatory was established, running costs to the University were recorded. The entry in the accounts of the Master of Works for 19 September 1760 shows:

1. To Mr Wilson – what he gave for towels for dusting the instruments 6. 1d.
2. Pd John Fisher for carrying lead from Mr Wilson's house to the Observatory 1. 0d.
3. A quire of paper for the Observatory 1. 0d.
4. 25 pens for Do. 8d.
5. ½ dozen screw nails for Do. 1-½d
6. 3 Batts for door of Do. 1. 6d.
7. ½ dozen more screw nails for Do. 1-½d.

Towards the end of the eighteenth century, the city developments presented an inspiring sight. Tobias Smollet,[49] in *The Expedition of Humphry Clinker*, described Glasgow as one of the prettiest cities in Europe. He also made reference to the Observatory as being 'well provided with astronomical instruments'. This special mention may have been made as Smollet had been a medical student at the College and was a friend of Macfarlane's brother; his description of the city may also have been swayed by a nostalgic bias. Murray[50] noted a visit to the Observatory by the Revd John Lettice, who was accompanied by Professor Patrick Wilson in 1792, and made particular reference to the reflecting telescope which had recently been provided by Sir William Herschel.

Upkeep of the grounds and land surrounding the Observatory was a continual problem. In August 1784, following recent flooding of the area, clean-up costs amounted to £4. Boney[51] has described a catalogue of costs for looking after the Observatory park in this period. Charges for March 1774 were made for 'clipping, digging the Hedge around the Observatory House'. For August 1784, the work carried out included 'cleaning the gravel about the house two times', 'clipping and digging the Hedge on the South side of the Observatory Ground' and '4 Carts dung for said Hedge and Cartage'.

In early December 1785, a Faculty committee of five met 'to consider the best methods for improving and disposing of the College Garden, the park around the Observatory, and to lay their views before the Faculty as soon as possible'. The resulting proposal involved large-scale ploughing, harrowing and dunging to enrich the ground for laying down with barley and grass. It was suggested that the scheme required fifty carts of dung and six carts of lime shells, with a cost of £6 for manual work of various types. Crop rotation was clearly intended to improve the quality of the soil, with a possibility of selling the grain for whisky-making. It is not clear how much of the plan was executed. On the income side, sale of grass cuttings from the Observatory banks regularly raised small sums (of the order of about 5s 6d to 11s 6d) and all of these are recorded in the University accounts. The grass sales were reviewed in May 1786 when two annual accounts were presented:

1784	Grass receipt	£10/15/2
	Balanced by accompt of work done	
1785	Grass receipt	£9/16/0
	Work done	9/15/10
	(only 2d due to the College)	

From these statements it can be seen that the sale of the 'produce' provides little or no return on the costs. Further details of the agricultural aspects of the Observatory park up to the time when the building was demolished and the area given over to student recreational activities can be found in Boney.[52] In around 1840, the area was described[53] in the following manner:

A fine park, interspersed with trees, stretches away behind the college towards the Gallowgate, and is admirably fitted for the recreation of the students. In summer it forms a most delightful promenade. An observatory is situated in the park, but as this has been found insufficient for astronomical purposes, a fine new erection has been built on an eminence in the western suburbs of the city, and will be stocked with the college instruments.

As related in Chapter 5, from the turn of the nineteenth century, astronomy teaching essentially disappeared; there was little or no activity at the Observatory and it was eventually demolished. Although Smith's maps[54] of the 1820s feature the Observatory, its operation was at a standstill by then. Plate 2 shows the extent of the College grounds in 1828. Mcfarlane Street, which was laid and opened[55] by the College in 1815, can be seen running from the Gallowgate towards the Observatory, but with St John's Church, built in 1819, in between. This building was directly on the north–south line and curtailed the possibility of undertaking transit observations in the southerly direction. Following the demolition of the Observatory in 1856, the land between its old site and the Hunterian Museum was used for a Highland Games in 1867, as can be seen in Fig. 3.11, but the area was in very poor condition. This can be clearly seen in Fig. 3.12, which presents the scene of the Old College around 1870, at the time when the site was abandoned. The path in this photograph along the northern boundary of the grounds, with the old Hunterian Museum to the left, was one of the walkways from the College to the Observatory. A telling description of the site was given by Ferguson[56] when he was a student, after the demise of the Observatory but prior to the College's move from the High Street to Gilmorehill. Professor

Figure 3.11 The activities of the Glasgow Highland Games of
1867 viewed from a location in front of where the Macfarlane
Observatory once stood. The old Hunterian Museum is to the right
of the tower, and Blackfriars Church just left of centre.

Ferguson's original article was re-read at a Glasgow University Club function
in 1930 and a relevant part of it is presented below.

The two greens were separated by the Molendinar Burn, which within the green
was covered over, except at a part where it was crossed by a bridge, and then
access to the stream was possible. Whether it could ever have been a haunt for
salmon, as has been said, maybe [sic] doubtful, unless there had been far more
water in it than I ever remember seeing. But in my time it had been put to base
uses and was no more than a drain of foul water, and if it contained any life at all,
it consisted of red fleshy lumps of red-like sewage worms, and an occasional rat. I
have seen them both.

The High Green was several feet higher than the other side, and it stretched
back to Hunter Street. On the south side was St John's Parish Church, on the
east side the old barracks and on the north side the Vennel. It is all gone now, the
greens are levelled up and locomotives run over the classic ground.

When I first saw it, the Observatory which had stood there before it was moved
to its present site, had been demolished, and heaps of stones and remains of walls
and foundations marked the place where it had stood. All the rest was bare and
deserted except for a gaunt, leafless tree. In this green there was no such thing as
athletics in those days; a football lay about which was occasionally kicked, and
on the High Green there was a hammer, an iron ball with a handle driven into a
hole through it, which one was tempted at times to compete in throwing over the

Figure 3.12 The pathway along the northern boundary of the Old College grounds provided one of the routes to the Observatory from the College, seen with its tower in the distance. On the left, the old abandoned Hunterian Museum stands within a desolate scene.

branches of the withered tree. But while one was making a cast a bell began to ring, and the place was far off to make a rush for the classroom necessary.

There is now nothing to show of Dow Hill and its Observatory. This site, where Professor Alexander Wilson made his momentous observations of sunspots (see Chapter 4), is now under a junction of railway lines as revealed in the map of Fig. 3.13, with its present condition shown in Fig. 3.14. Tangible references remain, however, in the form of local street signs, as shown in Fig. 3.15. Mcfarlane Street remains in the same location as shown in Plate 2 and Fig. 3.13, running from the Gallowgate directly towards where the Observatory once stood.

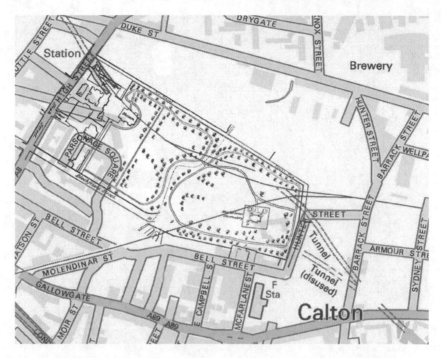

Figure 3.13 The map of the location of the Old College and its Observatory, superimposed on a more modern street map of the city. It can be seen that the fork of current railway lines, one from the north-west and another from the south-west, is very close to the location of Glasgow's first observatory. Mcfarlane Street runs northwards from the Gallowgate towards the site of the Observatory.

Figure 3.14 The location of the old Macfarlane Observatory as seen in 2007 through a gap between trees, where the tracks of two railway lines come together.

Figure 3.15 The current metal banners for Dow Hill (Dovehill) and McFarlane Street bear testimony to the locality around the Old College site. The latter street runs from Gallowgate directly towards the location of Glasgow's first observatory.

References

1. Coutts, J. (1909), *A History of the University of Glasgow (From its Foundation in 1451 to 1909)*, Glasgow: James MacLehose & Sons, p. 195 [GUL Education S271 1909-C – two copies; GUL Sp Coll Mu21-a.2; Sp Coll Bh12-e.14; GUL Sp Coll MacLehose 824].
2. Mackie, J. D. (1954), *The University of Glasgow, 1451–1951: A Short History*, Glasgow: Jackson, p. 217 [GUL Education S271 1954-M – six copies].
3. Gavine, David M. (1981), 'Astronomy in Scotland 1745–1900', unpublished PhD thesis, Open University, 55.
4. Coutts, *A History of the University of Glasgow*, p. 385.
5. Mackie, *The University of Glasgow*, p. 222; Murray, David (1927), *Memories of the Old College of Glasgow – Some Chapters in the History of the University*, Glasgow: Jackson, Wylie and Co., p. 50.
6. Gavine, 'Astronomy in Scotland 1745–1900', 55.
7. Boney, A. D. (1988), *The Lost Gardens of Glasgow University*, Bromley: Christopher Helm, p. 92 [GUL Botany A75.C7 – two copies].

8. Bryden, D. J. (1970), 'The Jamaican observatories of Colin Campbell, F.R.S. and Alexander Macfarlane, F.R.S.', *Notes and Records of the Royal Society*, 24, 261–72.
9. Smith, Robert (1738), *A Compleat System of Opticks, Vol. II*, Cambridge: Cambridge University Press, p. 341.
10. Glasgow University Manuscript 117/27408. Typewritten notes by Dr P. Swinbank, 15 September 1970 [Glasgow University Archives].
11. Sobel, Dava (1996), *Longitude*, London: Fourth Estate Ltd.
12. Glasgow University Manuscript 117/27403. Enventoire des Instruments Astronomiques Consistent en 14 Caisses ABCDEFGHIKLMNO dated 1756, compiled by Pierre Martel [Glasgow University Archives].
13. Gavine, 'Astronomy in Scotland 1745–1900', 91.
14. Smith, *A Compleat System of Opticks*.
15. Bryden, D. J. (1968), *James Short and His Telescopes*. Handbook and Catalogue of the Commemorative Bicentenary Exhibition, Royal Scottish Museum, Edinburgh.
16. University of Glasgow/Hunterian Museum Collections: Catalogue No: GLAHM 105684. Gregorian Telescope by James Short of Surry Street, London. Manufactured 1743. OA [overall measurement] when the tube is horizontal 530mm × 265mm × 440mm (3-inch diameter?). Stamped on eyepiece: 29/372 = 12; the first number is the model number, the second the serial number and the last the focal length in inches.
17 Bryden, 'The Jamaican observatories of Campbell and Macfarlane'.
18. Muirhead, J. P. (1859), *The Life of James Watt*, revised 2nd edition, London: John Murray, p. 41.
19. Mackie, *The University of Glasgow*, p. 218.
20. Muirhead, *The Life of James Watt*.
21. MacGregor, George (1881), *The History of Glasgow from the Earliest Period to the Present Time*, Glasgow: Thomas D. Morison; London: Hamilton, Adams, p. 356 [GUL History DX200 MACGR; GUL Sp Coll Mu23-a.17].
22. Muirhead, *The Life of James Watt*, p. 28.
23. Ibid., p. 225.
24. Mackie, *The University of Glasgow*, p. 218.
25. Turner, G. L. E. (1969), 'James Short FRS and his contribution to the construction of reflecting telescopes', *Notes and Records of the Royal Society*, 24: 1 (June 1969), 91–108.
26. Muirhead, *The Life of James Watt*, p. 43.
27. Coutts, *A History of the University of Glasgow*, p. 265.
28. *Glasgow Herald*, 3 May 1870, 'TUESDAY MORNING, May 3'.
29. Power, John (1999), 'Pipe organs in and around Glasgow', *The Organ*, 78: 307, 23–4.
30. Muirhead, *The Life of James Watt*, p. 41.
31. Emerson, Roger L. (2008), *Academic Patronage in the Scottish Enlightenment: Glasgow, Edinburgh, St Andrews*, Edinburgh: Edinburgh University Press, p. 134 [GUL Education B451.5].
32. Ibid., p. 136.

33. Ibid., p. 138.

34. Mackie, *The University of Glasgow*, p. 220.

35. Coutts, *A History of the University of Glasgow*, p. 320.

36. MacGregor, *The History of Glasgow*, p. 357.

37. Emerson, *Academic Patronage in the Scottish Enlightenment*, p. 139.

38. Coutts, *A History of the University of Glasgow*, p. 229.

39. Ibid., p. 229.

40. Mackie, *The University of Glasgow*, p. 222.

41. Glasgow University Manuscript 117/27400. Manuscript notes on a) The Macfarlane Observatory (1757) and the instruments contained in it [Glasgow University Archives].

42. Coutts, *A History of the University of Glasgow*, p. 229.

43. Murray, *Memories of the Old College of Glasgow*, p. 261.

44. King, Elizabeth Thomson (1909), *Lord Kelvin's Early Home: Being the Recollections of his Sister, the Late Mrs. Elizabeth King together with some family letters and a supplementary chapter by the editor*, London: Macmillan and Co., p. 95 [GUL LRA U11-c.13; GUL Sp Coll Mu8-g.11].

45. Wade, W. M. (1821), *The History of Glasgow, Ancient and Modern, with an Historical Introduction, and a Statistical Appendix*, p. 19 [GUL Sp Coll Mu26-f.46; GUL Sp Coll Bh11-e.15; GUL LRA R12-g.2].

46. Gibson, John (1777), *The History of Glasgow, from earliest accounts to the present time: with an account of the present state, of the different branches of commerce and manufactures now carried on in the city of Glasgow*, Glasgow: R. Chapman & A. Duncan [GUL History DX201 GIB; GUL Sp Coll Mu23-b.7; GUL Sp Coll BD16-e.12; GUL Sp Coll Bh11-a.24].

47. Fig. 3.9 shows, on the left-hand side, part of McArthur, John (1778), *Plan of the City of Glasgow, Gorbells and Caltoun From an actual survey by John McArthur*. The complete map is held by the Mitchell Library [GC 941.435 REI] and Glasgow University Library [GUL Maps C18:45 GLA15]. The right-hand side shows a picture from Jones, Nathaniel (1791), *Jones's directory; or, useful pocket companion, for the year 1791; containing alphabetical lists of the names and places of abode of the merchants, manufacturers, traders, and shopkeepers, in and about the city of Glasgow* [GUL Sp Coll Bh11-i.7].

48. Glasgow University Manuscript 117/27400.

49. Smollet, Tobias (1793), *The Expedition of Humphry Clinker*, vol. 2, London: J. Sibbald, p. 77.

50. Murray, *Memories of the Old College of Glasgow*, p. 262.

51. Boney, *The Lost Gardens of Glasgow University*, p. 213.

52. Ibid., pp. 215–17, 220–1, 230–1, 238.

53. *The Topographical, Statistical, and Historical Gazetteer of Scotland*, 1842. Vol 1, A–H, Glasgow: A. Fullarton & Co., pp. 656–7.

54. Smith, David (1822), *Map of the City of Glasgow and Suburbs. Originally published by Mr. Fleming in 1807 – with additional surveying for laying down the extension of the city & suburbs by David Smith to May 1821*, Glasgow: Alexander Findlay & William Turnbull [GUL Case Maps C18:45 GLA18]; Smith, David (1828), *Plan*

of the City of Glasgow and its environs with all the latest improvements accurately surveyed by Mr. David Smith, Glasgow: Wardlaw & Co. [GUL Case Maps C18:45 GLA20].

55. Glasgow University Manuscript 117/27400.
56. Aitken, Andrew (ed.) (1930), 'Glasgow Club Transactions – Session 1929–1930', in *Memories of the Old College in the High Street*, Glasgow: Aird & Coghill Ltd, pp. 29–47 [GUL Sp Coll Ferguson Af-y.41].

Chapter 4

The Wilsons

4.1 INTRODUCTION

The period of history covered in this chapter is the latter half of the eighteenth century. Following the contributions made by the Dick family to the field of natural philosophy, it was the Wilsons, father and son, who established astronomy as a significant contributor to teaching and research within the University. This chapter will cover these developments, but at the end, it will also mention a meteor event and a meteorite that affected some elements of the local population, revealing that astronomy also crosses the everyday life of the average person.

4.2 ALEXANDER WILSON – THE EARLY YEARS

It is always of interest to explore the background of an eminent person who makes a significant mark in science. The character of the first Professor of Astronomy in the University of Glasgow, and his life both prior to his appointment and afterwards, provides a fascinating tale of genius and of the good fortune of benefiting through personal contacts with people of influence. Such is the story of Alexander Wilson.

His parents were Clara Fairfoul and Patrick Wilson. When Alexander was born in 1714, his father was town clerk of St Andrews, but he sadly died when his son was a young child, after which Alexander was brought up by his mother. As a boy,[1] under his own inclinations, Alexander enjoyed technical drawing, modelling figures and engraving on copperplate, the latter serving him well in a later commercial pursuit. He was educated at the University of St Andrews, graduating with an MA on 8 May 1733. Subsequently, he was an apprentice to a surgeon, Dr George Martine, in the town. It was through the latter's influence that Alexander became interested in the art of glass-blowing for the production of thermometers which he calibrated with high-quality scales. Through his experiments on heat, Dr Martine also influenced another leading Scottish scientist, Joseph Black, already mentioned in Chapter 3 in relation to James Watt's activities. Black became Professor of Medicine at the University of Glasgow and Lecturer in Chemistry, establishing the notions of 'specific heat' and 'latent heat'. He also discovered carbon dioxide, noting that it was heavier than air.

In a biographical account of his life, Alexander's interest in several philosophical ideas are noted.[2] One direction of his thoughts was a scheme to reflect sunlight into a concentrated spot by means of a large collection of plain mirrors; he imagined that this had not been considered by any person before him. The same biography relates that, a few years later in France, de Buffon constructed such a magnificent array with success, so as to render the famous secret imputed to Archimedes of setting on fire the Roman galleys much less apocryphal than previously considered.

As many newly qualified young people did at the time, Wilson moved to London to seek employment in the medical profession, and he became assistant to a French surgeon, a political refugee, who welcomed him into his family circle. Among the Scottish colony in London, Wilson was introduced by David Gregory, Professor of Mathematics at St Andrews, to Dr Charles Stewart, the private physician of Archibald Campbell, Lord Islay, the brother and heir of the Duke of Argyll. Stewart introduced Wilson to His Lordship, who invited the young Scotsman to see some of the fine physical instruments and astronomical pieces in his collection. As part of this social interplay, Wilson was able to contribute by constructing thermometers of different kinds for Lord Islay and his friends, with more perfection and elegance than had hitherto been seen in London. This encounter with the eventual Duke of Argyll played a decisive part in opening the way for Wilson to become a professional astronomer, as related in Chapter 3.

Patrick Wilson,[3] Alexander's son, noted that this period was a happy and rewarding time for his father, as he was able to develop connections with people of a philosophical turn of mind whenever his medical duties permitted. A chance encounter with one of the advancing technologies of the time changed his life's direction, providing a profitable outlet for his latent genius. While visiting a letter foundry with a friend, who was there to purchase some printing types, he became captivated by the various contrivances that operated in the factory. His imagination was immediately triggered as to how the business of type founding might be improved, and his work in this enterprise was so renowned as to gain a biographical entry in Hansard's *Typographica*.[4]

He discussed his ideas with a friend named Bain, who also originated from St Andrews, and they decided to take practical steps to implement them. Wilson took the opportunity to communicate the notions of developing type founding to Lord Islay; he was pleased that his ideas met with the latter's approval and to receive his best wishes for success.

Wilson and Bain became partners, and took apartments for the preparatory steps to develop the business. Unexpected problems inhibited the progress they had initially expected. During their experiments they found the expense of living in London too great a barrier for opening a foundry and bringing their products to market. In 1739, they decided to return to St

Figure 4.1 The title page of the 1819 edition of *Specimen of Printing Types of Alexander Wilson & Sons*. The right-hand panel displays Greek fonts; the upper section is the 'Glasgow Homer' style used by the Foulis brothers in their production of the Greek classics.

Andrews. The following year, Wilson married Jean Sharp, the daughter of a St Andrews merchant. Jean and Alexander Wilson had three sons, who eventually joined their father in the business.

The partners decided to abandon their revolutionary schemes for setting up a foundry, which proved to be impossible to put into practice, and returned to more standard processes for producing type. Their infant letter foundry opened in 1742. Most of the printers in Scotland, based in either Edinburgh or Glasgow, depended on purchasing type from London, with the inconvenience of its transport. Orders accrued and, with this encouragement, they were soon able to expand their variety of fonts. After two years, with the prospect of extending sales to Ireland and North America, in 1744 they moved to Camlachie, a small village around a mile east of Glasgow. A few years later, Wilson was appointed as type founder to the University. In 1747, Bain left for Dublin to open a branch there. Soon after, he quit the Glasgow partnership, leaving Wilson the sole owner of the type foundry.

The quality of the type produced by the foundry was outstanding, and the finest of all was a Greek type (see Fig. 4.1). The Foulis Press, printers for Glasgow University and operated by the brothers Robert and Andrew Foulis, used Wilson's type, producing some of the finest and most beautiful books which no other press could match. Wilson provided a new type at an expense of time and labour that could not be recompensed by any profits from the sale of the types themselves. Such disinterested zeal for the honour of the University Press was well appreciated, and this was noted in the most glowing style in the preface of the famous Foulis edition of Homer's *Iliad* of 1756. Along with the *Odyssey*, published two years later, the production has been described as 'a landmark in the history of printing in Greek'. The large paper copy of this two-volume work was sold for £1 11s 6d, and the small version was priced at £1 1s. The recognition of Alexander Wilson by the Foulis brothers reads as follows:

> Omnes quidem tres regios Stephanorum characteres graecos expresserat jam apud nos, atque imitione accuratissima repraesent averat ALEXANDER WILSON A.M. egregius ille Typorum artifex, quem et hoc nomine adscripserat fibi Alma Mater. in his autem grandioris formae characteribus Stephaniansis, id unum desiderari quodammodo videbatur, scilicet, si resita ferre posset, ut, falvâ tamen illa solidae magnitudinis specie quâ delectantur omnes, existeret una simul elegantiae quiddam, magis, atque venustatis. rogatus est igitur ille aritifex, ut, in hoc assequendo, solartiam suam, quâ quidem pollet maximâ, strenul exercet. quod et lubenter aggressus est, et ad votum usque videtur consectus vir ad varias ingenuas artes augendas natus.

Although not directly related to our main theme, it is interesting to note the place of the printing press within the University precincts. According to Mackie,[5] it was a tradition in Glasgow that there should be a close connection between the College and the printing press, but the system in Glasgow had fallen into a bad way in the hands of various local tradesmen, all of whom were connected with the College, some having rooms there. A new era dawned when Robert and Andrew Foulis, sons of a Glasgow maltman, who had turned their energies to bookselling, obtained permission to carry on their trade in the north-west corner of the Inner Quadrangle, selling books in a small room on the ground floor. Soon afterwards, they erected a printing press in a room overhead. In 1750, finding that they needed more room, they acquired part of the building in the neighbouring Shuttle Street. They may have owed their privilege to the fact that Robert Foulis was brother-in-law to Professor Moor, but they did not misuse it. They began to print in 1741 and, in 1743, Robert was made Printer to the University; it is not clear when his brother, Andrew, became a partner. In that year, they produced the first book in Glasgow to be printed in Greek, and, in the next two decades, they issued the 'immaculate' Horace, wonderful folios of the *Iliad* and the *Odyssey*, and other classics. For the technical excellence of the work, they were themselves responsible; they were at pains to have the best paper, good ink and good type in cooperation with Alexander Wilson, as mentioned above. In the matter of proofreading, they had the help of Muirhead and Moor. They benefited from the patronage of William Hunter, who in 1766 ordered a complete set of their publications, and they enjoyed the friendship of many professors into whose company they were willingly accepted.

Following his appointment as the first Regius Professor of Practical Astronomy, with concurrent residence in the College, Professor Wilson found it impracticable to carry on his outside business conveniently without a founding house near to hand. In March 1762, he made representation to conduct these operations within the College precinct. According to Coutts,[6] following a motion proposed by Adam Smith it was agreed that the University should build a foundry in a convenient place, at a cost of not more than £40, and that Wilson should pay a reasonable rent and agree that, should the building become useless before the University was compensated for the outlay, he should pay such compensation as might be awarded by arbiters. The building was erected at a cost of about £59 in the little garden adjoining the Physick Garden, not far from the chemistry laboratory. The site of the foundry is marked on a map provided by Murray.[7] The yearly rent was fixed at £3 15s; a proposal by Moor that Wilson and his son should occupy the building rent-free, but should pay 7.5 per cent on the outlay and hand over the foundry in good condition for carrying on the business if they ceased to use it for type founding, was rejected. In 1769, Wilson applied for an extension to the building; as the University had erected it at a cost of £57 14s, an additional rent of £3 9s 3d was imposed. It may be noted[8] that the operation of the foundry steadily poisoned the soil of the adjacent Physick Garden with its volatile fumes of lead, antimony and tin. By 1803, the Garden was so barren that the Faculty decided to dispense with it and develop another site, but this was delayed for various reasons until 1818. Jointly with the Royal Botanic Institute, the University opened up a botanic garden on the Sandyford Estate lying between Argyle Street and Sauchiehall Street. The area is now built over but the enterprise was the forerunner of the present Botanic Garden further out from the city. Despite Wilson's other time-consuming scholarly activities, the firm he ran in partnership with his three sons continued to operate throughout his life. He published *A Specimen of some of the Printing Types Cast in the Foundry of Alexander Wilson & Sons*[9] in 1772, providing a fine example of the capabilities of the firm. In fact, Alexander Wilson & Sons continued to be a thriving business after the death of its founder, and the headquarters moved from Glasgow to London in 1834.

What were Wilson's other scholarly activities prior to his accepting the position within the University? One remarkable enterprise was through a social connection with Thomas Melvill, who was known for his mathematical prowess and who studied divinity in 1748 and 1749 at the Glasgow College, also attending lectures on natural philosophy given by Professor Dick. Melvill was a regular visitor to the Wilsons' bustling household at Camlachie. In one of several philosophical schemes on which they collaborated, Wilson proposed[10] investigating the temperature of the higher atmosphere by raising a number of paper kites, one above the other on the same line, with some of his thermometers appended to those reaching the highest levels. The topmost kite ascended to amazing heights, disappearing at times among the white summer clouds. The pressure of the breeze interplaying on the kite elements made it too difficult for the line to be controlled by a single person, and the scheme needed some specially designed machinery for the experiments to be conducted. To obtain the temperature information, the thermometers were secured to the line by bushy strings of paper which acted as fuses. After reaching the desired heights, the thermometers were allowed to fall, released by the slow burning of their attachment to the kite line.

Although the equipment used rose to cloud level, the experiments were conducted in fine, dry weather, and so no effects whatsoever of an electric nature were recorded. Such analysis of lightning flashes was undertaken by Dr Benjamin Franklin two years later. Melvill, while in Switzerland in April 1753, wrote with enthusiasm to Wilson on the prospect of undertaking research along the lines of Franklin's experiments, using their combined kites system. Unfortunately Melvill died of tuberculosis in Geneva later the same year at the age of twenty-seven, and this research was abandoned. For the record, Melvill[11] performed many experiments exploring the nature of light and may be considered as the originator of spectroscopy. He discovered that the yellow light seen in flames containing incandescent sea salt had a sharp and unique monochromatic colour, this later being referred to as the 'sodium line'. He also contributed to the understanding of stellar aberration, as will be mentioned again later.

In 1752, the Wilson family moved to Glasgow and, around five years later, Alexander invented hydrostatical glass bubbles, or specific gravity beads, for determining the strengths of spirits (see Fig. 4.2 and Cat. No: GLAHM 105661 in the Appendix to Chapter 1). His designs were more accurate and commodious than devices used by distillers and merchants in the West Indies. A philosophical and literary society, comprising professors and their friends, used to meet weekly within the Glasgow College and, in the winter of 1757, Wilson presented a discourse on his devices. He demonstrated how the sealed glass bubbles containing air could be used to measure small differences in the specific gravity of fluids of the same kind at a particular temperature by floating them and selecting the one that became stationary within the liquid without sinking. The specific gravity beads he produced were of high quality and sensitivity, and he made packaged sets commercially. Wilson's invention was copied by other local glass-blowers and also put on the market.

In the following year, 1758, Wilson presented another discourse to the same society on the motion of pendulums. On this occasion, he exhibited a spring clock from a small compass which beat at the rate of seconds by means of a pendulum he had contrived upon the principle of the balance,

Figure 4.2 A set of Wilson's specific gravity beads, known as philosophical bubbles, made by James Brown of Glasgow, c. 1780.

whose centres of oscillation and motion were very close to one another. At one of the trials, it performed so well as to be correct to one second in forty hours when compared with a very accurate astronomical clock – possibly the Shelton purchased by the younger Robert Dick. These experiments showed Wilson's ingenuity in using simple and cheap machinery to serve a purpose normally fulfilled by more complicated and expensive systems.

Soon after, he produced a thermometer with a capillary bore of elliptical cross-section, this giving better visibility for the reading with a broader mercury column, and perfected this design after many trials. At about the same time, he mused on how he might apply a graduated thermometer to measure the temperature of the boiling point of water affected by atmospheric pressure, with the instrument contrived to act as a 'marine barometer'.

Some of the activities of Alexander Wilson have recently come to light via a notebook,[12] with entries from 1735 to 1753 covering his time in London and in Camlachie. Descriptions are given of the materials used to produce blue or red tinctures in 'spirit of wine' thermometers; details of some of the thermometers made in Camlachie and sold are noted, a typical price being one guinea. Brief comments can be found on the fabrication of metal mirrors for telescopes, with a note on experiments involving the addition of arsenic to the amalgam, which unfortunately produced brittle casts. Towards the middle section, there is a drawing of an orrery, or Jovilabe, depicting the motion of Jupiter's satellites around the parent body, with details of all the required gear ratios. King[13] has provided a general history of the development of Jovilabes.

4.3 ALEXANDER WILSON THE ASTRONOMER

After the receipt of the Macfarlane bequest, the building of the first observatory by the University and the machinations of the establishment of the Regius Chair of Practical Astronomy (see Chapter 3), with the Royal Warrant dated 11 January 1760, Alexander Wilson took up the prestigious appointment. After giving an inaugural exposition on comets, he was inducted on 17 February 1760. He held the post until his resignation in 1784. A caricature of a portrait of him is presented in Fig. 4.3, this based on the colour portrait displayed in Plate 4.

As described in the previous chapter, the University had originally proposed the appointment of an Assistant to the Chair of Natural Philosophy who should act as an astronomical observer, and though what was done went beyond this notion, the title of Practical Astronomy given to the new chair seemed to emphasise that it was a position that would involve observational work. It was not considered that the Professor was obliged to lecture to students, as that was not done at Greenwich or elsewhere in Great Britain, but

Figure 4.3   A caricature of a portrait of Alexander Wilson (1714–86), Professor of Astronomy at Glasgow University from 1760 to 1783.

since Wilson was willing to do so, he was authorised to give instruction in the three following branches to students who applied to him: (1) the application of spherical trigonometry to the solution of astronomical problems; (2) the construction of astronomical tables and how to use them to make calculations; and (3) the construction of astronomical instruments and methods of using them to make observations. His courses were not requisites for graduation, but physical astronomy was taught, with obligatory requirements for natural philosophy classes, for all students under Professor John Anderson (see Chapter 3).

In addition to type founding carried out in buildings within the College, reference to business ventures for making scientific instruments with James Watt was made in a lecture given by Professor R. V. Jones[14] and reprinted in 2008. Patrick Wilson was noted as being the partner, but the dates provided are more appropriate to his father, Alexander. Although Patrick maintained a friendship with Watt and the two corresponded on many occasions, there is no direct evidence of firm business ventures between either of the two Wilsons and Watt.

Records in the Faculty minutes show that Wilson was soon busy establishing astronomical observations but that there were interruptions. In the entry for 4 February 1761, a problem is described whereby:

> Mr. Wilson reported that the Observatory had been sometime ago broken into in the night, and that several attempts of the same kind had been made since, and he is authorised to get a man to sleep in the Observatory at a shilling per week or thereby, as he can agree with one in whom he can confide.

Modern-day experience shows that such problems of intrusion, theft and vandalism have continued, but University security is no longer of the 'sleep-in' type.

Among some of Wilson's pursuits around this time were a continuation of experiments he had undertaken earlier to cast materials for telescope mirrors and to improve the polishing process for producing their optical surfaces. He conducted trials with a variety of speculum metals to produce reflecting surfaces carrying a curved profile in the form of a parabola, so improving image quality. The mirrors were of small diameter and focal length, although he had designs to make them much larger. Apparently he was often heard to rue the fact that no patron seemed amenable to offering support for the construction of large telescopes that would further discoveries related to the Moon or planets, or to exploring the heavens. Again, according to his biography,[15] if his own means had been less circumscribed, he would himself have attempted something of this kind. The successes of William Herschel had shown that the suggestion of producing larger mirrors was not just a romantic dream. With little or no direct experience of astronomy, Wilson turned his skills to observational studies. Made using an 18-inch telescope, his observations[16] included those of the first Jovian satellite over an interval of seven years, the solar eclipse of 31 March 1769 and an occultation of 6 November 1769. By comparing the timings of the eclipses of Jupiter's satellite with those observed at Greenwich, chiefly by Maskelyne, and those made by James Short at his private observatory in Surry Street in The Strand, some twenty-six seconds west of Greenwich, he aimed to deduce the longitude of the Observatory at Glasgow. Although all the sets of timings are listed, only four relate to the same eclipse being observed at two sites. The difference of meridians between Greenwich and Glasgow was not calculated; the application of simple arithmetic on his data shows that any such determination would not have been very accurate.

One of Wilson's instrumental developments also seemed worthy of publication at the time. He proposed[17] an improvement for eyepiece cross-wires within transit telescopes using a flattened silver filament, with the smallest cross-section being presented to the eye. His description of their preparation, with the strands pressed between two polished steel plates, indicates that the original wire was two-thousandths of an inch thick.

Wilson's collected papers contained considerable amiable correspondence with the Astronomer Royal, Nevil Maskelyne, and it may be noted that all his publications in the *Philosophical Transactions of London* were communicated by the latter.

4.4 TRANSIT OF VENUS

On 6 June 1761, Wilson observed the Transit of Venus and, for the second of the pair of such events on 3 June 1769, he organised a group of observers at the College. The latter transit was forecast in the *Glasgow Courant* of 1 June (see Fig. 4.4), the announcement apparently made independently of Professor Wilson. In his report[18] of the occasion, he noted that he had placed a request

> in the local news-paper begging the inhabitants, in cases where it would not be inconvenient, to put out their fires from three o'clock that afternoon till sunset; the politeness of the inhabitants of Glasgow, in complying with the request, was far greater than could well be expected, insomuch that there was not a spire of smoke to be perceived in that quarter from which the observations could be incommoded.

Wilson chose an observation site for himself at a house some distance from the Observatory, although within sight of it, to be more free from the smoke of the town. He took two telescopes by Short with him, one of which had a focal length of 18 inches. Remarkably, the following day there was a solar eclipse, and, in his table of timings[19] related to the Transit of Venus, as presented in Fig. 4.5, Wilson makes reference to the eclipse with observations made using a Short telescope with a focal length of 9 inches. This instrument may well be the second Short telescope he had used the day before. The focal length suggests that it may be one located within the Glasgow

As Saturday evening prov'd favourable here for obſerving the tranſit of Venus, we could have wiſhed it had been in our power, to have procured a particular accoúnt of this uncommon phenomenon, from the gentlemen ſkilled in Aſtronomy, but as no ſuch account is hitherto ſent us we ſhall be obliged to any gentleman who will be ſo kind as to favour us with the obſervations he may have made on the tranſit, and ſhall take the firſt opportunity of giving them to the public.

The following Table (taken from Mr. Martin's Inſtitutions of Aſtronomical Calculations juſt publiſhed) exhibits all the Tranſits of Venus from the year 1631 to 2160, in cluſive, with the interval of years between each two, that the regularity and order in which they return, may be the better obſerved.

Anno Dom.	Month.	Interval of Years.
1631	December	
1639	December	8
1761	June	122
1769	June	8
1874	December	105
1882	December	8
2004	June	122
2012	June	8
2117	December	105
2125	December	8
2247	June	122
2255	June	8
2360	December	105

As Saturday evening prov'd favourable here for observing the transit of Venus, we could have wished it had been in our power, to have procured a particular account of this uncommon phenomenon, from the gentlemen skilled in Astronomy, but as no such account is hitherto sent us we shall be obliged to any gentleman who will be so kind as to favour us with the observations he may have made on the transit, and shall take the first opportunity of giving them to the public.

The following Table (taken from Mr. Martin's Institutions of Astronomical Calculations just published) exhibits all the Transits of Venus from the year 1631 to 2160, inclusive, with the interval of years between each two, that the regularity and order in which they return, may be the better observed.

Anno Dom.	Month	Interval of Years.
1631	December	
1639	December	8
1761	June	122
1769	June	8
1874	December	105
1882	December	8
2004	June	122
2012	June	8
2117	December	105
2125	December	8
2247	June	122
2255	June	8
2360	December	105

Figure 4.4 The *Glasgow Courant* of 1 June 1769 had recourse to Benjamin Martin's *Institutions of Astronomical Calculations* for the prediction of the Transit of Venus of 3 June, and all future similar events through to AD 2360. The contemporary ones of 2004 and 2012 are listed. The article has been transcribed for clarity. Benjamin Martin was an eighteenth-century London populariser of astronomy.

University Hunterian Museum, since this collection contains two with focal lengths close to 9 inches (see GLAHM 113782 and 105681 in the Appendix to Chapter 1).

Three other instruments were employed at the Observatory. A solar image of about 6 inches in diameter was projected onto a screen using a Dollond refractor 29 inches in focal length, under the supervision of Drs Williamson and Reid; this equipment was adjacent to the Shelton clock which had been regulated by stellar transit observations. A Dr Irvine watched from the south window using a telescope with a focal length of 13 feet, together with Alexander Wilson's son, Patrick, who used a Short reflector with a focal length of 12 inches. The reference to its focal length suggests that this is likely to be the earliest of the three James Short telescopes (see GLAHM 105684 in the Appendix to Chapter 1), and a picture of it is displayed in Plate 5. Both

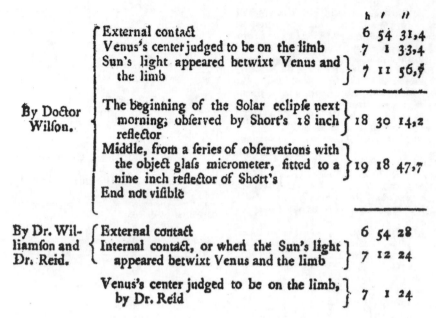

		h	′	″
	External contact	6	54	31,4
	Venus's center judged to be on the limb	7	1	33,4
	Sun's light appeared betwixt Venus and the limb	7	11	56,5
By Doctor Wilson.	The beginning of the Solar eclipse next morning; observed by Short's 18 inch reflector	18	30	14,2
	Middle, from a series of observations with the object glass micrometer, fitted to a nine inch reflector of Short's	19	18	47,7
	End not visible			
By Dr. Williamson and Dr. Reid.	External contact	6	54	28
	Internal contact, or when the Sun's light appeared betwixt Venus and the limb	7	12	24
	Venus's center judged to be on the limb, by Dr. Reid	7	1	24

Figure 4.5   A table of timings relating to the Transit of Venus of 3 June 1769 and the solar eclipse of the following day showed that Alexander Wilson used both the Short 18-inch and 9-inch telescopes, the latter probably being one in the safekeeping of the Glasgow University Hunterian Museum.

of these observing teams employed smoked glasses to allow direct viewing of the event. Within the observing room, an assistant stood by the Shelton clock, purchased by Professor Dick, and used a clapper stick to beat out the seconds, so that the progress of the Transit of Venus could be timed. Equipment from natural philosophy laboratories was also used by John Anderson (see Chapter 3), and assistants used both reflectors and refractors to watch from the College steeple, with their clock regulated by signals from the nearby Observatory.

Wilson commented on the quality of the viewed images, referring to a 'considerable tremor', now termed 'poor seeing', which is the effect caused by atmospheric turbulence, adding difficulty to the process of measuring the apparent diameter of Venus when projected on the solar disc. He noted that:

> After the center [sic] of Venus had passed the Sun's limb, she appeared to us not to be circular, but oblong, the longest diameter being that which passed through the Sun's center. As the internal contact approached, Venus appeared to us to adhere to the Sun's limb, by a dark protuberance or neck, both the length and breadth of which varied every moment by a constant undulation: neither did this neck break off instantaneously, but changed colour from black to dusky brown, till at last the interval betwixt Venus and the Sun's limb appeared quite clear.

Reports on the behaviour of the image at the times of internal and external ingress of Venus were given by Wilson's son, Patrick, and by colleagues Drs Williamson and Reid. As for Mr Anderson and his assistants in the College steeple, he noted that all of them 'were uncertain about the timing of the external contact, owing to the state of the atmosphere, and a tremor given to the steeple by the wind; but none of their observations varied, above three seconds, from my own'.

As a footnote to his paper discussing the Transit of Venus, Wilson mentions that during the course of 1769, he saw the Aurora Borealis several times and that it formed itself into an arch twenty or thirty degrees above the horizon, with the vertex to the west of north.

4.5 OBSERVATIONS OF SUNSPOTS

The work for which Alexander Wilson is most renowned relates to the appearance of sunspots viewed when they are close to the solar limb. In November 1769, he made observations of a sunspot of extraordinary size, but the reports and interpretation were delayed until 1774.[20] As a result of what he recorded and how he interpreted what he had seen, the term 'Wilson effect' became linked to the change in appearance of a spot as, under the effects of solar rotation, it approaches the solar limb. In recounting his original observations in a second paper,[21] Wilson noted that his discovery was first reported in the *London Chronicle*, although Pedersen[22] has drawn attention to the fact that this article in the 1 December 1770 issue was written anonymously, with the author designated as 'X'; from reference to other material in the article, it is obvious that 'X' was Alexander Wilson. It was also noted in the *Chronicle* that the 'discovery of the nature of solar spots' had been presented to the Literary Society of Glasgow the previous year.

The presence of the particular spot of November 1769 had been intimated to him in a letter from a friend in London. At first, he was prevented from observing it until the weather cleared on 22 November, when he used a Gregorian telescope, noted as having a focal length of 26 inches and a magnification of ×112; Bryden[23] has referred to this telescope as being made by Short and having a focal length of 24 inches. In the late afternoon, he made a record of the spot when it was close to the Sun's western limb. On the following day, he had more time to examine the solar image, and noticed a remarkable change (see Fig. 4.6). The description of his observations, taken from the *Philosophical Transactions of the Royal Society*,[24] is as follows:

the umbra, which before was equally broad all round the nucleus, appeared much contracted *on that part which lay towards the center* [sic] *of the disc*, whilst the other parts of it remained nearly of their former dimensions.

This change of the umbra seemed somewhat extraordinary, as it was the very reverse of what I expected from the motion of the spot towards the limb. But next day, at 10 o'clock, I had another observation, and discovered changes, which were still more unexpected. The distance of the spot from the limb was now about 24″. By this time, the contracted side of the umbra above mentioned had entirely vanished; and the figure of the nucleus was now remarkably changed, from what it had been the preceding day. This alteration of the figure appeared evidently to have taken place upon that side which had now lost the umbra, the breadth of the nucleus being thereby more suddenly impaired than it ought to have been, by the motion of the spot across the disc.

Regarding these circumstances as new, I began to consider, what might be the cause of them. One of two things seemed necessarily to be the case; either, that they were owing to some physical alteration or wasting of the spot, and of that part of it where the deficiency of the umbra was observed; or else, that they were owing to the nearer approach of the spot to the limb, by the sun's rotation on his axis.

The last of these two ideas had no sooner struck me, than I began to suspect, that the central part, or nucleus of this spot, was beneath the level of the sun's spherical surface; and that the shady zone or umbra, which surrounded it, might be nothing else but the shelving sides of the luminous matter of the sun, reaching from its surface, in every direction, down to the nucleus: for, upon this supposition, I perceived, that a just account could be given of the changes, of the umbra and of the figure of the nucleus, above described.

The opinion therefore, which I ventured to form from what I had seen this day, was, that this spot might, probably, be a vast EXCAVATION in the luminous matter of the sun: the nucleus, commonly so called, being the bottom, and the umbra the shelving sides of the excavation: and, moreover, that the umbra, next the center of the disc, although out of my view, did still however exist, and was rendered invisible by its present position only; and further, that the sudden alterations, now discernible in the figure of the nucleus, were occasioned by some part of it also being hid, by the interposition of the edge of the excavation, between the nucleus and the eye.

These views, which now presented themselves, I remember to have communicated, that afternoon, to my son; when I then told him, that, if they were well founded, there will be room to verify them, if the spot should again return upon the opposite side of the disc. I was however uncertain, if it would last so long upon the sun's body, as to be again visible after the time of half a revolution; a circumstance which I wished to take place, as I was aware, that my present observations might justly be deemed insufficient, for establishing so singular an opinion concerning the nature of this spot: and that, not withstanding all which I had seen, we might still imagine, that these changes were produced by certain physical alterations of the spot itself.

These considerations made me attentively wait its return. At last, on December 11th, I again discovered it, on the opposite side of the disc, it having by that time advanced a little way from the eastern limb, being distant from it 1′ 30″. And now I could only perceive three sides of the umbra, namely, the upper and under sides, and that towards the limb, which was this side that formerly had vanished. The side towards the center of the disc was not as yet visible: which I concluded, upon the same grounds as formerly, that it was hid from my sight, by its averted position only, and that, after the spot had advanced a little further, it would make its appearance. Accordingly, the next day, being December 12th, at ten o'clock, it came into view, and I saw it distinctly, though narrower than the other sides. After this, my observations were interrupted, by unfavourable weather, till the 17th, when the spot had passed the center [*sic*] of the disc, the umbra now appearing to surround the nucleus equally . . .

All the foregoing appearances, when taken together, and when duly considered, seem to prove in the most convincing manner, that the nucleus of the spot was considerably beneath the level of the sun's spherical surface.

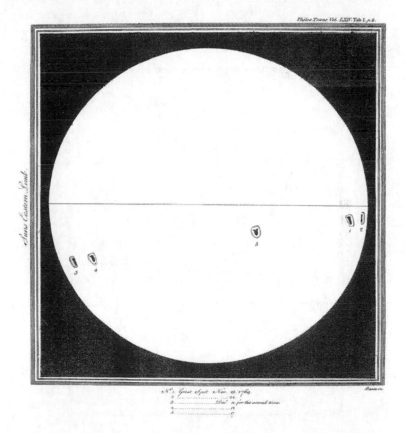

Figure 4.6 The large sunspot of November/December 1769 as drawn by Alexander Wilson. Images 1 and 2 are of 23 and 24 November respectively; images 3 and 4 are of the same spot as seen on 11 and 12 December, following its reappearance on the eastern limb. On 17 December the spot was recorded as image 5, near to the centre of the solar disc.

'The Nature of Sunspots' was announced by the Royal Society of Copenhagen as the subject of a prize essay. This induced Wilson to submit a paper, written in Latin, containing an account of his observations and of the conclusions drawn from them. In return he obtained the honourable distinction of a gold medal of nearly sixteen guineas' intrinsic value, that had on its reverse the figure of 'Truth' pendent in the air, holding a wreath in one hand and in the other a perspective glass, and the motto 'VERITATI LVCIFERÆ'. This token was presented to the Hunterian Museum[25] in June 1818 by his son, Andrew Wilson. The importance of Wilson's work was reported in Edinburgh[26] shortly after the publication of his landmark paper abstracted above.

By applying some simple geometry to his observations, as shown in Fig. 4.7, Wilson concluded that the nucleus of this large spot was 'not less than a semidiameter of the earth, below the level of the sun's spherical surface'. Although Wilson suggested that all spots might have a cavernous nature,

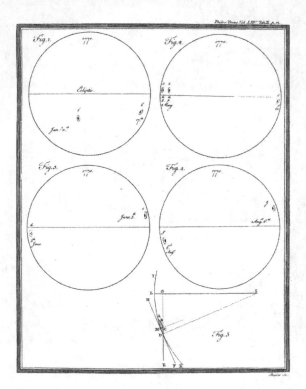

Figure 4.7 Drawings of sunspots made by Alexander Wilson in 1770; Fig. 4 (the bottom-right circle) is noteworthy, as on 5 August it records two spots, each displaying the Wilson effect but in the form of mirror images as they appear on opposite limbs of the Sun. Fig. 5 at the bottom shows the geometry used by Wilson to determine the depth of a spot cavity.

he tempered his proposal with some caution, encouraging others with good telescopes to make their own observations. He also presented the findings of previous observations by Scheiner and Hevelius regarding the general behaviour of spots, and how they related to the observations that he made on their evolution with their comings, goings and fragmentation. He commented that the important factor in his discovery of their depressions was the result of using a good telescope, with earlier observers tending to use a simple camera obscura. Wilson was very interested in the physical cause behind the observed effect, and he speculated that the Sun had structure, with a hotter outer layer comprising an elastic luminous material covering a dark globe which emitted little or no light. This model was accepted as true for around a century, and Wilson's theory that sunspots were depressions was 'proved' in 1861 when the first stereoscopic photographs of the Sun were taken. Only after 1900 was Wilson's idea shown to be inaccurate, when in 1908 Hale discovered that the apparent depressions were caused by strong magnetic fields that inhibited the convection of hot gas to the solar photosphere.

Wilson's interpretation of the cavernous nature of sunspots was not accepted by everyone, and in particular not by Jérôme Lalande in France, who wrote a critical paper in 1776, published in 1779, arguing that sunspots were the tops of mountains showing above a liquid surface. Wilson felt it necessary to respond in 1783 by writing a paper[27] in his defence. Its introduction was written with a mastery of diplomacy. In considering his further observations of the phenomenon, he obviated the objections, maintaining the reality of his propositions. At one point he says that:

> the spots are *cavities* or *depressions* in that immediately resplendent substance which invests the body of the sun to a certain depth; that the dark nucleus of the spot is at the bottom of this evacuation, which commonly extends downwards to a space equal to the semi-diameter of our globe; that the shady or dusky zone which surrounds the nucleus, is nothing but the sloping sides of the excavation reaching from the sun's general surface downward to the nucleus or bottom.

This was demonstrated by a strict induction drawn from following the phases of the spots as they transverse the Sun's disc.

In furthering his thesis, he described the making of a model Sun with a cavity on its surface, which he was able to view at a distance with a telescope. From the sizes of solar spots, and the changes in their apparent shape according to their closeness to the limb, Wilson estimated cavity depths at ~4,000 miles, suggesting that it might be the staggering scale of such features that caused the sceptical reaction to his proposal. His argument was that relative to the dimensions of the Sun and its overall sphericity, these cavities were essentially insignificant.

As described by Patrick Wilson,[28] it is of interest to note that in 1676, Flamsteed had very nearly anticipated Wilson's discovery when he reported on an asymmetry of the penumbra of a spot close to the solar limb. When

Alexander Wilson saw the great spot on 23 November 1769, the geometry was much the same as Flamsteed had described. However, Wilson had the advantage of being able to see the spot the following day, noting the change in appearance resulting from the aspect change caused by the Sun's rotation. Again, the notion of an *excavation* or *depression* was confirmed by subsequent observations of the same spot. It was only very much later that Wilson became aware of Flamsteed's writings and was pleased to learn of the description made 100 years before his own now famous observations.

The interpretation of changes to the 'dished' appearance of a spot according to its aspect at the solar limb remains a topic for discussion to this day. A summary presented by Bray and Loughhead[29] in 1964 was not optimistic as to there being a true physical depression in the photosphere, as they were unable to reveal any dip in the solar circumference profile for a spot which was exactly on the terminator. They maintained that insight was gained by considering that there are essentially differences in the opacity of the plasma materials in the volumes around the magnetic active zone. By simply applying distinct absorption coefficients $\kappa_1, \kappa_2, \kappa_3$ to the photospheric, penumbral and umbral material, the differences in the apparent penumbral widths across a symmetrical spot are readily explained. In a simplistic model of the situation, the photosphere/penumbra and penumbra/umbra boundaries may be defined as being the projections of points at which the line of sight has unit optical thickness, and they show that by simple geometry, the apparent width of the penumbra on the side of the spot remote from the limb decreases more rapidly than a simple geometrical foreshortening law would imply (see Fig. 4.8). They concluded that the dominant factor in producing the Wilson effect is the much greater transparency of the umbra, relative to the penumbra, in comparison with the photosphere.

An interesting effect of the opacity is that it partially occludes the dark interior of a spot seen at large solar longitudes relative to the surrounding photosphere, making any spot more difficult to detect at these locations. When examining the longitude distributions of more than 25,000 sunspots automatically detected in images from the Michelson Doppler Imager instrument on the SOHO satellite, the inclusion of the Wilson effect was required in the models to mimic the statistics. With Monte Carlo modelling of spot geometry and evolution, the 'depth' of sunspots was determined to be around 1,000km below the otherwise spherical surface. It is perhaps more than a coincidence that this work[30] was undertaken at Glasgow University, some 240 years after the phenomenon was recognised locally and termed the 'Wilson effect'.

4.6 SOLAR MOTION AND UNIVERSAL GRAVITY

Patrick Wilson[31] mentioned that in January 1777, while having an evening discussion with his father, their attention turned to a query, among many

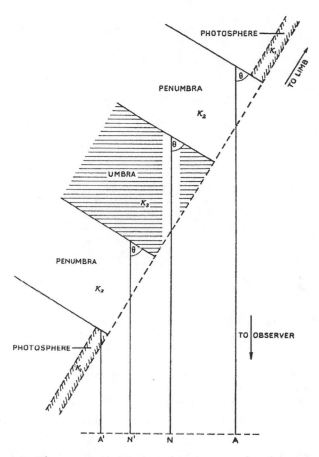

Figure 4.8 The points A', N', A and N correspond to lines of sight for which the optical depth is unity, with absorption coefficients κ_1, κ_2, κ_3 being assigned to the photosphere, penumbra and umbra respectively. For a circular spot with $\kappa_1 > \kappa_2 > \kappa_3$, the apparent distance AN across the penumbra nearer to the limb is greater than the apparent distance A' N' across the penumbra nearer to the centre of the solar disc.

others, set by Sir Isaac Newton at the end of his treatise on optics, namely: 'What hinders the fixed stars from falling upon each other?'

On reflecting upon this matter, they quickly came to the opinion that, if a similar question had been applied to the components of the Solar System, the answer would be readily available: that stability results from a balance of 'periodical motion' with gravity. In a like manner, Alexander Wilson reasoned that the same principle might apply to the stars, and that the order of stability, even of the Universe and every individual system comprehended in it, might depend upon periodical motion around some grand centre of general gravitation. An anonymous paper,[32] attributed to Alexander Wilson,

THOUGHTS

ON

GENERAL GRAVITATION,

And VIEWS thence arising

As to the STATE of the UNIVERSE.

———*What hinders the fixed Stars from falling upon one another?*
Newton's Optics, 3d Edit. p. 344.

LONDON:
PRINTED FOR T. CADELL IN THE STRAND.
MDCCLXXVII.

THOUGHTS

ON

GENERAL GRAVITATION.

THE laws of motion which regard all bodies, are laid down by Sir Isaac Newton in four general propositions. The first three he lays down as axioms, conceiving that they had already been sufficiently proved by Galileo, Wren, Hugens, Wallace, and others, and only points out the chief heads of the induction, which leads to them. The fourth law, commonly called the Law of Gravitation, he proves by a large and particular induction founded upon the phenomena of our planetary system. There appear to be other powers in Nature somewhat analogous to that of gravity, but which extend to less distances,

B.   and

Figure 4.9 The frontispiece and leading page of Wilson's article on the proposition that the stars are bound by gravitation and balanced by their orbital motions.

was published in 1777 on this proposition; it could be construed by the use of 'we' that it might have been jointly authored with Patrick, although the latter never laid claim to this. The frontispiece and first page of the work are depicted in Fig. 4.9. The outer cover shows that the article was for sale at a cost of 1s.

Considering the Moon orbiting the Earth and the Jovian system of planets, Wilson wrote:

As any of the former is but an epitome of that greater system to which they belong, may not this in turn be but a faint representation of that GRAND SYSTEM of the UNIVERSE, round whose centre, this Solar System of ours, and an inconceivable multitude of others like to it, do in reality revolve according to the Law of Gravitation?

This conception, as well as appearing warranted by the nature of gravity, was also advanced with regard to the discoveries of Halley referred to as the 'proper motions of the fixed stars'. It was given impetus particularly through the opinions of Dr Maskelyne, as noted by Patrick Wilson:[33] 'That probably all the stars are continually changing their places by some slow and peculiar motions throughout the mundane space'.

As if foreseeing the eventual achievements related to the structures of the stellar Universe, Wilson[34] concluded: 'Let posterity therefore determine how far the observed laws of these celestial motions shall favour these hints of One Grand Universal System'. Were it not for the periodical motion of the stars about a common centre, 'the whole glory of Nature would terminate in one universal ruin. But a supposition so injurious to the wisdom and to the over-ruling power of the Deity must be rejected as derogatory in the highest degree.'

Herschel,[35] in a renowned paper on the subject of proper motions of stars and related to the proper motion of the Sun and Solar System, read to the Royal Society on 6 March 1783, noted at the end that he had recently received a short tract from Dr Wilson, entitled 'Thoughts on general Gravitation, and Views thence arising as to the state of the Universe'; this was the article written by Wilson six years earlier on the possibility of there being a general motion of the stars and a solar motion. A few days earlier, on 3 March 1783, Herschel had written[36] to Wilson addressing him as 'a Gentleman so dear to all lovers of science and to Astronomers in particular'. Herschel continued:

I was apprehensive that what I wrote on the motion of the Solar System might be thought too much out of the way to deserve the notice of Astronomers; but since I have seen the contents of the valuable tract you have sent me, I am not without hopes that what I have said will be received, by a few at least, with no disapprobation; and if you should be one of that number, I shall think myself particularly flattered. You will perhaps like to hear a short report of that paper as it may be a good while before a copy of it can come to you.

From my last observations I give an account of a great number of changes amongst the fixed stars; such as – stars that are lost – that have changed their magnitude – that have moved from their former place – new ones come to visible, &c. Hence I conclude that it is highly probable that every star is more or less in motion. By analogy I apply this to the Sun as one of the fixed stars. I deliver afterwards a method that will easiest enable us to detect the direction and quantity of the solar motion and apply it then to observations that have already been made, and facts that are known; whence at last I draw the conclusion that our Sun is now actually moving with a velocity greater than that which our Earth has in her annual orbit, towards a point of the heavens in the northern hemisphere not far from λ or ρ Herculis. I have not seen what Mr De la Lande has said upon the subject, but will endeavour to see it. I ought to mention also that I have used the theory of attraction in support of my Hypotheses of the motion of the Solar System. And as I find the Tract you have favoured me with rests on the same foundation, I shall not fail to add a note, or Postscript to my paper (if it should be honoured with an impression in the *Transactions*) to pay the proper respect of quotation to the Tract as well as to De la Lande's Idea; tho' as to the latter, it does not appear to me at present in what manner the rotatory motion upon an axis is connected with progressive movement.

Your most obed. serv.
WM. HERSCHEL.

On the wider scene, Wilson was not alone in thinking about the mechanical behaviour of the stars. Some twenty-five years before, in 1750, the Durham astronomer Thomas Wright[37] had written of 'the projectile, or centrifugal Force, which not only preserves [the stars] in their Orbits, but prevents them from rushing altogether, by the common universal Law of Gravity'. Wright also talked about the proper motions of the stars and of the appearance of the Milky Way in terms of the optical effect of our immersion in a flat structure comprising stars. This idea was taken up and expanded by Immanuel Kant. Another of Wright's ideas, which is also often attributed to Kant, was that many faint nebulae are actually incredibly distant stellar systems similar to our own galaxy.

4.7 A COLD WINTER

Alexander Wilson and his son Patrick took temperature readings during the cold winters in Glasgow, conducting several experiments on the nature of hoar frost and the evaporation of snow. Alexander Wilson's curiosity[38] on these matters was triggered by the fact that the water in his bedside decanter was frozen when he awoke on the morning of 3 January 1768. With his thermometers to hand, he measured the outside temperature as 5 °F (−15 °C). He surmised that it might have been colder during the night. He then made arrangements to monitor the temperature over the next twenty-four hours, noting that it fell as low as −2 °F before midnight the following night, while earlier he had taken a reading 10 inches below the fallen snow at 30 °F. It was also noted that the intermittent formation of fog seemed to go hand in hand with minute variations of temperature. Experiments were also conducted on the evaporation of ice which he had formed on broken pieces of speculum by breathing on them. At the end of this first paper on the subject, he stated: 'Some particular reasons have occurred, which will hinder me from transmitting to you the paper on the solar spots, till some time next winter, by which time I shall have finished everything I have to say on that subject.'

4.8 ALEXANDER WILSON – CONCLUSION

As well as being a practical person with keen observational skills, Alexander Wilson was a true philosopher. His subtlety of expression in his writings reveals this quality. His son, Patrick, characterised his father's understanding of the progess of science by writing:[39]

> It sometimes struck him when looking over the progress of our philosophical discovery that many things of high moment appear to have lain long wrapped up in embryo by not employing ourselves more frequently in what may be called 'a direct search', and in filling up with more attention and boldness the list of desiderata. Between the last step and the accomplishment of a profound discovery, he

conceived that the transition might sometimes be made with no great effort of invention merely by sifting carefully such principles as are already known and familiar to us, and availing ourselves of them to their full extent.

In summing up his character, Patrick commented[40] that he

was amiable to an uncommon degree. The cast of his temper, though uniformly cheerful and serene, was yet meek and humble, and his affections flowed in the warmest current immediately from the heart. His looks, as well as his conversation and demeanour, constantly indicated a soul full of innocence and benignity in harmony with itself, and aspiring to be so with all around it.

Alexander Wilson always shied away from self-promotion; his modesty was self-evident as his first announcement related to the observation of sunspots and his proposal related to stellar motions under the influence of gravity were both made anonymously.

Although Alexander Wilson was not an official 'teaching' professor, according to Mackie[41] he played a considerable part in University affairs, frequently supporting Anderson (see Chapter 3) in the ongoing internal controversies; he helped to entertain distinguished visitors such as Boswell and Paoli (1771) and, as Vice-Rector, had a casting vote in the rectorial election of 1776. Wilson was awarded an honorary degree by the University of St Andrews on 6 August 1763, and was a founding member of the Royal Society of Edinburgh in 1783. As mentioned earlier, he was awarded the Gold Medal of the Royal Danish Academy of Copenhagen in 1771 for his discovery related to sunspots and their saucer-shaped depressions in the photosphere.

Short biographies of Alexander Wilson have been written by several authors, such as Armitage[42] and Garstang.[43] According to the latter, Wilson's memory is perpetuated by a lunar crater that Schröter named after him. The standard coordinates are $\xi=-0.238$, $\eta=-0.936$, or latitude 69°.2 S, longitude 42°.4 W, and it has been described as a grand crater, 70 kilometres in diameter, with lofty walls and containing other features catalogued on selenographic maps with roman letters. On the floor is a low central ridge, and there is a smaller crater on the inner west slope. On the outer north-west slope is a peculiar depression, K, with interior craterlets. Of these, L is the largest, but a craterlet M, on the north crest, is still larger. North of M is a crater H, with a smooth floor and a row of craterlets on its outer south slope. To the south-west are the craters A, Z and C, with crater F on the west. On the north-east is the named crater, Kircher, and on the south, beyond some craters, is Legentil. A selenographic map is shown in Fig. 4.10 to help identification, together with a photograph of the region.

In 1784, Alexander Wilson resigned from his University post, and lived in retirement among his family until he passed away on 16 October 1786. A transcript of his testament can be found on the internet.[44] The type-founding business was taken over and continued under the direction of his

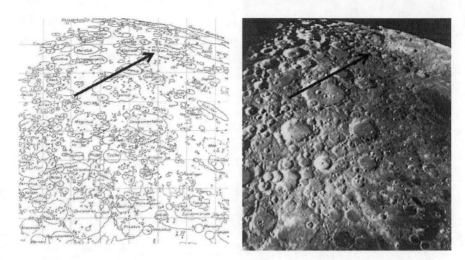

Figure 4.10   A lunar map and photograph covering the south pole region
and including the crater named after Professor Alexander Wilson.

sons. According to a 1791 business directory,[45] 'Andrew Wilsone' was a
'Type-founder – College'; also, 'Andrew Foulis' was a 'printere Shuttle ftr –
lodgings College'.

In 1774, during Alexander Wilson's tenure of the Chair of Astronomy, a
dissertation[46] was presented locally by a John Fleming. What is both remarkable and anomalous in relation to the date is that the discussion of the relative distances of the Sun and Moon is based on an Earth-centred Universe,
as depicted in one of the diagrams illustrating the argument.

4.9 PATRICK WILSON

In 1782, Alexander Wilson's request to have his middle son, Patrick,
installed in the Chair of Astronomy was at first rejected by the Crown,
the submission considered as a nepotistic bad precedent. However, the
Faculty had permitted Alexander to employ Patrick as his assistant at the
Observatory, and Patrick had cared for the instruments and taught there.
In 1783, Patrick Wilson was elected Clerk to the Senate. He was re-elected
each year until 1799, with the exception of 1795, when the Senate minutes
record that Patrick Cumin was elected in his absence. On 28 July 1784, the
Crown finally accepted a re-submission, and Patrick was appointed as his
father's successor on the latter's resignation. His formal trial bore the title
De Stellarum Fixarum Parallaxe et Distantia; it included a discourse on the
possibility of directly determining the velocity of light within a refracting
medium by measuring stellar aberration using a telescope with a water-filled
tube, rather than air (see below). A profile of him in the form of a medallion
by Tassie is shown in Fig. 4.11.

Figure 4.11 A Tassie medallion of Patrick Wilson. He followed in his father's footsteps as Professor of Astronomy.

On 23 October 1798, Patrick informed the Faculty that his ill health had caused him to consider retiring, although Emerson[47] believed that the underlying reason was related to a wish to relocate his business to London. He also wanted to nominate the successor to his Regius Chair. Though asked to reconsider his tenure, his mind was made up, as conveyed in a letter[48] to the Duke of Portland that included a letter of resignation. At his suggestion, the Faculty sent a signed memorial[49] to the Duke, via Henry Dundas, recommending the appointment of Thomas Jackson who had taught natural philosophy. After a few months' delay, Patrick Wilson relinquished the Chair in 1799. The political machinations of the episode have been presented by Emerson,[50] and further comment on the developments, and on the appointment of the third holder of the Chair, will be made in the following chapter.

Wilson moved to London, first to 16 Sackville Street, Piccadilly, then, in around 1806, to Windmill Hill, Hampstead; at the time of his death in 1811, he was at 29 Kensington Square, Kensington. In London, he was able to enjoy the company of the Herschels and other friends distinguished in pursuits congenial to his own. Patrick was a welcomed guest of the Herschels at Slough, where the two astronomers amicably observed sunspots together. According to Muirhead,[51] on Wilson's leaving Glasgow, 'as a mark of his high regard' for the learned body 'so long and so intimately known to him, and in testimony of his desire for the future prosperity of the office of Professor of Practical Astronomy in the said College which was first filled by his ever-honoured father', he endowed[52] the University of Glasgow with some of his instruments and a fund of £1,000, for the purpose of purchasing astronomical books and apparatus, and occasionally bestowing a gold medal prize on meritorious students of his favourite science. These monies were later referred to as the Wilson Fund.

Gavine[53] has noted that, in 1798, Patrick Wilson purchased refracting and reflecting telescopes from London, as well as a magic lantern with twelve astronomical slides, so there must have been some element of teaching at this time in Glasgow. It is likely that one of his Herschelian telescopes bought for £200 is now in the possession of the National Museum of Scotland, Edinburgh, having previously been kept at Jordanhill House in Glasgow (see Morrison-Law).[54]

At the time of Wilson's retirement, the inventory of equipment, with item values, included: a Herschel telescope, a Dollond achromat (£18 10s), a small loud-beating clock for the west room (£2), a Hadley sextant from Dollond (£2 2s), a Russell lunar globe (£6 10s), an old celestial globe (10s), a circular instrument, sextant and azimuth compass by Troughton (£51 11s), achromatic objectives and eyepieces (£50 19s), an astronomical circle 'as soon as Troughton can execute it' (£68 5s) and a small telescope (no price). Total value including insurance was £413 18s.

In the fifteen years that Wilson held the Chair, he performed very little in the way of direct observational astronomy. In following his father's thinking on the motion of the Solar System indicating that there was a similarity between the general motions of stars, Patrick considered undertaking telescope experiments associated with the velocity of light and stellar aberration. In a paper[55] he wrote prior to taking the Chair, he proposed an experiment related to the effect of stellar aberration but with the aim of directly determining the speed of light within dense but transparent materials, rather than calculating it from the 'sine formula' associated with refraction. He claimed that his proposed experiment would offer 'very strong additional evidence' for Newton's optical principles. Stellar aberration, as discovered by Bradley, results from the fact that light takes a finite time to travel from the entry point of a telescope's aperture to the focal point, where the stellar image and its position are recorded. For a given star, as a result of the Earth's motion about the Sun, the light path to the focal plane changes according to the time of year, so causing the star's apparent position to execute an ellipse with a scale of several seconds of arc, depending on the stellar direction relative to the Earth's orbital plane. Wilson asked how the 'constant' of stellar aberration might be affected by replacing the air in the telescope with some refracting material, such as water. His discussion related essentially to the speed of light in the refracting material and on whether absolute or relative frames of reference are used for the measurements. He suggested that the result should be independent of the style of telescope used; any difference in the aberration measurements would abrogate the notion of there being velocity changes in the medium of the telescope which were relatable to the sine law of refraction. From Wilson's paper,[56] it might be construed that he thought the velocity was greater in a more refracting medium, rather than the inverse; this was in keeping with the Newtonian corpuscular notion of light, rather than Huygens' undulatory theory. Apparently there was a small dispute over the priority of Wilson's proposal, and when the topic came up at a Royal Society Club dinner, according to a biography of Cavendish,[57] 'Mr Cavendish put in, that he did not think it a matter of any consequence . . . as nothing seemed likely to be determined by that method'.

According to Pedersen,[58] the article announcing the solar Wilson effect in the *London Chronicle* of 1 December 1770, and mentioned earlier, also noted that the son of the gentleman who had made the discovery related to the 'nature of sunspots' had also presented to the Glasgow Literary Society a dissertation concerning Dr Bradley's theory of aberration of the fixed stars, this being some thirteen years prior to Patrick Wilson's later paper.[59] The article in the *Chronicle* notes that there was a material error in the application of Bradley's theory in the determination of the velocity of light, related to its velocity in the vitreous humour within the eyeball not being taken into account, the measurements being made by and on the surface of the retina, rather than at the telescope's focal plane itself. Pedersen refers to strong similarities between Wilson's lecture and a letter written by Thomas Melvill to Bradley seventeen years previously. Whether Wilson knew of Melvill's thinking and of the letter, and why it was not mentioned when he wrote on stellar aberration, is open to conjecture. As mentioned earlier, Melvill was a family friend and might have discussed such ideas within the Wilson household. Patrick Wilson's colloquium in Glasgow took place some fifteen years after Melvill's premature death (see earlier in this chapter). The developments of Wilson's thinking between 1770 and his paper of 1782,[60] and the importance of his 'thought experiment' related to a water-filled telescope and to the concepts associated with the velocity of light, have been summarised by Pedersen.[61] The notion of experimenting with water-filled telescopes to investigate the nature of light was also aired by Boscovich[62] in 1766. Wilson's contribution, firmly rooted in the corpuscular emission theory of light, was too readily dismissed by Cavendish; it played an important role in the later considerations of Arago and Fresnel, as wave theory became established. Arguments on the matter resurfaced many times in the nineteenth century; these related to 'aether drag', and it was only through experiments carried out by Airy[63] in 1871 using a water-filled telescope that the matter was finally put to rest. Wilson's proposition that there should be no change on the effect of aberration according to the light passing through refracting material within a telescope was vindicated.

4.10 INVESTIGATING DEW POINT

Following his father's interest in investigating the temperatures of snow on the ground and in the air above it, and the behaviour of the formation of

hoar frost, Patrick Wilson submitted two further papers[64] on the extreme
low temperatures occurring in Glasgow in January 1780 and January 1781,
and on experiments conducted at the Observatory there. His work was
briefly noted[65] in Edinburgh circles. In the first of the papers presented
to the Royal Society, he comments that his father did not take part in the
observations as he was confined to his room with ill health. At the time of
writing the first paper,[66] in February 1780, he intimated to Maskelyne that
his father was well again and wished to be remembered to him. Again temperatures as low as 0 °F were recorded. A large variety of experiments were
conducted, one of the aims being to elucidate how and why cold air appears
to deposit hoar frost on some but not all materials. Again, in the first of
Wilson's papers he noted that 'a figure of an unicorn in stone, which stands
within the college, had resisted the attacks of the air all to the tip of his horn,
which accordingly was the only part distinguished by a patch of hoar-frost'.
The monument referred to is now located on the west side of the University
building at Gilmorehill and is part of the 'Lion and Unicorn' staircase moved
from the Old College. An image of the unicorn's head is displayed in Fig.
4.12, although it cannot be guaranteed that the delicate spiralled horn is the
original referred to by Wilson.

In the second paper,[67] the temperatures at different depths in the snow
were given; the coldest was at the surface of the snow where a temperature of

Figure 4.12   The unicorn on the western side of the main
University Building at Gilmorehill.

−13 °F was recorded. On the night of 21 January 1781, it was noted that there was a 'lively aurora borealis, most part of it of bright red, which formed a crown near to the zenith: but it mostly vanished about three o'clock'.

In his essay on the theory associated with dew, Tomlinson[68] presented Patrick Wilson's observations, commenting on their historical importance within the subject of the 'dew point' and noting that Wilson's work had been very much ignored and neglected by subsequent experimentalists. Although Wilson appeared to have missed out on making any statement of advance on the understanding of the subject, Tomlinson was very much impressed by his observations and on the way he reported them. He noted that:

> With these important results before him, it is wonderful to me how Wilson could persist in the idea that they depended on an 'excess of cold'; that our winters are so short, and the cold not usually severe enough, for the prosecution of 'so dark a subject', which he therefore hands over to the philosophers of more rigorous climes. The fact is again and again brought before him, that it is 'in the clearest and stillest nights only that the cold at the surface of the boards was observed to be most remarkable', and yet he allows the great discovery to escape him, that the deposit of moisture is a consequence, not a cause, of this cold. He discovers the highly important and suggestive fact, that different bodies freely exposed to the clear sky cool at different rates, and yet this great fact bears for him no fruit. He makes the equally suggestive discovery that the refrigeration is checked by screens, even where the material is so flimsy as gauze; and this is equally barren. Truly we can only wonder that a man having done so much should have stopped short – should have ploughed the land, sown the seed, watched the growing crop, and then have left others to possess themselves of the harvest. What is so surprising is, that Wilson should not have continued his experiments and observations at different seasons, and that such men as Black, Crawford, and Irvine, should not have seen in these results something to suggest the laying out of thermometers on other nights after the frost had disappeared.
>
> Another remarkable fact is, that Wilson's experiments should have been unknown to subsequent inquirers in the same field, the first of whom that has now to be mentioned is Mr. James Six, F.R.S., the inventor of a well known register-thermometer, an account of which, and of experiments made with it, was published in the 'Philosophical Transactions', and also in the Monthly Review.

Although the subject matter might be considered somewhat whimsical, or even bizarre, Patrick Wilson[69] reported in 1795 on the behaviour of 'lighted wicks when swimming in a bason of oil'. Perhaps an even more bizarre record of 1785 relates to his dog, which he described[70] as having a 'self-preservating instinct'. Apparently choking on some foreign substance, the dog began to eat cinders from the fireplace and eventually was sick, releasing an undigested hare's pelt. Of more importance, while in Glasgow he experimented[71] with etching and engraving glass plates with the purpose of using them, instead of copper plates, for printing images with exceedingly fine detail. Part of the incentive for his investigations was to improve the quality of banknotes which, alarmingly, he noted, were frequently forged, especially those of the Bank of England.

4.11 A FRIENDSHIP WITH WILLIAM HERSCHEL

Patrick Wilson had a long and fruitful friendship with Sir William Herschel; this was one of the issues that influenced him to move to London. As noted by Clark,[72] at some point he dispatched a letter to Herschel asking him to explain why he had claimed the discovery of sunspots as openings when Wilson senior had published this same conclusion a decade earlier. Apparently, Herschel destroyed the letter, as was his custom with communications that distressed him, but he did reply, the record being transcribed by his sister, Caroline, into the family's letters book. Herschel implied that he had known of Wilson's paper on the subject, but now he had read it, he 'avowedly disclaimed every merit as a first discoverer'. He then explained that he had neglected to survey the recent literature on sunspots because he wanted to avoid a fight with the German astronomer Johann Hieronymus Schröter, who, he confided, 'had written a tedious treatise on the solar spots'.

It is noted in Clerk[73] that in March 1789, Herschel wrote to Patrick Wilson saying:

> I have finished a second speculum to my new twenty-foot, very much superior to my first, and am now reviewing the heavens with it. This will be a work of some years; but it is to me far from laborious, that it is attended with utmost delight.

Wilson's admiration for William Herschel is evident in a letter of 3 November 1789,[74] written to Dr James Lind, a tutor at Eton, in which he expressed it in flowing terms and talked about 'the sublimity of Herschel's genius upon what he has devised, and is attempting, by means of *large apertures*; though no person admires more than I do the discoveries he has made within our own system'.

According to Coutts,[75] in June 1792, Professor Wilson informed the Faculty that William Herschel of Slough, the celebrated astronomer, was to visit Glasgow early the following month, and out of respect to him and his friend, John Komarzewski, a dinner was arranged in the forehall of the College, to which such other gentlemen as the committee thought proper might also be invited. At the same time the Senate resolved to recognise his eminence as a practical astronomer and the sublime discoveries made in the heavens by means of his highly improved telescopes by conferring upon him the degree of LLD, and ordered a diploma to be prepared for presentation to him when he came to Glasgow.

The friendship obviously continued when Wilson moved to London, and he was able to be involved with the great scientific investigations of the time. It can be claimed that he helped Herschel with his investigations into which

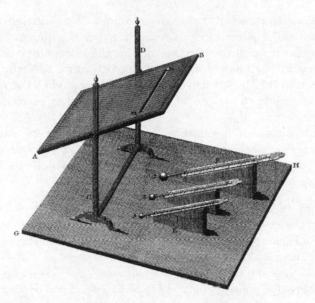

Figure 4.13 The arrangement of thermometers in Herschel's experiments which led to the discovery of infra-red radiation. The two smaller devices were lent to Herschel by Patrick Wilson, and were probably made by his father.

colours within the spectrum carried 'heat', the experiments leading to the discovery of infra-red radiation. From his experiments in Glasgow on the nature of hoar frost, Wilson had a good collection of sensitive thermometers which had been manufactured by his father, Alexander. Two of these were loaned to Herschel to form an array of three for placement behind a masking board covering a dispersed spectrum of the light of the Sun.

In describing his experimental arrangement supporting the three thermometers, Herschel[76] noted that that the first of the three

> was rather too large for great sensibility. No. 2 and 3 were two excellent thermometers, which my highly esteemed friend, Dr. WILSON, late Professor of Astronomy at Glasgow, had lent me for the purpose: their balls being very small, made them of exquisite sensibility . . . By repeated trials, I found that Dr. WILSON'S No. 2 and mine, always agreed in shewing the temperature of the place where I examined them, when the change was not very sudden; but mine would require ten minutes to take a change, which the other would shew in five. No. 3 never differed much from No. 2.

The arrangement of the three thermometers is depicted in Fig. 4.13. It seems likely that Alexander Wilson's thermometers, probably originally used in Glasgow and loaned by Patrick Wilson to Herschel, were key to the discovery of infra-red radiation, as noted by Jones.[77]

In addition, Herschel also discovered ultra-violet radiation by exploring the response of silver nitrate soaked in paper at different locations in a dispersed spectrum of the Sun, noting that the maximum discolouration occurred beyond the visible limit of the blue part of the spectrum. This fact was discussed in the Herschelian circle of friends, including Patrick Wilson. Herschel's experiments were noted by James Lind and, in one letter, Wilson wrote to him about how their 'excellent friend' Dr Herschel's last paper on the construction of the heavens contained matter very interesting to astronomers. Part of the letter has been transcribed by Goulding;[78] he has emphasised the words related to the discovery of ultra-violet radiation as in the following excerpt of Wilson's original letter:

> namely, the effect of the sun's emanations in changing the colour of paper or vellum moistened with the nitrate of silver from white to gray-brown and nearly black . . . It is remarkable that the less refrangible rays of the solar spectrum have very little effect in changing the colours – and that the blue and violet change it powerfully – but still exceedingly more remarkably that beyond the violet rays, and in the invisible boundary there acts something which powerfully also operates in changing the colours. It is not improbable that the sun's emanations may be more compounded than hitherto had been imagined, and that the principle upon which such effects depend may be quite a different form of Heat or Light . . . [and might be] revealed to us by the refraction of the prism.

According to Goulding,[79] the verse drama *Prometheus Unbound* by Percy Bysshe Shelley has long been recognised as a work in which extensive use is made of astronomical imagery as a metaphor for revolutionary political ideas. Goulding's thesis promotes the notion that James Lind, Shelley's mentor at Eton, was a direct conduit for much of the poet's astronomical knowledge. As well as being an accomplished astronomer himself, Lind was a friend, neighbour and correspondent of Sir William Herschel. Excerpts from Shelley's verse are reinterpreted as incorporating Herschelian visions of a huge cosmic life cycle at work in the Universe, encapsulating the organic process of formation of all things, from the greatest stars to the smallest particles of matter. The quotation above from Wilson's letter is used by Goulding as an example of how the concept of there being an 'ultra-violet' presence might alter our overall perspective of the underlying picture being portrayed in lines 206–35 of Act IV of Shelley's work.

Patrick Wilson supported Herschel when his famous 40-foot telescope was the object of derogatory comments by Lalande in 1806. Lalande remained sceptical of its merits although he had not seen it for himself, a fresh outbreak of war preventing an intended visit to Slough. In a letter[80] to Herschel, Wilson wrote:

> I don't know if as yet you have met with De La Lande's *History of Astronomy* for the year 1806 . . . There is a paragraph concerning you and the 40-feet Telescope,

evidently calculated to impress the belief of the *total* failure of your noble
Instrument and resting the proof on his correspondence with yourself. The structure of the whole paragraph appears to me very base and an outrange against all
decorums which govern men who stand upon their good character. I have often
wished that, for former provocations of a similar nature, you had denounced the
Haters of Merit, in the face of Europe, as unworthy of your correspondence.

Herschel ignored his friend's forceful advocacy and maintained a dignified
silence. His brother, Dietrich, took up the call and wrote an article defending
the telescope.

For the whole of his life, while in Glasgow as well as later in London,
Patrick Wilson remained in contact with James Watt, who in turn was also
involved in Herschelian society. One of Watt's 'inventions' was a micrometer (see Chapter 3, p. 59), and Wilson was supportive in that he thought that
mention of it should be made. On 24 July 1778, Wilson wrote to Watt,[81]
noting that the invention was little known and was several years ahead of
other developers' contributions. If he were to assert this priority with any
publication, Wilson conveyed that it would give him great pleasure to contribute evidence in his favour.

In 1781, Wilson came into possession of one of Watt's page-copying
machines, and, after practising with it, wrote to Watt[82] offering a suggestion
on how best to prepare the paper prior to using the machine and commenting on a potential problem with the breaking of its rollers, with the intention
of improving success with future sales.

James Watt also began to develop a machine for copying 3D sculptures,
and, in a letter to Wilson[83] dated 7 March 1811, in which he described its
progress, his initial comments are:

> I well remember your most excellent father's and your kindness to me at a period
> when I did not consider myself in meriting them, and I hope I shall always remem-
> ber them with gratitude while I live. It is, as you observe, a long time since —more
> than half a century. Let us be thankful we are spared to tell the tale.
>
> We were much gratified by Dr. and Mrs. Herschel's call upon us; our regret was
> that they could not stay longer. It gives us much pleasure to hear that they have got
> over the winter so well, and that the Doctor's ardour of research continues. Long
> may it do so; – without a hobby-horse, what is life?

Wilson replied[84] from 29 Kensington Square on 11 May 1811. He died on
31 December of that year at Kensington. Following his death, a letter of
condolence,[85] dated January 1812, was written to Andrew Wilson, his elder
brother, by James Watt. It reads as follows:

> It was with extreme concern I heard of my dear friend your brother's death.
>
> On these subjects I can offer no other consolations than what are derived from
> religion: they have only gone before us a little while, in that path we all must tread,
> and we should be thankful they were spared so long to their friends and the world.

4.12 METEORS AND METEORITES

As also happens today, reports of fireballs and meteors regularly appeared in newspapers. An early record[86] refers to an occurrence seen in Glasgow on 25 November 1752. The strange description reads:

> We had an account from Glasgow, that on Nov. 25, about four in the afternoon, a remarkable meteor, consisting of a large ball of fire and a long tail, passed over the place. Its direction was from north-east to south-west; and after having, for a short space of time, exhibited in its tail the various colours of the rainbow in the most beautiful manner, it seemed to expand, and burst into a thousand sparks of fire; it was immediately followed by a great shower of hail.

On 18 August 1783, at the time when the Chair was changing hands from father to son, a magnificent meteor was again seen in Glasgow and, indeed, across the whole of the country, although no mention of it appears in the local Observatory records. There were several reports of it made by local laymen, one of which[87] is of particular interest in that it gives an insight into other social activities in the city. Comments were made on its significance to the science of meteorology, and the report described the discussion made by more professional people. A part of it reads as follows:

> It was during the time when one of these early family hops was given to us that the great fiery meteor of the 18th of August 1783 made its appearance. This meteor was remarkable on two accounts; first, because it was the largest and most brilliant meteor ever taken notice of in the annals of Scotland; and in the second place, it may be said to have laid the foundation of the great science of meteorology; for, before the appearance of this great luminous body, if you learned men had turned their attention to the subject . . .
>
> At a quarter-past nine o'clock, while a country-dance was upon the floor, and while the late Robert Aitken, Esq. (of the Bank of Scotland), was leading down the dance, our room all of a sudden was illuminated, as is if a thousand gas-burners had suddenly burst forth, and after a period in which a person might quickly have counted six, had as suddenly been extinguished! It was not like a rapid and instantaneous flash of lightning, which disappears almost in an indivisible point of time, for there was a fleeting delay of light, while in a twinkling the dance was stopped and every person in the room hurriedly rushed to the windows to see what this mighty burst of light could have been; but to behold it was gone! and nothing outward could now be seen, except dark objects glimmering for the murky twilight. As I stood near the centre of the room, I unfortunately did not see this great ball of fire, but by its light, it appeared to me to have passed over the city in the direction seemingly in a line from Cuninghame's house (now the Royal Exchange) to the Slaughter-house (or Court-house on the Green). In its passage southward it was seen to burst into two parts, the discs of each of these parts appearing as large as the original ball seen in Glasgow. At the time when it thus burst, eight brilliant luminous fragments were thrown off from it, besides numerous lesser sparkles.

Nine years later, in August 1792, another brilliant meteor was recorded.[88] The report read:

> On Wednesday night, about eleven o'clock, a luminous meteor, of apparently great magnitude, and of a globular form, was seen from Glasgow, in the northern hemisphere. In its progress to the South-east, it assumed an appearance somewhat resembling a Comet, the tail of which was visible for some seconds, after the body was hid by the horizon.

Of much greater significance, some twelve years later a meteorite was seen to fall on the northern outskirts of Glasgow, in a quarry near High Possil. It is the first recorded fall in Scotland, although three other events have been reported, or older fragments found. A fragment of the High Possil fall, weighing about 150g, is held in the Hunterian Museum at Glasgow University. The major constituents are similar to those of a basalt: orthopyroxene, olivine, plagioclase feldspar and diopside. About 9 per cent of the meteorite consists of nickel–iron alloys, with traces of other minerals such as troilite, whitlockite, chromite and copper. A picture of it is shown in Plate 6, and images of sections of it under the microscope can be found on the internet,[89] together with its history subsequent to its impact in Glasgow on 5 April 1804.

An account of the fall was presented in the *Herald and Advertiser* on 30 April 1804; the article commented that the 'phenomenon in the neighbourhood . . . is at present the subject of much conversation and Philosophical Enquiry':

> Three men at work in a field at Possil, about three miles north from Glasgow, in the forenoon of Thursday the 5th curt. were alarmed with a singular noise, which continued, they say, for about two minutes, seeming to proceed from the south-east to the north-west. At first, it appeared to resemble four reports from the firing of cannon, afterwards, the sound of a bell, or rather of a gong, with a violently whizzing noise; and lastly they heard a sound, as if some hard body struck, with very great force, the surface of the earth.
>
> On the same day, in the forenoon, six men were at work in the Possil quarry, thirty feet below the surface of the ground, and there too an uncommon noise was heard, which, it is said, seemed at first to proceed from the firing of some cannon; but afterwards, the sound of hard substances hurling downwards over stones, and continuing in whole, for the space of a minute.
>
> By others who were at the quarry, *viz.* the overseer of the quarry and a man who was upon a tree, to whom he was giving directions, the noise is described as continuing about two minutes, appearing as if it began in the west, and passed around by the south towards the east; . . . as if three or four cannon had been fired off, about the great bridge which conducts the Forth and Clyde canal over the river Kelvin, at the distance of a mile and a half westward from the quarry; and afterwards, as a violent rushing, whizzing noise.
>
> Along with these last people, there were two boys, one of ten, and the other of

four years old, and a dog; the dog, on hearing the noise, ran home, seemingly in a great fright. The overseer, during the continuance of the noise, on looking up to the atmosphere, observed in it a misty commotion, which occasioned in him a considerable alarm, when he called out to the man on the tree, 'Come down, I think there is some judgement coming upon us,' and says that the man on the tree had scarcely got upon the ground, when something struck with great force, in a drain made for turning off water in the time of, or after rain, about ninety yards distance, splashing mud and water for about twenty feet around. The elder boy, led by the noise to look up to the atmosphere, says that he observed the appearance of smoke in it, with something of a reddish colour moving rapidly through the air, from the west till it fell on the ground. The younger boy, at the instant before the stroke against the earth was heard, called out 'Oh! such a reek!' [a foul smell] and says that he saw an appearance of smoke near the place where the body fell on the ground.

The overseer immediately ran up to the place where the splashing was observed, when he saw a hole made at the bottom of the drain. In that place a small stream of water, perhaps about a quarter of an inch deep, was running over a gentle declivity, and no spring is near it. The hole was filling with water, and about six inches of it remained still empty. The overseer having made bare his arm, thrust his hand and arm into the hole, which he judges to have been almost perpendicular, the bottom being perhaps a little inclined to the east, and the upper part to the west; at the bottom of the hole, he felt something hard, which he could not move with his hand. The hole was then cleared out, with a shovel and mattock, from an expectation that a cannon ball might be found, but nothing was observed except the natural stratum of soil, and a soft sandy rock upon which it lay, and two pieces of stone that had penetrated a few inches through the rock.

The pieces of stone, he took to be whinstone, and thinks that they were eighteen inches below the bottom of the drain, and that the hole was about fifteen inches in diameter. He was not sensible of any particular heat in the water, or in the pieces of stone, nor of any uncommon smell in the latter, although he applied them to his nostrils. He says that the one piece of stone was about two inches long; that the other piece was about six inches long, four inches broad, and four inches thick, blunted at the edges and end; that the fractures of these pieces exactly coincided; that he does not know whether the fracture was caused by the violence of the fall, or by the mattock; and that he never saw any such stone about the quarry.

Some days later, when the particulars which have been narrated, became known; a careful search was made for these pieces of stone, which had been disregarded, and the first mentioned piece was soon found; but the largest piece having been used as a block in the quarry, and having fallen among rubbish could not be discovered. Some days after, a fragment of it was detected. The two fragments recovered make the two extremes of the stone: on the surface, they are pretty smooth, and of a black colour; but internally they have a greyish appearance. The intermediate part larger than both seems, as yet, to be lost.

At the village of High Possil which is within a quarter of a mile of the place where the stone fell, the noise gave much alarm to those who were in the open air; and there, it seems, they thought that the sound proceeded from south east to north-west, agreeably to the report of the three men first mentioned.

Two men at work within a hundred yards of the house of Possil were alarmed by the noise; they thought it over their heads, and that it resembled the report of a cannon, six times repeated, at equal intervals, with a confused uncommon sound of ten minutes duration; the noise seemed to begin in the north, and to turn round by the west, south and east, to the north.

The 5th curt. was cold and cloudy; a little more cloudy in the north-east than in the other quarters. It would gratify many who have heard the above circumstances, to learn through the Glasgow Newspapers, whether any remarkable noise was observed in the neighbouring parts of the country, on the forenoon of the 5th curt. that had any resemblance to what was remarked near Possil.

April 30th 1804.

Much of the above report can also be found, with a little more material, in the *Philosophical Magazine* in an article entitled 'Another stone from the clouds'.[90] The event was investigated by Robert Crawford, the proprietor of the quarry, and Dr Freer, Dr Jeffry, Professor Davidson and Professor M'Turk of the University, but with no reference to the current Professor of Astronomy, Dr Couper, the date being very soon after his appointment. Mr Crawford remarked that both fragments had a fishy, fetid smell when he first examined them. Apparently the noise of the explosion was heard in Falkirk, some 24 miles east of Possil. A small monument, close to the original impact point, was dedicated in 2005 to celebrate the bicentenary of the event (see Plate 6).

Meteors seen over Glasgow have continued to be reported spasmodically, as that of 1816 bears testament. A report of this event[91] reads:

METEOR. On Tuesday evening, a very brilliant meteor was visible here for several hours. Soon after sun-set the horizon towards the north became very luminous, seemingly with electric light, or Aurora Borealis. About seven o'clock, a band of light seemed to detach itself from the mass, and very soon formed a nearly semi-circular arch, or band, of brilliant white light, about three or four degrees broad, well defined, apparently tapering near its extremities to points in the horizon, on one side nearly E.N.E., and the other W.S.W., or perhaps nearly S.W. It rose on the east side a few degrees east of the Pleiades in Taurus, and, passing though Andromeda, nearly filled the space between the stars Epsilon and Zeta in the Swan, and entering the Milky Way at a very acute angle, coincided with it for a great number of degrees. Between the constellations Aquila and Lyra it passed so as to be nearly equidistant from the bright stars Altair and Lyra, whence, continuing very much in coincidence with the Milky Way, it descended to the horizon on the west side. It continued nearly stationary, and of almost equal brightness, till near ten o'clock, when it began gradually to disappear. Its appearance during the whole time it was visible was very much the same that would have been produced by a cylinder of electric light at a great height in the atmosphere, stretching along a line of great and indefinite extent from the E.N.E. to the S.W., having its axis in a straight line, and being throughout of equal diameter.

On 3 November 1872, a magnificent bolide was observed in Glasgow by a Robert McClure. His description[92] notes that it was seen in Auriga about 10° east of Capella and 'glided' across the sky, shining with a brilliant green light and exhibiting a pear-shaped disc one-third of the semi-diameter of the full Moon. At the middle of its path, almost due north, its velocity abated and its colour changed to a whitish-blue. The meteor, accompanied by a diminutive red tail and followed by a train of sparks, gradually approached the horizon and disappeared behind a low cloud in the NNW.

References

1. Patrick Wilson (1789), 'Biographical Account of Alexander Wilson, M.D. late Professor of Practical Astronomy in Glasgow. By the late Patrick Wilson, A. M. Professor of Practical Astronomy in the University of Glasgow', *Transactions of the Royal Society of Edinburgh*, 10, 279. This memoir was read at the Royal Society in 1789, but was not published at the time. It was found among Patrick Wilson's papers, and then published in 1826.
2. Ibid., 281.
3. Ibid., 282.
4. Hansard, T. C. (1825), *Typographia: An Historical Sketch of the Origin and Progress of the Art of Printing*, London: Baldwin, Cradock and Joy.
5. Mackie, J. D. (1954), *The University of Glasgow, 1451–1951: A Short History*, Glasgow: Jackson, p. 227 [GUL Education S271 1954-M – six copies].
6. Coutts, J. (1909), *A History of the University of Glasgow (From its Foundation in 1451 to 1909)*, Glasgow: James MacLehose & Sons, p. 263 [GUL Education S271 1909-C – two copies; GUL Sp Coll Mu21-a.2; Sp Coll Bh12-e.14; GUL Sp Coll MacLehose 824].
7. Murray, David (1927), *Memories of the Old College of Glasgow – Some Chapters in the History of the University*, Glasgow: Jackson, Wylie and Co., p. 185.
8. Boney, A. D. (1992), personal communication.
9. Wilson, A. (Dr) & Sons (1772), *A specimen of some of the printing types cast in the foundery of Doctor A. Wilson and Sons*. Glasgow: College of Glasgow. This early version contains a small range of roman and italic typefaces, with print style examples using passages of the Bible [GUL E-Book]. Other editions available are:
   (1783), *A specimen of printing types . . . cast in the letter foundery of Dr. Alex Wilson and Sons*, Glasgow: Alexander Wilson and Sons [GUL Sp Coll e245].
   (1789), *A specimen of printing types cast in the letter foundery of Alexander Wilson and Sons*, Glasgow: Alexander Wilson and Sons. Contains a wide range of fonts including Saxon, Hebrew and Classical Greek [GUL E-Book].

(1815), *A specimen of modern cut printing types, by Alex. Wilson and Sons, letter found-ers, Glasgow*, Glasgow: Alexander Wilson and Sons [GUL Sp Coll Mu25-y.46].
(1819), *Specimen of printing types, by Alexander Wilson and sons, letter founders, Glasgow*, Glasgow: Alexander Wilson and Sons [GUL Sp Coll Mu-y.47].
(1834), *Specimen of modern printing types, cast at the letter foundry of Alex. Wilson and Sons, Glasgow [And Supplement]*, Glasgow: Alexander Wilson and Sons [GUL Sp Coll BD18-a.1 vol. 1, BD18-a.2 vol. 2].

10. Wilson, 'Biographical Account of Alexander Wilson, M.D.', 284.
11. Melvill, Thomas (1752), 'Observations on light and colours'. Read before the Medical Society of Edinburgh on 3 January and 7 February 1752. A convenient presentation of the more important sections of this work can be found in the *Journal of the Royal Astronomical Society of Canada*, 8, 231–72, with a prefatory note by J. A. Brashear at pp. 229–30.
12. Wilson, Alexander, Notebook entries from 1735 to 1753. Donated by Helen E. Babington Smith (2009) [GUL Sp Coll Ms 574].
13. King, H. C. (1978), *Geared to the Stars*, Bristol: Adam Hilger.
14. Jones, R. V. (2008), 'Physical science in the eighteenth century', *Journal of Eighteenth-Century Studies*, 1: 2, 73–88.
15. Wilson, 'Biographical Account of Alexander Wilson, M.D.', 290.
16. Wilson, Dr (1769), 'LVI. Eclipses of Jupiter's First Satellite, with an Eighteen Inch Reflector of Mr Short's. Observed by Dr Wilson at the Glasgow Observatory', *Philosophical Transactions*, 59, 402–3.
17. Wilson, Alexander (1774), 'XI. An Improvement proposed in the cross Wires of Telescopes, by Dr. Wilson, of Glasgow. In a Letter to the Astronomer Royal', *Philosophical Transactions*, 64, 105–7.
18. Wilson, Alexander (1769), 'XLIII. Observations of the Transit of Venus over the Sun, contained in a Letter to the Reverend Nevil Maskelyne, Astronomer Royal, From Dr. Alexander Wilson, Professor of Astronomy in the University of Glasgow', *Philosophical Transactions*, 59, 333–8.
19. Ibid., 336.
20. Wilson, Alexander (1774), 'I. Observations on the Solar Spots. By Alexander Wilson, M. D. Professor of Practical Astronomy in the University of Glasgow. Communicated by the Rev. Nevil Maskelyne, Astronomer Royal', *Philosophical Transactions*, 64, 1–30.
21. Wilson, Alexander (1783), 'VII. An Answer to the Objections Stated by M. De la Lande, in the Memoirs of the French Academy for the Year 1776, against the Solar Spots Being Excavations in the Luminous Matter of the Sun, Together with a Short Examination of the Views Entertained by Him upon That Subject. By Alexander Wilson, M. D. Professor of Practical Astronomy in the University of Glasgow; Communicated by Nevil Maskelyne, D.D. F.R.S. and Astronomer Royal', *Philosophical Transactions*, 73, 144–68.
22. Pedersen, K. M. (2000), 'Water-Filled Telescopes and the Pre-History of Fresnel's Ether Dragging', *Archives for the History of the Exact Sciences*, 54, 499–564.
23. Bryden, D. J. (1968), *James Short and His Telescopes*, Handbook and Catalogue of the Commemorative Bicentenary Exhibition, Royal Scottish Museum Edinburgh.

24. Wilson, 'I. Observations on the Solar Spots', 7–10.
25. Gold medal given by the King of Denmark in the year 1768 to the late Alex. Wilson LL.D., Prof. of Astronomy in the University of Glasgow as a testimony of the high sense entertained in that Kingdom of the merit of his observations upon the spots of the Sun. Deposited in the Hunterian Museum by his son Andrew Wilson, Glasgow. Donation list of the Hunterian Museum, June 1818. Marks: OBV: Bust Christian VII, right Leg: CHRISTIANUS VII.DG.REX. DAN.NORV.VAND.GOTH (around); I.H.WOOLF.F (below bust) REV: Nude female holding wreath and horn in clouds and rays of Sun Leg: VERITATI LVCIFERAE (above); P.G. (base); ALLECTAE. / MDCCLXVIII. (exergue) [GLAHM 35381].
26. Wilson, Alexander (1780), 'Observations of Solar Spots', *Medical and Philosophical Commentaries by a Society in Edinburgh – Collected and Published by Andrew Duncan*, 3, 46–50.
27. Wilson, 'VII. An Answer to the Objections Stated by M. De la Lande'.
28. Wilson, 'Biographical Account of Alexander Wilson, M.D.', 292.
29. Bray, R. J. and Loughhead, R. E. (1964), *Sunspots*, London: Chapman & Hall Ltd.
30. Watson, F., Fletcher, L., Dalla, S. and Marshall, S. (2009), 'Modelling the longitudinal asymmetry in sunspot emergence: The role of the Wilson depression', *Solar Physics*, 260, 5–19.
31. Wilson, 'Biographical Account of Alexander Wilson, M.D.', 294.
32. Wilson, Alexander (1777), *Thoughts on General Gravitation and Views Thence Arising as to the State of the Universe*, London: n.p. [GUL Sp Coll Hunterian Ae.5.6 – the cover page has a handwritten inscription 'From the author' at the top; GUL E-Book].
33. Wilson (1789), 'Biographical Account of Alexander Wilson, M.D.', 295.
34. Wilson, *Thoughts on General Gravitation*.
35. Herschel, William (1783), 'On the Proper Motion of the Sun and Solar System; With an Account of Several Changes That Have Happened among Fixed Stars since the Time of Mr. Flamstead', *Philosophical Transactions*, 73, 247–83.
36. Lubbock, Constance (1933), *The Herschel Chronicle*, Cambridge: Cambridge University Press, p. 186.
37. Wright, Thomas (1750), *An Original Theory of New Hypothesis of the Universe*, London: n.p. [GUL Sp Coll BD5-a.21].
38. Wilson, Alexander (1771), 'XXXVIII. An Account of the remarkable Cold observed at Glasgow, in the Month of January, 1768; in a Letter from Mr. Alexander Wilson, Professor of Astronomy at Glasgow, to the Rev. Mr. Nevil Maskelyne, B. D. F. R. S. and Astronomer Royal', *Philosophical Transactions*, 61, 326–31.
39. Wilson, 'Biographical Account of Alexander Wilson, M.D.', 295.
40. Ibid., 297.
41. Mackie, *The University of Glasgow, 1451–1951*, p. 222.
42. Armitage, A. (1950), 'Alexander Wilson, M. D. – A University Astronomer of Eighteenth Century Scotland', *Popular Astronomy*, 58, 388–94.

43. Garstang, R. H. (1964), 'Alexander Wilson, 1714–1786', *Journal of the British Astronomical Association*, 74, 201–3.

44. Testament of Alexander Wilson, CC9/7/73, p. 266. Available at: www.scotland-speople.gov.uk/content/images/famousscots/fstranscript69.htm

45. Jones, Nathaniel (1791), *Jones's directory; or, useful pocket companion, for the year 1791; containing alphabetical lists of the names and places of abode of the merchants, manufacturers, traders, and shopkeepers, in and about the city of Glasgow* [GUL Sp Coll Bh11-i.7].

46. Fleming, John (1974), 'A Dissertation upon the Sun and Moons proportional and Relative distance from the Earth. As exhibited By the New Mathematical Instruments Constructed for that purpose By John Fleming Inquirer into the Arts 1774'. Several astronomical drawings are included. At the end of the introduction is the signature 'John Fleming' and the date and place: 'Nook of Kilbride near Glasgow May: 13th. 1774' [GUL MS Hunter 623 (S.8.5)].

47. Emerson, Roger L. (2008), *Academic Patronage in the Scottish Enlightenment: Glasgow, Edinburgh, St Andrews*, Edinburgh: Edinburgh University Press, p. 194 [GUL Education B451.5].

48. Copy of letter to the Duke of Portland. Memorial of Patrick Wilson by which he intimated to the Faculty his determination to resign, together with a copy of his resignation letter [GUL Sp Coll MS Murray 663/18/4].

49. Copy of letter signed by Patrick Wilson, Professor of Astronomy, to Henry Dundas, 1st Viscount Melville, London: 1799. The letter concerns a Memorial of Thomas Jackson, whom Wilson had in mind as his successor when he retired [GUL Sp Coll MS Murray 663/18/5].

50. Emerson, *Academic Patronage in the Scottish Enlightenment*.

51. Muirhead, J. P. (1854), *The Origin and Progress of The Mechanical Inventions of James Watt, Vol. II. Extracts from Correspondence*, London: John Murray, p. 107.

52. The Maitland Club (1850), *Deeds instituting Bursaries, Scholarships and other Foundations in the College and University of Glasgow*, Glasgow: n.p., pp. 253–8.

53. Gavine, David M. (1981), 'Astronomy in Scotland 1745–1900', unpublished PhD thesis, Open University, 62.

54. Morrison-Law, A. D. (2007), *Making Scientific Instruments in the Industrial Revolution*, Farnham: Ashgate Publishing. Fig. 5.5 on p. 131 shows the interior of Jordanhill House, with a telescope by William Herschel, once owned by Patrick Wilson.

55. Wilson, Patrick (1782), 'VII. An Experiment proposed for determining, by the Aberration of the fixed Stars, whether the Rays of Light, in pervading different Media, change their Velocity in every Medium whose refractive Density is known. By Patrick Wilson, A. M. Assistant to Alexander Wilson, M. D. Professor of Practical Astronomy in the University of Glasgow; communicated by the Rev. Nevil Maskelyne, D. D. F. R. S. Astronomer Royal', *Philosophical Transactions*, 72, 58–70.

56. Ibid., 58–70.

57. Jungnickel, C. and McCormmach, R. (1996), *Cavendish*, Philadelphia, PA: The American Philosophical Society.

58. Pedersen, 'Water-Filled Telescopes', p. 513.
59. Wilson, 'VII. An Experiment'.
60. Ibid.
61. Pedersen, 'Water-Filled Telescopes'.
62. Krajnović, Davor (2011), 'A Jesuit Anglophile: Rogerius Boscovich in England', *News & Reviews in Astronomy and Geophysics*, 52, 6.16–6.20.
63. Airy, Sir George Biddell (1871), 'On the supposed alteration in the amount of Astronomical Aberration of Light, produced by the passage of the Light through a considerable thickness of Refracting Medium', *Proceedings of the Royal Society of London*, 20, 35–9.
64. Wilson, Patrick (1780), 'XXVI. An Account of a most extraordinary Degree of Cold at Glasgow in January last; together with some new experiments and Observations on the comparative Temperature of Hoar-frost and the Air near to it, made at the Macfarlane Observatory belonging to the College. In a Letter from Patrick Wilson, M. A. to the Rev. Nevil Maskelyne, D. D. F. R. S. and Astronomer Royal', *Philosophical Transactions*, 70, 451–73; Wilson, Patrick (1781), 'XXVI. Farther Experiments on Cold. Made at the Macfarlane Observatory Belonging to Glasgow College. In a Letter from Patrick Wilson, M. A. to the Rev. Nevil Maskelyne, D. D. F. R. S. and Astronomer Royal', *Philosophical Transactions*, 71, 386–94.
65. (1785), *Medical Commentaries For the Years 1783–84 – Collected and Published by Andrew Duncan*, 9, 424–5.
66. Wilson, 'XXVI. An Account of a most extraordinary Degree of Cold at Glasgow'.
67. Wilson, 'XXVI. Farther Experiments on Cold'.
68. Tomlinson, Charles (1863), *Experimental Essays – III History of the Modern Theory of Dew*, London: Virtue Brothers & Co.
69. Wilson, Patrick (1795), 'An Account of certain Motions which small lighted Wicks acquire when swimming in a Bason of Oil; together with Observations upon the Phenomena tending to explain the Principles upon which such Motions depend. Glasgow College, April 28, 1795', *A Journal of Natural Philosophy, Chemistry and the Arts – Illustrated with Engravings* (1798), vol. II, 167–72.
70. Wilson, Patrick (1783), Extract of a letter from Patrick Wilson. *Medical Commentaries for the Year 1780 – Collected and Published by Andrew Duncan*, 7, 422–6.
71. Wilson, Patrick (1798), Copy of a Letter from Professor Wilson, of Glasgow, on the Art of multiplying Copies of engraved Plates and Stamps in relief. Glasgow College, 23rd March, 1798. *A Journal of Natural Philosophy, Chemistry and the Arts – Illustrated with Engravings* (1799), vol. II, 60–3.
72. Clark, S. (2007), *The Sun Kings: The Unexpected Tragedy of Richard Carrington and the Tale of How Modern Astronomy Began*, Princeton, NJ: Princeton University Press.
73. Clerk, Agnes M. (1895), *The Herschels and Modern Astronomy*, London: Cassell and Co., p. 36 [GLA LRA U11-f.18].
74. Muirhead, J. P. (1854), *The Origin and Progress of The Mechanical Inventions of James Watt, Vol. II. Extracts from Correspondence*, London: John Murray, p. 229.

75. Coutts, *A History of the University of Glasgow*, p. 306.
76. Herschel, W. (1800), 'Investigation of the Powers of the Prismatic Colours to Heat and Illuminate Objects; With Remarks, That Prove the Different Refrangibility of Radiant Heat. To Which is Added, an Inquiry into the Method of Viewing the Sun Advantageously with Telescopes of Large Apertures and High Magnifying Powers', *Philosophical Transactions of the Royal Society of London*, 90, 255–83.
77. Jones, R. V. (1978), 'Through music to the stars', *Notes and Records of the Royal Society of London*, 33: 1, 37–56.
78. Goulding, C. (2006), 'Shelley's Cosmological Sublime: William Herschel, James Lind and "The Multitudinous Orb"', *The Review of English Studies*, new series, 57: 232, 783–92.
79. Ibid., 783–92.
80. Lubbock, *The Herschel Chronicle*, p. 313.
81. Muirhead, *The Origin and Progress of The Mechanical Inventions of James Watt*, p. 107.
82. Ibid., p. 125.
83. Ibid., p. 329.
84. Ibid., p. 333.
85. Ibid., p. 336.
86. The Monthly Chronologer (1752), *The London Magazine, or Gentleman's Monthly Intelligencer*, 21, 574.
87. Senex, Aliquis, J. B., etc. (1884), *Glasgow past and present: illustrated in Dean of Guild court reports and in the reminiscences and communications of Senex*, ed. David Robertson, Vol. 2 1773–1865, Glasgow: n.p., p. 194 [GUL History DX200 SEN; GUL Sp Coll Mu23-b.21-24; GUL Sp Coll Bh11-a.25-27].
88. *The Times*, 31 August 1792, p. 4. The article originally appeared in a Scottish newspaper on 25 August 1792. This reference is taken from the website of Mark Bostick and listed as NPA 08-31-1792. Available at: http://www.mail-archive.com/meteorite-list@meteoritecentral.com/msg28070.html
89. Faithfull, John (2000), 'The High Possil Meteorite'. Available at: http://www.hmag.gla.ac.uk/john/Huntmin/hposs.htm
90. 'Another stone from the clouds' (1804), *The Philosophical Magazine*, 18, 371–4.
91. *The Times*, 3 October 1816, p. 4. This reference is taken from the website of Mark Bostick and listed as NPA 10-03-1816 – Glasgow Meteor Report. Available at: http://six.pairlist.net/pipermail/meteorite-list/2005-January/007576.html
92. McClure, Robert (1872), 'Brilliant meteors', *Nature*, 7, 28–9.

Chapter 5

The Early Nineteenth Century

5.1 INTRODUCTION

After the turn of the nineteenth century, the early years found natural philosophy and astronomy at Glasgow University to be in the doldrums. Two appointments were made to the Chair of Astronomy, but the incumbents made little impact on the development of teaching or research in the subject. This was all to change remarkably, however, when John Pringle Nichol became the fifth holder of the Chair in 1836. Among the townsfolk of Glasgow there was a thirst to be involved in the development of science, and there were enterprises of great notoriety on the astronomy front, with the building of two observatories beyond the confines of the University.

5.2 WILLIAM MEIKLEHAM

According to Coutts,[1] the problems within the University regarding physics and astronomy began on 13 April 1796, when James Brown, Minister of Denino (in Fife) and assistant to Nicholas Vilant, Professor of Mathematics at the University of St Andrews, became Professor of Natural Philosophy in succession to Anderson. The appointment was one of the most unfortunate ever made. Having taught for one session, Brown declared himself to be in ill health and recommended that a John Leslie of Largo should conduct the class for the next session. Leslie did not accept the invitation, saying that a large annuity had recently been settled upon him. Thomas Jackson, a young MA who had won the Gartmore Prize at the end of the previous session, was appointed. He taught for two sessions, receiving the thanks of the Faculty for the ability and diligence with which he had acquitted himself.

At this stage, the appointment of the next Professor of Astronomy became an issue. Patrick Wilson declared his intention to retire and, although he was asked to reconsider, he adhered to his decision. He postponed his retirement for a few months while his successor was sought. At Wilson's suggestion,[2] the Faculty sent a memorial to the Chancellor for presentation to the Home Secretary, the Duke of Portland, recommending Thomas Jackson, who had taught the natural philosophy class during the earlier part of Brown's absence. The Chancellor declared that he could not recommend Jackson, and expressed a hope that the Faculty would not expect him to present to the Home Secretary a memorial with which he did not concur. The Faculty then asked Henry Dundas to transmit a memorial to Portland; Dundas did so, at the same time informing the Faculty that he considered it impossible for Portland to recommend anyone who did not have the support of the Chancellor, more especially as the Faculty were themselves not unanimous. All this time there was no actual vacancy, for Wilson had not yet placed his resignation in the hands of the Faculty; although he had sent it to the Chancellor, it had been returned to him. Details of the interchanges and arguments related to the delay in accepting Wilson's resignation and to the promotion of Jackson as his replacement have been presented by Emerson,[3] and some of the original related letters[4] are available in Glasgow University Library. The affair was deemed to reflect the differences of two camps, one based on a professorial lobby, which thought it best to preserve the teaching and observational work of the subject, and the other with a more political base, guiding the appointments of the Crown. Jackson, a worthy and prom-ising young man, rather suffered from the unskilful handling of the affair, but ten years later, he was appointed Professor of Natural Philosophy at St Andrews University, an office he filled with credit and success until 1837. In August 1799, the King issued a commission to William Meikleham to be Professor of Practical Astronomy and Observer, and, on 29 October 1799, Meikleham was admitted to office. For his trial he presented a discourse entitled *De Methods Longitudinam determinandi.* He held the Chair for four years only until he was promoted in 1803 to the post of Professor of Natural Philosophy. Mrs King,[5] Lord Kelvin's sister, noted in 1835 that Professor Meikleham 'was a good-natured, fat, little hunchback with a very red face; and he had a fat, little curly-haired black dog called Jura, that always toddled behind him'. A portrait of Professor Meikleham is shown in Fig. 5.1. Other notes and anecdotes related to him can be found in Murray.[6]

Prior to this latter development, efforts were made to induce Brown to reside in Glasgow and, at least in some part, to undertake the teaching of natural philosophy, but he avoided doing so. Meikleham had been appointed to teach for him. In May 1800, Brown proposed that Lockhart Muirhead, then acting as librarian, should be appointed as his assistant and successor, Brown retaining £150 of the annual emoluments of the Chair during his lifetime. The Faculty did not approve this, and, on 19 June, Meikleham was appointed to teach for another session. This arrangement continued for some time and, in November 1801, the Faculty allowed Brown to be absent until the end of March 1803, in the hope that his health might be restored, announcing that if this was not the case, they would have to endeavour to come to a permanent arrangement.

In April 1803, the Faculty resolved to procure reports from eminent medical men, appointed by themselves, on the state of Brown's health, given that he had not attempted to teach since 1797, nor even to reside in Glasgow.

Figure 5.1 A portrait of William Meikleham, the third Regius Professor of Astronomy (1799–1803), who later became Professor of Natural Philosophy.

Brown then began to make suggestions as to how the problem might be resolved, but those first made were considered inadmissible. Reports on his health were procured from Dr Munro and Dr Wardrop, but, as Brown agreed to resign on an annuity, they were not opened. Though Brown did woefully little as a professor, he made a very good bargain for himself, obtaining an annual allowance of £165 for life, £100 from the emoluments of the Chair of Natural Philosophy and the rest paid by the Faculty. He resigned on these terms in August 1803, and Meikleham was transferred from the Astronomy Chair to that of Natural Philosophy. It seems most remarkable that Brown taught the natural philosophy class for a single session but held the professorship for seven years and the annuity for thirty-three.

As time went on, Meikleham became exasperated with the arrangement under which £100 was withdrawn from his emoluments to make up part of the annuity allowed to the inert Dr Brown and, in June 1812, he proposed that after 1 November 1813, he should be relieved of this burden, and that the whole annuity of £165 should be taken from College funds. This suggested arrangement was carried by a majority. Meikleham was elected Clerk of Faculty in 1829 and 1830.

Natural philosophy continued in an inauspicious way for some years. Meikleham, however, did incorporate astronomy into its curriculum, as is clear from the notes of a William Sommerville who was a student in 1818–19. His manuscript[7] shows that the lectures included topics such as:

(i) What is the Evidence on which we pronounce that the Earth revolves on her axis? (ii) On the Laws of Kepler, (iii) On the explanation which the theory of Gravitation affords of the phenomena of the Tides, (iv) On the Laws of the Propagation of Light, (v) On the Evidence on which we pronounce the Velocity of Light to be Ascertained, (vi) On the circumstances which distinguish planets from those that were formerly known.

The latter topic shows that recent asteroid discoveries were being discussed; the notes continue with the phrase 'Of the four Planets – Ceres, Pallas, Juno and Vesta were all discovered since the commencement of the nineteenth century . . .'. Around this time, a few instruments were purchased for natural philosophy, including a Brewster's 'improved' telescope (£16 16s), a 3-inch telescope (£30 9s) from Robert Morton of Edinburgh and a pair of achromats from Adie.

Meikleham remained in office until his death in 1846 but for some years after 1838 his increasing age meant that he was unable to take his class. He was seized with illness around the time of New Year 1839, and after the class had been left untaught for a fortnight, the Faculty requested that Dr Thomson and Dr Nichol divide the work of teaching it between them, to which they agreed. The following October, Meikleham, still an invalid, was empowered to employ Nichol to assist him in the session about to commence. A year later, David Thomson of Trinity College, Cambridge, was appointed to teach the class, and he remained for four or five years before being appointed Professor at King's College, Aberdeen. Finally, in the last year of Meikleham's life, the Revd John Cunningham held the reins.

On the demise of Meikleham in 1846, John Nichol, then Professor of Astronomy, proposed an arrangement to bring teaching into line with recent discovery. Mackie[8] noted that he probably hoped to succeed to the Chair, but the Faculty, deciding that any Professor appointed must have knowledge of all the recent developments in the subject, elected William Thomson, son of the Professor of Mathematics, who had entered the University of Glasgow at the age of ten and was now a man of twenty-two and a fellow of Peterhouse College. The University gained the services of one who was to cast lustre upon the Chair of Natural Philosophy for fifty-three years, and who, by the title of 1st Baron Kelvin, or Lord Kelvin, is revered throughout the world. He at once put his stamp on defining the position of natural philosophy in the degree course. He changed what was a fairly stagnant situation into one that completely revolutionised the status of physical science in Glasgow, with worldwide repercussions. The Faculty were persuaded to spend money on apparatus, and, in 1855, Thomson established the first laboratory in experimental philosophy in which students could work for themselves. Lord Kelvin himself made several contributions to astronomy, perhaps the most important being those of calculating the time scales associated with the cooling of the Earth and of considering the process of energy generation within the Sun as a result of gravitational contraction, showing that this mechanism was totally inadequate to keep it radiating over its known lifespan.

5.3 JAMES COUPER

After Meikleham's transfer to natural philosophy, in August 1803, the King appointed the Revd James Couper, minister of Baldernock, to be the fourth Professor of Practical Astronomy and Observer in the University. According to Coutts,[9] Robert Findlay, Professor of Divinity and Dean of the Faculty, recalled the resolution adopted in 1797, when Mylne was elected Professor of Moral Philosophy, in which a condition applied whereby he could not be admitted until he had resigned his charge as one of the ministers at Paisley. Findlay, therefore, with the concurrence of George Jardine, Professor of Logic and Rhetoric, and Mylne, dissented from appointing a day for the admission of Couper until the latter had resigned his pastoral charge at Baldernock. Despite this, Couper was admitted on 24 October 1803. His formal trial was entitled *De Systemate Mundi*. Couper soon retired from the ministry at Baldernock, but he taught classes in astronomy only in the early part of his incumbency. In four years the class numbers dropped from twenty-one to fifteen, and then to two, with the blame being put on student poverty, the natural philosophy course covering the same ground and the state of the observing conditions at the Macfarlane building. Eventually Couper gave up observing too, his time taken up by his additional post as College Librarian. The fact that a splendid new observatory appeared in the city in 1810, initially with significant financial support and equipped with the latest technology, must also have affected his attitude to his position at the University.

The physical conditions at the Macfarlane Observatory were made progressively worse by the growth of the city and the concomitant smoke pollution. This was exacerbated by the erection of St John's Church in 1819. The building was substantive and, located as it was immediately south of the Upper Green, it constricted observations. An 1818 map of Glasgow, presented as a frontispiece of Chapman's book[10] and reproduced in Fig. 5.3, and that by Smith[11] of 1828 (see Plate 2) show the location of the new church, directly between the Observatory and the north end of Mcfarlane Street, newly opened in 1815; it would have directly compromised observations of the skies to the south. The tower was said to stand on the meridian line of the transit instrument. According to Murray,[12] the church, and most of the other new buildings, stood on land belonging to the University, but when the land was sold, no restrictive conditions seem to have been made regarding its future use, nor regarding the nature of the buildings that might be erected on it. As time went on, the usefulness of the Observatory was greatly eroded.

UNIVERSITY	UNIVERSITY OF GLASGOW
Dr Couper will begin his ANNUAL COURSE of LECTURES on	MATHEMATICS AND GEOGRAPHY
ASTRONOMY	THE MATHEMATICAL CLASSES will
TO-MORROW EVENING at eight o'clock. The Popular Course on Mondays an Fridays – The Mathematical Lectures on Wednesdays. Tickets to be had at Brash & Reid's Shop, or at Dr. Couper's house. COLLEGE; 10th Nov 1808.	Open on Tuesday the 6th of November, the SENIOR CLASS at *Ten* - and the JUNIOR at *Twelve o'clock* A COURSE of POPULAR LECTURES, on — GEOGRAPHY and ASTRONOMY, will commence on MONDAY the 19th November, at *Three o'clock.*

Figure 5.2 The left-hand image is a representation of an advertisement in the *Glasgow Herald* of 11 November 1808, giving details of Dr Couper's proposed lectures at the University. The right-hand image is a similar advertisement of 2 November 1832, probably placed by Dr Couper.

The story, if there is one, of Couper's contribution to Glasgow astronomy is an enigma, although he did play some part in the operation of the Garnethill Observatory, as outlined below. The whereabouts of any portrait or other image of him have proved to be elusive. The instruments were kept in good order, however, and the students were instructed in their use. There are references in the Senate minutes[13] of 1805 and 1813 to the cleaning of a clock, probably the Shelton, belonging to the Macfarlane Observatory, and this work would have been instigated by Couper. He advertised popular lectures in astronomy and mathematics (see Fig. 5.2) in the *Glasgow Herald* in November 1808, but what response this received is unknown. Again in 1832, in the same newspaper, he advertised popular lectures and a course on spherical trigonometry. He argued that interest might have returned to its former level if there had been a good, well-sited observatory in the local area. With an annual salary of £220 from the supplementary fund, plus the £50 from the Treasury and the availability of a house, the position of Regius Professor of Astronomy was an office of honour encumbered with relatively few duties.

Professor Couper certainly helped in procuring positions for his sons, as related by Coutts.[14] In 1809, his son James was appointed Keeper of the Hunterian Museum, but Couper himself was requested to take charge until his son took up the post, and to assist by acquainting himself with the collection. There is little trace of James's involvement, and the elder James Couper probably continued as Keeper from that time. In 1820, he proposed that his second son, William, be joint Keeper with him, and the latter was duly appointed on an annual salary of £70. Dr William Couper did have considerable skill in mineralogy and he lectured on its associated topics. In his later ill health, Professor Nichol (see next chapter) lectured for him on geology and crystallography.

Professor James Couper died on 7 January 1836. The instruments were still useable at this time as it may be noted that soon after John Pringle

Nichol had been appointed to the Chair of Astronomy, he was able to instruct William Thomson in the art of transit measurements, and the required telescope must then have been in good order.

During Couper's incumbency, there was a noteworthy episode in 1810 with respect to a 'battle' in the grounds of the Macfarlane Observatory. Area maps show that the College grounds were intersected by the Molendinar Burn (see, for example, Fig. 3.8). The Observatory stood on the eastern part of what was known as the High Green; it was reached by the bridge over the burn and a winding path beyond. There was a securely locked gate on the bridge to ensure the privacy of the Observatory, and students were not normally allowed access to this part of the grounds. As described by Murray,[15] one morning in November 1810, some students who were gathered in the western section of the College grounds were astonished to see a considerable number of soldiers of the 71st Regiment (the Highland Light Infantry, or Glasgow Regiment) from the barracks careering about to the east in the sacred enclosure around the Observatory. This was taken as an insult to the University, and the students resolved to expel the intruders. A large number of them – *togati et non-togati* – assembled under the leadership of a William Couper (1793–1857), then an Arts student but later President of the Faculty of Physicians and Surgeons of Glasgow, and Professor of Natural History. They speedily broke open the gate, crossed the bridge, rushed up the hill and attacked the Highlanders. Reinforcements were gained from the barracks, and the students were driven back up the slope to the Hunterian Museum. The fracas degenerated into a battle in which stones were thrown and several windows in the museum were broken. In the meantime the professors had communicated with the regiment officers, some of whom were dispatched to the scene, and the soldiers were called off.

5.4 GARNETHILL OBSERVATORY

A new observatory was erected in the city in 1810 by the Glasgow Society for Promoting Astronomical Science, a body founded in 1808 by a number of gentlemen who received a 'Seal of Cause' from the Magistrate and Council of Glasgow, allowing it to trade within the burgh system of guilds and incorporations. The Society acquired a site on Garnethill, which Murray[16] describes as being part of this area on the south side of Hill Street, between Scott Street and Thistle Street, the latter now renamed Garnet Street. Prior to the establishment of the Observatory, the site[17] was wasteland, although it was ploughed for growing crops of strawberries on the sunny slopes. Chapman's map[18] (see Fig. 5.3) shows the footprint of the building isolated from any other developments at the western edge of the city; only a few years later, Garnethill was a prime area of urban growth. This map also clearly shows the rectangular footprint of the older Macfarlane Observatory in the east side of the city. During this period, Glasgow boasted two observatories a little more

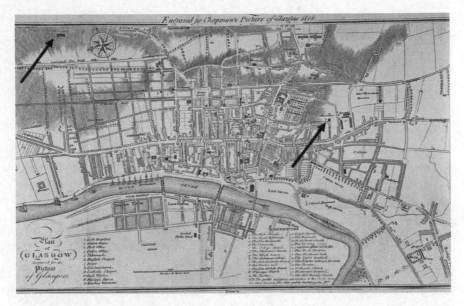

Figure 5.3 Chapman's 1818 engraving shows the locations, about one mile apart, of the two extant observatories in Glasgow. Garnethill is on the upper left, while the Macfarlane Observatory is centre right. Macfarlane Street (see Figure 3.13 and Plate 2) can be seen running from Gallowgate to the Macfarlane Observatory in the College grounds.

Figure 5.4 Smith's maps of 1822 and 1828 show the detail of the footprint and location of the Garnethill Observatory. Soon after its opening in 1810, the surrounding area fell victim to the rapid urban sprawl, as the two images graphically illustrate.

than a mile apart; that of the University was essentially defunct, however, and the new one at Garnethill had a very short lifespan. In addition to the Observatory's financial problems, urban growth was proceeding at such a lively pace that, within just a few years, the Observatory became surrounded by prestigious properties, all contributing to local smoke pollution. Fig. 5.4 shows two maps of the area separated by an interval of fewer than ten years.

Descriptions of the building and its costly equipment have been given by Chapman,[19] MacGregor[20] and McUre.[21] Drawings showing the features of the Observatory and its surrounding gardens are displayed in Figs 5.5 and 5.6.

Figure 5.5 A sketch of the Garnethill Observatory depicting its neo-Egyptian style of architecture.

Figure 5.6   A postcard of a drawing of Garnethill Observatory.

Garnethill Observatory was in the Egyptian style of architecture and was divided into three compartments. The central part formed a scientific observatory for the purpose of watching and recording celestial phenomena in order to promote the general interests of science, and was crowned by a rotatable cupola. It contained three massive 20-feet-high stone pedestals; one supported a sidereal clock, another an alt-azimuth instrument and the third a large mural circle by Troughton. The eastern section of the building was given over to more popular pursuits and was furnished with every instrument capable of blending instruction with amusement. According to Chapman,[22] a camera-obscura apparatus, on a great scale, gave a 'vivid representation of the surrounding landscape, comprising living, moving and even varying panorama, surpassing by far the finest works of the pencil, as nature is superior to art'. Telescopes of different kinds were available to 'display the magnificent host of heaven', while Dolland's largest ever solar microscope revealed the other extremes of the insect world. A brief description of some of the delights of the available slides for this instrument is given by Wade.[23] It was reasonably expected that a considerable revenue would have been derived from this part of the establishment, to which the public would be indiscriminately admitted. The third section was dedicated to the accommodation of the subscribers. It was furnished with maps, charts and globes of the largest dimensions and finest execution, and also provided a library of valuable treatises on astronomy, navigation and commerce. The meteorological instruments, destined to indicate with accuracy the existing state of, and approaching changes in, the weather, were also arranged in this room. An adjoining terrace held a Herschel telescope, 10 feet in length and 10 inches in diameter. On the roof of the building was a second Herschelian telescope, 14 feet long. The subscribers alone had access to this area containing the delicate scientific instruments. According to Cleland, 'This valuable institution which is exceeded only by Greenwich Observatory, has been honoured by the approbation of the most eminent astronomers of the country. Dr. Herschel, who has repeatedly visited the observatory, has been liberal in his approbation.'[24]

To establish the enterprise, a sum of £3,000 was required, and this was raised by selling 150 shares of £20 each. The shares were made heritable and deemed to be transferable property; their high price was to ensure that the subscribers 'would be as select as possible'. The construction, superintendence and management of the institution was vested in the proprietors, or in the committee appointed by them at a general meeting. In the foundation stone were deposited some newspapers, several medals and a plate bearing the inscription depicted in Fig. 5.7.

According to Morton-Gledhill,[25] the Garnethill Observatory could probably have been the first neo-Egyptian building in Glasgow, being an early example of a fashion that had first appeared in France and England. The Glasgow Necropolis, the largest of its kind, is the most spectacular example

MAY XI, MDCCCX
GEORGE III. L YEAR
THIS BUILDING, SUGGESTED BY THE LOVE OF SCIENCE,
ERECTED BY INDIVIDUAL SUBSCRIPTION
INTENDED TO PROMOTE THE STUDY OF ASTRONOMY, AND TO
RECORD OBSERVATIONS THE MOST INTERESTING,
IT IS HOPED WILL LONG REMAIN
APPROPRIATED TO ITS DESTINED OBJECT,
AND A MAGNIFICENT MONUMENT
OF THE SCIENTIFIC TASTE, AND PUBLIC SPIRIT, OF
THE CITY OF GLASGOW,
AND OF THE PRESENT TIMES.

Figure 5.7   The inscription on the plaque, set in the foundations of
Garnethill Observatory in 1810, expresses the sentiments of the age.

of the Egyptian revival in Scotland, but these burial grounds were developed
later, between 1832 and 1867, when the revival was at its height. The choice
of style for the Observatory of 1810 appears to be quite simple. The architect
was one Thomas Webster (1773–1844) of London. The Egyptian revival was
still ongoing in England at this time, so it is likely that he believed himself to
be designing a building at the very height of fashion, whereas locally he was
some years ahead of his time. It is Webster's only known work.

The Herschelian telescope mentioned above is listed in the inventory
of the major instruments at Garnethill, as provided by Howse.[26] In addition, there were two pieces by Troughton – an alt-azimuth instrument and
a mural circle – two clocks, an Arnold and a Hardy, both from London,
and a second, larger telescope made by Herschel, 14 inches in diameter and
with a focal length of 14 feet.

Dr Andrew Ure, MD (1778–1857), a lecturer in natural philosophy at
Anderson's Institution, took an active part in the establishment of the
Observatory, and was installed as the first Superintendent and Observer,
occupying an official residence in the building from 1809 to 1813. Dr Ure
was a man of great versatility but is best known as a chemist. One of his
admissions was that he secretly conducted experiments on the cadaver
of an executed murderer. He appears to have been vain, overbearing and
aggressive. He figures in *Northern Sketches*[27] as 'Dr Transit', and the character presented is nearer the truth than some of the others portrayed in that
volume. He has been described as 'a young man high in the estimation of the
world, but far higher in his own opinion . . . Believing he knows as much as
any man, he has the air of being perfectly satisfied on that head and envying
nobody.'

After acting as the first Superintendent and Observer from 1809 to 1813,
Dr Ure was succeeded by James Cross, former Professor of Mathematics,
also at Anderson's Institution. He was appointed to teach astronomy and

navigation and held office in the years 1814–15. Popular lectures were given in the Institution by a Dr Thomas Garnett. He was succeeded by Dr George Birbeck, who added a course of geography and astronomy. The Observatory was very popular for a few years, but the interest of the members of the Society and that of the public gradually waned, and it eventually fell into disuse. Although the Observatory had good equipment, there were inadequate funds to run it, and, towards the end of 1812, the Directors offered to sell it to Glasgow University. The proposal had the support of Couper, Meikleham and others, but there were inhibiting conditions that the University could not accept. Further efforts were made in 1819 and 1821 to induce the University to purchase with the restrictive issues removed, but the overtures fell on deaf ears. By this time, Garnethill and the adjoining land was being built upon; the University was of the opinion that the site had become unsuitable and reluctantly declined the offer. The Observatory's operation went into its final decline and an announcement[28] came on 24 February 1824 that it would be sold to the highest bidder. The equipment was sold, and it may be noted that the building, although it still remained in 1836, was gone by 1847, together with the pious wish expressed in its foundation stone.

It is interesting to note that in 1827, the 14-inch Herschelian telescope had found its way, via London and the Board of Longitude, to the Royal Observatory at the Cape of Good Hope. The inventory for the Cape Observatory, given by Howse,[29] cites the telescope as being in South Africa in 1827, but states that it had a diameter of 13 inches. Warner[30] has traced the history of the mirror from its manufacture to its present location in the safekeeping of the Museum of the South African Astronomical Observatory in Cape Town, and a picture of it is displayed in Fig. 5.8. It was cast by William Herschel on 14 April 1810, using a mixture of five parts tin to twelve parts copper, and it proved to be of excellent quality. It weighed fifty-three and a half pounds. In a letter sent from Glasgow on 2 September 1811 to his sister, Herschel noted that its diameter was 14 inches, but with 13 of polished surface. While installing it, he observed the Great Comet of 1811 through it using a power of around 110: 'the Moon was up, and the atmosphere was hazy . . .'.

A few days later, Herschel was in Newcastle upon Tyne, and on 10 September he wrote:

Dear Sister

I had not time at Glasgow to write another letter and we have been detained on account of the erection of the 14 feet telescope. It could not be put up till the place was properly prepared for it which was September 2. In the evening of that day I saw the Comet through it which you and our brother [Alexander] have probably observed every night . . .

Clerk[31] noted that Herschel travelled to Scotland in the summer of 1810 and received the Freedom of the City of Glasgow. This does not tally with

Figure 5.8 The 14-inch mirror, originally from the Herschelian telescope at Garnethill, now in the museum of the South African Astronomical Observatory in Cape Town.

the official list of recipients listed from 1800, as Herschel's name is not included. It may be noted, however, that the celebrated French astronomer François Jean Dominique Arago was awarded this honour on 20 September 1834. It is not clear who championed Arago, but he was much impressed by the Industrial Revolution and its implications for the city, and he wrote in admiring terms of James Watt. The citation of the award reads:

> Dominique François Jean Arago Member of the Chamber of Deputies, Astronomer-Royal, and Secretary of the Institute of France, FRS, L and E: 'In testimony of their admiration of his high talents and eminent scientific and literary attainments, and particularly of his successful exertions to extend the boundaries of astronomical science'.

By 1822, the Society, with its Observatory, was in the process of winding down its activities and was in correspondence with the Board of Longitude relating to the sale of the Herschelian telescope. These records have been explored by Warner,[32] who noted that the writer from Glasgow was an Auduir Trumpletin, although the surname was difficult to decipher; it may be noted, however, that an Andrew Templeton acted as Treasurer to the Society over several different periods and the signature may have been his. A letter to an unknown addressee at the Board of Longitude dated 14 June 1822 reads:

Sir,

In a letter I have from my friend Dr. Ure he mentions you had been inquiring at him respecting the 14 Feet Telescope belonging to the Glasgow Observatory with the view of purchasing it for the Board of Longitude, and desiring me to inform you the price at which the Society would be disposed to part with it. I have not had an opportunity of meeting the Directors of this subject, but think they would let it go for £300 or so, and pay the expenses of transmitting it to London. The condition of the instrument is as good as when new when we paid Dr. Herschell [*sic*] I think 400 for it, besides a considerable expense into bringing it down and fitting it up here. I shall be glad to hear from you when I will immediately have a communication with the Director and receive authority to treat with you.

> I am respectfully
> Your most obed. Ser^v
> Auduir Trumpletin

Nearly four years later, this was followed by a further letter to 'J. F. W. Herschell [*sic*] Esq., Secretary of the Royal Society', and dated 'Glasgow 10 March 1826'. It read:

Sir,

When Dr. Ure was in London about three years ago, he had a conversation with some Gentlemen connected with the Board of Longitude (Dr. Young particularly I think) respecting the fourteen feet Telescope belonging to the Glasgow Observatory, which was made under the directions of the late Dr. Herschel. It is now intended that this valuable Instrument should be disposed of and from what passed with Dr. Ure I understand a wish was expressed by some of the Gentlemen that if it was for sale, it might be offered to them for the use of some of their Establishments. As I learned that you are connected with some of the Members of this Board, and I know I presume the value of this Instrument, I take leave to advise you of this circumstance and to say that if the Board have any desire to be possessed of it, we will be happy to treat with them.

The Instrument is in perfect preservation and in no respect inferior in condition than when it left the Maker's hands.

May I request you would be at the trouble of procuring information as to the views of the Board respecting this, and favour me with the result at your earliest convenience.

For the Glasgow Astronomical Society, I Have the Honour to be

> Sir. Your very obd. Servant,
> Audir Trumpletin

A deal seems to have been done, and a letter of 7 November 1826 to Dr Young from John Barrow, Secretary of the Admiralty, records the safe arrival of the telescope in London and gives a direction for the Board of Longitude to purchase it for 300 guineas. Whether there was a delay in the payment being made or in the dispatch of a receipt for the monies is uncertain, but acknowledgement was not made until March 1827. It reads as follows:

> To Dr. Young,
> Glasgow 10 March 1827
> Sir,
> I have the honour to acknowledge receipt of the Navy Bill inclosed by you for Three hundred and fifteen pounds being the price of the Telescope sold by the Astronomical Society here to the Board of Longitude and the same is settled accordingly.
>
>> I have the Honour to be
>> Your very obedient Servt
>> Auduir Trumpletin
>> For Glasgow Astronomical Society

The date of shipment to the Cape is not known, but it missed the main consignment arriving on 19 November 1826 for establishing the Observatory. The instrument was not unpacked immediately and languished until the appearance of Halley's Comet in 1835 when it was used by Thomas Maclear, Her Majesty's Astronomer at the Cape. Diary references to its use in 1835–6, and again in 1843, have been collated by Warner,[33] who mentions that in 1843, it was probably the first telescope used to attempt to detect polarization in the light of a comet. This claim is challengeable, however, as Arago[34] made such observations of the comets of 1819 and 1835; this was noted by Grant.[35] The mirror was repolished by J. F. W. Herschel in 1838, before he returned to England. After 1853, there is no published reference to its use, although it occasionally appeared in later inventories of the Cape Observatory. It was rediscovered by Dr Ian Glass in 1987 under the McLean Telescope, and the mirror, as previously mentioned, is now on display (see Fig. 5.8) in the museum of the South African Astronomical Observatory in Cape Town.

It is especially noteworthy that around 1840, a group of people in Manchester were proposing to establish an observatory, referred to as the Royal Lancashire Observatory. An enquiry was made of the Astronomer Royal, Sir George Biddell Airy, as to equipment that might be of use for making observations. Part of his reply to a John Davies, dated 31 January 1840, is presented in Fig. 5.9. Good positive advice was given on instruments considered useful, but there was a dire warning about the difficulty of running such an institution, and Airy cited Glasgow as an example. After the introductory niceties, the letter notes that:

> In several instances of similar institutions, no difficulty has been experienced in the first construction of the Observatory, purchase of instruments, etc. The difficulty which has soon begun to show itself is in the maintaining of it. You are doubtless aware that some years ago an Observatory erected in a similar way at Glasgow and furnished (I believe) with excellent instruments was dismantled and

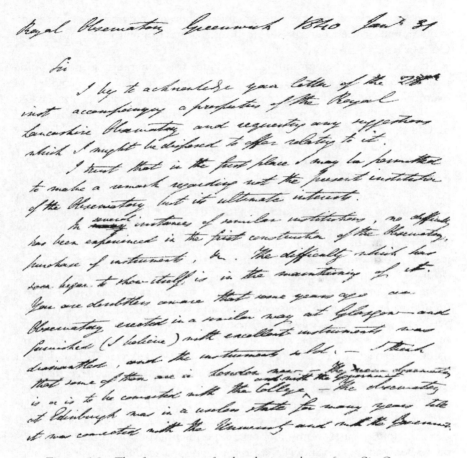

Figure 5.9 The first section of a handwritten letter from Sir George Biddell Airy (Astronomer a Royal 1835–81) draws attention to the problems of establishing and running an observatory. Dated 31 January 1840, it is in response to an enquiry by a John Davies of Manchester concerning the setting up of a Royal Lancashire Observatory.

the instruments sold. I think that some of these are in London now. The new Observatory is or is to be connected with the College.

The first-mentioned observatory was the one located at Garnethill, while the 'new Observatory' refers to a later enterprise at Horselethill, also known as Dowanhill (described in Chapter 6).

From the turn of the nineteenth century, and even before this time, the thirst of the public at large for keeping abreast with scientific developments is self-evident and readily seen within the general social activity of the city. Many individuals contributed to promoting an understanding of science and engineering. Educational courses of all kinds were burgeoning at this time.

There were several people who taught astronomy beyond the University walls, some of their enterprises no doubt competing with the courses that Couper tried to promote.

In most cases, the communicated science was general, astronomy going hand in hand with geography; the second image in Fig. 5.10 provides an example of this trend. Aspects of geography teaching have been discussed by Withers,[36] who has provided a list of twenty lecturers who operated in Glasgow from 1755 to 1830, with many of the courses advertised, or reported on, in the *Glasgow Herald*. Several of the activities specifically mention astronomy as being a topic within the presented subjects. For example, Robert Dobson 'opened a school for accounts, arithmetic, algebra, geography with globes and astronomy' in the Old Coffee House Land at Glasgow Cross, operating from 1755 to 1761. Later, between 1790 and 1801, Robert Lothian is noted as being a teacher at St Andrew's Entry, Saltmarket, and latterly at the Trongate, of geography, mathematics, military mathematics and astronomy 'with a set of improved machinery'; he is recorded as offering geography with globes and navigation, and a ladies' geography class, using a planetarium, orrery, armillary and sphere, as well as 'other instruments made by himself'. Between 1808 and 1816, James Denholm (see Fig. 5.10) is reported to have given a course of seventy lectures 'including the use of instruments' and 'transparent apparatus' with additional demonstrations using terrestrial and celestial globes and an orrery. Between 1820 and 1832, Alexander Watt gave geography lectures at Anderson's College in 1830, referring to himself as 'Professor'. He taught geography at 43 Dunlop Street (1820) and 72 Buchanan Street (1825). Fig. 5.10 includes a copy of a circular for winter courses, commencing on 23 October 1826, advertising geography, the principles of astronomy, history, experimental philosophy and chemistry to be given by Alexander Watt in the Glasgow Academy, a private school in the city (not to be confused with the present institution of that same name). Classes were available for ladies and gentlemen; those for the ladies were held in the afternoons while those for the men took place in the evenings after they had finished work. Educational aids were available in the form of star globes. Fig. 5.11 shows that a Mr Mackie included astronomy in his courses on mathematics and geography from 1825 to 1830. In 1835, John Gullan gave a course of twenty-five lectures on geography, geology and astronomy to the Gorbals Popular Institution for the Diffusion of Science.

On a much grander scale, a Thomas Longstaff, employed by the Glasgow Mechanical Institution, lectured on astronomy in 1825, using an orrery about 10 feet in diameter, at the Trades Hall and Glasgow's Theatre Royal at its earlier location; the presentations at the latter venue were accompanied by music (see Fig. 5.12). A few years later, in November 1829, under the auspices of the city authorities, an event was presented in the Assembly Rooms in Ingram Street, and a copy of its associated promotion is shown in Fig. 5.13. The lecturer, R. E. Lloyd, toured Britain presenting his show in most

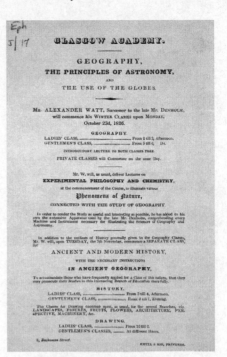

Glasgow Academy
———
Geography
The Principles of Astronomy
and
Use of the Globes
Mr Denholm will, as usual, begin his CLASSES
for the ensuing session upon MONDAY the 21st
OCTOBER inst.
The Class for Ladies from 5 till 6 afternoon
The Gentleman's Class from 7 to 8 in the evening
The Classes for DRAWING and PAINTING in all their
branches and for different departments of Mathematics
and Science, including PRACTICAL GEOMETRY,
DOCTRINES of LOGARITHMS, TRIGONOMETRY,
MENSURATION of HEIGHTS and DISTANCES and
SUPERFILES and SOLIDS, LAND SURVEYING,
NAVIGATION, FORTIFICATION, &c remain open at
the ordinary hours
GLASGOW ACADEMY
12th Oct 1816

Figure 5.10 The left-hand image is a representation of an advertisement in the *Glasgow Herald* of 14 October 1816, providing details of astronomical classes to be given by Mr Denholm, who had previously given courses for a number of years while the Garnethill Observatory was at its height of popularity. By October 1826 a Mr Alexander Watt had taken over these classes, as the image of the pamphlet advertising the courses available at Glasgow Academy shows. Note that all these courses had interlinks with geography and were independent of the University.

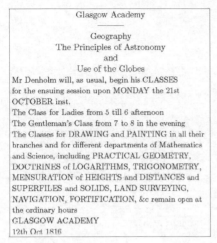

MATHEMATICS AND GEOGRAPHY
MR. MACKIE
LECTURER ON MATHEMATICS, GLASGOW MECHANICS'
INSTITUTION

Respectfully announces that on TUESDAY the 5th
April, he will commence a GEOGRAPHICAL
and ASTRONOMICAL CLASS, for Young Gentlemen at three o'clock P.M.
At present, various Classes are open for Theoretical and
Practical Arithmetic, Algebra, Geometry, Land-Survey-
ing, Navigation, and other branches of Practical Mathematics.
10 COCHRAN STREET
March 30th, 1825.

Figure 5.11 The left-hand image displays an advertisement in the *Glasgow Herald* of 1 April 1825 announcing lectures on astronomy by a Mr Mackie. More than four years later, in November 1829, he was still active as the flier presented in the right-hand image shows.

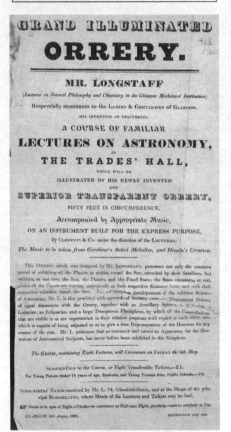

THEATRE ROYAL, QUEEN STREET

—

MR LONGSTAFF

Respectfully informs the Ladies and Gentlemen of
Glasgow, that he purposes delivering THREE
LECTURES on the CONSTRUCTION of the UNI-
VERSE in the Theatre on the Evenings of Wednesday,
Thursday, and Friday the 10th, 11th and 12th instant, at
half-past eight o'clock. They will be illustrated by his
Grand Transparent *Orrery*, together with a variety
of other magnificent Scenes, accompanied by appropriate
Music.
Admission, Boxes 2s 6d; Pit 1s 6d; Gallery 1s —
Tickets to be had of Mr L, and at Booksellers' Shops,
Observatory, August 6th, 1825.

Figure 5.12   The upper element is a representation of an advertisement in the
Glasgow Herald of 5 August 1825 for an orrery show to be given by Thomas
Longstaff at Glasgow's Theatre Royal. A flier announcing six lectures in the
Theatre Royal is presented on the bottom right; the advertisement for eight lectures
in the Trades' Hall in May 1825 is depicted on the bottom left of the figure.

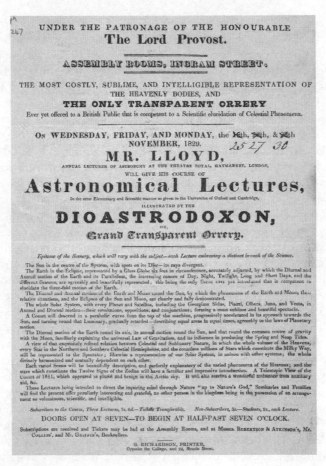

Figure 5.13 A flier for three lectures on astronomy, illustrated by a dioastrodoxon, to be given by a Mr Lloyd in the Assembly Rooms in Ingram Street in November 1829.

of the major cities, with his *Dioastrodoxon*, or Grand Transparent Orrery, some 21 feet in diameter. Some details of this spectacular circus have been collected by King.[37]

5.5 FULTON'S ORRERY

A biography of Fulton and the story of his wonderful orrery is given in a threepenny pamphlet[38] of 1859. An account of the history of this exquisite piece of machinery has also been provided by Grogans.[39]

The son of a shoemaker from near Fenwick in North Ayrshire, John Fulton (1800–53) had been given an interest in astronomy and navigation by his father. By natural inclination, he developed a passion for orreries, having read about their designs in various books and encyclopaedias. Fulton outlined his plans for his orrery in 1823 and spent three years undertaking the various calculations on the different gearing mechanisms, the design comprising 175 gears with a total of 10,300 teeth and over 200 moving parts. Making the patterns for the gear wheels took a further seven years. Its design is based on that of Pearson's improved orrery, but it is much larger. Originally its height from the ground was ~ 4 feet 6 inches to the top of the sun ball; the arm carrying Uranus is ~ 4 feet 4 inches long. In addition to the models and movements for the then known planets, there are radial arms to hold four asteroids. Further details of the dimensions of the mechanisms, pulleys and silk cords can be found in King.[40]

The exhibit was completed in 1833 and had its first display in Glasgow, which lasted three months during the winter of 1833–4. After receiving local commendation, Fulton submitted it to the Society of Arts of Scotland in Edinburgh and, in 1835, he was awarded the Society's Silver Medal, which had a value of ten sovereigns. The orrery was pronounced as the most perfect that had ever been constructed. It then travelled to Perth, Aberdeen, around the towns of Ayrshire and then to Dumfries. In England, it was exhibited in Carlisle, Liverpool, Manchester, Stockport and then in London and out to Windsor. While in the Gallery of Practical Science in London, the building in which it was housed was damaged by a severe hailstorm and the exhibit was packed up and returned to Scotland in 1852.

Fulton died in London in 1853, but the orrery continued to be exhibited by his brother, Thomas, in Argyle Street, carrying an admission charge of a shilling. The *Glasgow Examiner* of 1 January 1859 noted that it had been partially destroyed in an accident but successfully repaired by Thomas, and was again on show in Buchanan Street. A delightful description[41] of it appeared in the *Glasgow Herald* in 1863 when it was on display at 12 St Vincent Street. Later, a number of Glasgow citizens, impressed by the instrument's educational value, contributed £400 to buy it from the Fulton family and presented it to the city's Corporation. It was exhibited in the City Industrial Museum in the 1870s, and then in various branch museums, but was not given the proper care and attention it deserved. Having been partially restored several times, it now resides in the Kelvingrove Art Gallery and Museum in Glasgow. Images related to Fulton's Orrery can be found on the Glasgow Museums Photo Library website.[42] Its current display in the Kelvingrove Art Gallery and Museum is depicted in Plate 7, together with an advertisement of a earlier display in Argyle Street.

5.6 CONCLUSION

In summary, the first thirty-five years of the nineteenth century saw little activity in astronomy within the University. This was offset to some degree by the activities of citizens who had a thirst for knowledge, wishing to be enlightened by the then current developments in the subject. These passions continued, and were harnessed by the appearance of a brilliant star in the form of the fifth Regius Professor to be installed within the University. Although he did little in the way of direct observational work, he had vision and the rhetoric to carry his followers with him, altering the local astronomical landscape. His name was John Pringle Nichol, and his story is told in the following chapter.

References

1. Coutts, J. (1909), *A History of the University of Glasgow (From its Foundation in 1451 to 1909)*, Glasgow: James MacLehose & Sons, p. 321 [GUL Education S271 1909-C – two copies; GUL Sp Coll Mu21-a.2; GUL Sp Coll Bh12-e.14; GUL Sp Coll MacLehose 824].
2. Ibid., p. 319.
3. Emerson, Roger L. (2008), *Academic Patronage in the Scottish Enlightenment: Glasgow, Edinburgh, St Andrews*, Edinburgh: Edinburgh University Press, p. 194 [GUL Education B451.5].
4. The following references relate to letters in the Murray Collection.
   Copy of Letter to the Duke of Portland – Memorial of Patrick Wilson – intimated to the Faculty his determination to resign – Copy of resignation letter [GUL Sp Coll MS Murray 663/18/4].
   Letter from Patrick Wilson, Professor of Astronomy, to Henry Dundas, 1st Viscount Melville. London: 1799. The letter concerns Thomas Jackson, who Wilson had in mind as his successor when he retired. Memorial in Favour of Thomas Jackson – Signed by Patrick Wilson [GUL Sp Coll MS Murray 663/18/5].
   Letter from Downing Street 10th April 1799 – To the Principal Davidson and George Jardine, Clerk of Faculty noting the impossibility for his Grace (Duke of Portland) to recommend any person to fill the Chair of the Professor of Astronomy [GUL Sp Coll MS Murray 663/18/6]. See also GUL Sp Coll MS Murray 663/18/7; GUL Sp Coll MS Murray 663/18/8.
5. King, Elizabeth Thomson (1909), *Lord Kelvin's Early Home: Being the Recollections of his Sister, the Late Mrs. Elizabeth King together with some family letters and a supplementary chapter by the editor*, London: Macmillan and Co., p. 121 [GUL LRA U11-c.13; GUL Sp Coll Mu8-g.11].
6. Murray, David (1927), *Memories of the Old College of Glasgow – Some Chapters in the History of the University*, Glasgow: Jackson, Wylie and Co., p. 263.
7. Somerville, William (1818–19), 'Exercises in Natural Philosophy, prescribed by Dr. Meikleham and executed by William Sommerville' [GUL MS Gen 608].
8. Mackie, J. D. (1954), *The University of Glasgow, 1451–1951: A Short History*, Glasgow: Jackson, p. 260 [GUL Education S271 1954-M – six copies].
9. Coutts, *A History of the University of Glasgow*, p. 352.

10. Chapman, Robert (1820), *The topographical picture of Glasgow in its ancient and modern state; with sketches of a tour to the lakes and romantic scenery in the shires of Dumbarton, Argyll, and Perth, and to the Falls of Clyde*, Glasgow: R. Chapman [GUL Sp Coll Mu1-f.25].

11. Smith, David (1828), *Plan of the City of Glasgow and its environs with all the latest improvements accurately surveyed by Mr. David Smith*, Glasgow: Wardlaw & Co. [GUL Case Maps C18:45 GLA20].

12. Murray, *Memories of the Old College of Glasgow*, p. 264.

13. GUA 26696 (p. 369) [10 October 1805], GUA 26697 (p. 410) [15 January 1813] – notes of the accounts for the cleaning of the Observatory clock, Glasgow University Archives.

14. Coutts, *A History of the University of Glasgow*, p. 353.

15. Murray, *Memories of the Old College of Glasgow*, p. 492.

16. Ibid., p. 264.

17. Glasgow University Manuscript 117/27407. c. 1922. Extract from St Aloysius College Magazine with article on Garnethill Observatory by Fr Joseph Bullen SJ, Glasgow University Archives.

18. Chapman, *The topographical picture of Glasgow*.

19. Ibid., p. 166.

20. MacGregor, George (1881), *The History of Glasgow from the Earliest Period to the Present Time*, Glasgow: Thomas D. Morison; London: Hamilton Adams, p. 392 [GUL History DX200 MACGR; GUL Sp Coll Mu23-a.17].

21. McUre, John (1871?–2), *A View of the City of Glasgow*, Glasgow: John Tweed, vol. 3, p. 658 [GUL History DX200 MACUR3].

22. Chapman, *The topographical picture of Glasgow*, p. 166.

23. Wade, W. M. (1822), *A Tour of Modern and Peep into Ancient Glasgow*, Glasgow: n.p. [GUL Sp Coll Mu26-f.47].

24. Cleland, James (1840), *The Rise and Progress of the City of Glasgow: Comprising an Account of its Ancient and Modern History, its Trade, Manufactures, Commerce, and other Concerns*, Glasgow: John Smith, p. 187 [GUL Sp Coll Mu22-b.9].

25. Morton-Gledhill, R. I. (1988), 'The architecture of astronomy in the British Isles: A general study', *Vistas in Astronomy*, 32, 235–83.

26. Howse, D. (1986), 'The Greenwich list of observatories – a world list of astronomical observatories, instruments and clocks 1670–1850', *Journal for the History of Astronomy*, 17, 1–100.

27. Smith, Leonard [pseudonym for John Finlay] (1810?), *Northern Sketches, or, Characters of G*******, London: J. Dick. Five copies are available in the Glasgow University Library: GUL Sp Coll Mu1-e.32; GUL Sp Coll Mu1-e.33; GUL Sp Coll Mu1-f.46; GUL Sp Coll BD16-h.36; GUL Sp Coll Bf72-k.20.

28. Glasgow University Manuscript 117/27407.

29. Howse, 'The Greenwich list of observatories'.

30. Warner, B. (1987), 'The William Herschel 14-foot telescope', *Monthly Notes of the South African Astronomical Society*, 46, 158–63 [Board of Longitude Papers – Public Record Office: PRO 28].

31. Clerk, Agnes M. (1895), *The Herschels and Modern Astronomy*, London: Cassell and Co., p. 49 [GLA LRA U11-f.18].
32. Warner, 'The William Herschel 14-foot telescope', 158–63.
33. Ibid., 158–63.
34. Arago, D. F. J. (1855), *Popular Astronomy Vol. I – The Comets*, London: Longman, Brown, Green and Longman, p. 629.
35. Grant, R. (1852), *History of Physical Astronomy*, London: Henry G. Bohn, p. 313.
36. Withers, C. W. J. (1998), 'Towards a history of geography in the public sphere', *History of Science*, 36, 45–78.
37. King, H. C. (1978), *Geared to the Stars*, Bristol: Adam Hilger, pp. 314–17.
38. Rose, Thomas and Walker, William (1859), *Fulton's Orrery Applied and Explained with a Memoir of John Fulton*, Glasgow: George Gallie [GUL Sp Coll Mu56-i.4; GUL Sp Coll BG57-e.12].
39. Grogans, R. M. (1978), 'Fulton's Orrery', *Journal of the British Astronomical Association*, 88, 277–80.
40. King, H. C. (1964), *Exploration of the Universe: The Story of Astronomy*, London: Secker & Warburg, p. 337 [GUL Astronomy A11 1964-K].
41. The *Herald*, 21 Jan 1863, 'Fulton's Orrery'.
42. Glasgow Museums Photo Gallery website. Available at: http://www.glasgow. gov.uk/en/visitors/museumsgalleries/photolibrary.htm; see images T.2002.9; T.2002.9_01 and T.2002.9_03.

Chapter 6

A Professor of Eloquence

6.1 INTRODUCTION

Following the death of Professor Couper in the first week of 1836, for astronomy to return with a prominent profile within the University a man with charisma and inspiration was required. As it turned out, that person was immediately to hand. John Pringle Nichol (1804–59) had displayed his talents as a student at King's College in Aberdeen, achieving Honours in mathematics and physics. A verbal portrait, including the details of his family origins, marriages and forays into political economy, is contained in an essay by MacLehose[1] and in the *Oxford Dictionary of National Biography*.[2] It may be noted that after ordination and his brief career as a minister, a change in Nichol's theological views saw him pursue a career in education. In his early years, before he came to Glasgow, he was an inspiring teacher, writing several articles on the philosophy of education and political economy. Later, he frequently lectured at public events on the platform of 'education' and 'the cause of oppressed nationalities'.

His first wife, a Miss Tullis of Auchmuty, Fife, was the eldest daughter of the proprietor of the *Fife Herald*. Nichol's early political essays appeared in that newspaper, with himself acting as editor. There were two children, both later associated with Glasgow University. His son, John, became Professor of English Language at Glasgow after a time at Balliol College, Oxford; his daughter, Agnes, married William Jack, Professor of Mathematics at Glasgow. After his first wife's death in 1850, Nichol married Elizabeth Pease of Darlington on 6 July 1853, at an independent chapel. The Pease family were antagonistic to the union, as they thought Nichol had designs on Elizabeth's wealth, and there were also religious difficulties, as he was not a Quaker. She had taken an active part[3] in the movement to abolish slavery, and later became well known in her own right as a philanthropist.

The key stepping stones in Nichol's career can be seen from a collection of testimonials[4] which appear in a timeline. Over the period from 1819 to 1823, the various letters record his illustrious progress through undergraduate courses in mathematics, natural philosophy and chemistry at Aberdeen. In his undergraduate career, his teaching abilities had already been noted as, at the age of seventeen, he had successfully filled a vacancy to be in charge

V. *From Nassau W. Senior and James Mill,*
Esquires, London.

17th January 1836.

WE certify that when the Chair of Political Economy
in the Collége de France was vacant by the death of M.
Say, and it was understood that the French Government
were disposed to appoint a foreigner to that office, we,
being applied to privately to recommend a gentleman
whom we considered to be qualified for it, recommended
the Rev. J. P. NICHOL.

NASSAU W. SENIOR.

J. MILL.

Figure 6.1 A recommendation on behalf of J. P. Nichol for
the Chair of Political Economy in the Collège de France.

of the parish school at Dun during one of the vacations. By October 1823, his references declared that he was a fit person to teach mathematics within Cupar Academy. Four years later, he moved to Montrose to become Rector of Montrose Academy. At the time of his application, it may be noted that he had taught French in addition to mathematics, and the Revd Birrel commented that he had also given a 'Popular Course in Astronomy'. Around 1834, he applied for a post as a mathematics teacher at the High School of Edinburgh. Whether he took this position is not known, as he appeared to have suffered some illness which limited his activities. He was certainly in and around Edinburgh at this time, as he delivered public lectures through the Philosophical Society there, which attracted large audiences. His skills in this arena were noted by everyone who attended. In both platform presentations and his writings, he was a master of rhetoric.

In January 1836, Nichol was promoted by James Mill and Nassau W. Senior, the historian of India, to take the Chair of Political Economy at the Collège de France (see Fig. 6.1) when it became vacant after the death of Jean Baptiste Say, but he declined the nomination. Timed as it was, such a submission by these two gentlemen no doubt would have given more power to Nichol's application for the Astronomy Chair at Glasgow. In addition, for the latter, he also carried a portfolio of testimonials, including formal letters of support from Professor Henderson, Professor of Practical Astronomy in Edinburgh, Sir Thomas MakDougall Brisbane and Sir David Brewster. Sir Thomas, who was from Largs, where he had his own private observatory, was famed for establishing an observatory in Parramatta, near Sydney, Australia. He continued his scientific work on his return home to Makerstoun, near Kelso, and he had political influence within the University. Sir David was also a famous and influential scientist, known especially for the

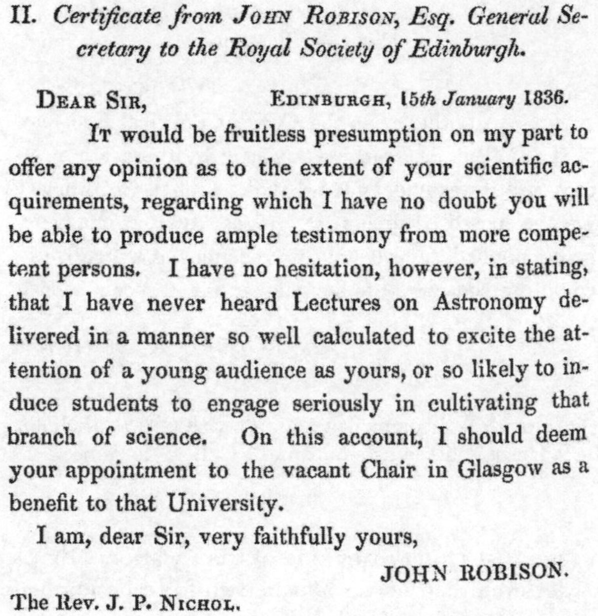

II. *Certificate from* JOHN ROBISON, *Esq. General Secretary to the Royal Society of Edinburgh.*

DEAR SIR,   EDINBURGH, 15th January 1836.

It would be fruitless presumption on my part to
offer any opinion as to the extent of your scientific ac-
quirements, regarding which I have no doubt you will
be able to produce ample testimony from more competent persons.   I have no hesitation, however, in stating,
that I have never heard Lectures on Astronomy delivered in a manner so well calculated to excite the attention of a young audience as yours, or so likely to induce students to engage seriously in cultivating that
branch of science.   On this account, I should deem
your appointment to the vacant Chair in Glasgow as a
benefit to that University.

I am, dear Sir, very faithfully yours,

JOHN ROBISON.

The Rev. J. P. NICHOL.

Figure 6.2   The letter of support from John Robison, General Secretary of
the Royal Society of Edinburgh, emphasises Nichol's eloquent lecturing style.

establishment of laws, now carrying his name, related to optical reflections,
refractions and polarization; he was Principal of St Andrews University
(1837–59) and later of Edinburgh University (1859–68). Included with other
testimonials, that of John Robison, General Secretary of the Royal Society
of Edinburgh (see Fig. 6.2), emphasised Nichol's gifts as a popular lecturer.
An alternative candidate, one known for his great literary contributions, was
Thomas Carlyle, but he was not truly a man of science and was probably
not as organised as Nichol in submitting his application. At thirty-two, tall
and handsome, with a gift for oratory and a reputation throughout Scotland
as a brilliant populariser of astronomy, he was an excellent choice for the
post as the fifth Regius Professor of Practical Astronomy. His appointment
was advanced by Sir Robert Peel, who mooted a knighthood at some stage,
although Nichol declined this. In 1857, he was encouraged by the local
Liberal Party in Glasgow to stand for Parliament, but he did not follow the
call, instead throwing his weight behind a Walter Buchanan.

Incidentally, the James Mill mentioned above was the father of John
Stuart Mill (1806–73), one of the greatest Victorian liberal thinkers, who
popularised the principles of philosophical empiricism and utilitarianism.
He became sympathetic to socialism and was a strong advocate of women's
rights and political and social reforms such as proportional representation,

trade unions and farm cooperatives. While at Montrose Academy, Nichol began a lifelong correspondence with Mill, in which they discussed questions on matters of philosophy and economics which were further developed in Mill's published works.

On 6 February 1836, King William IV appointed John Pringle Nichol to the Regius Chair. His trial subject was entitled *De finibus usu statuque praesenti Astronomiae practicae*. Technically Nichol's position was that of observer rather than teacher, and it is doubtful whether he was actually required to teach. Nichol was concerned that the teaching of practical science was somewhat lacking in the University, and he wanted to address this. The mathematics syllabus in Scottish universities was insufficiently substantive to allow great advances in theoretical astronomy, so the emphasis needed to be placed on practical work, as far as Scotland's climate would allow, but without duplicating the astrophysics undertaken at Greenwich and Edinburgh, that is physical and extra-meridional observations. With the founding of the Astronomical Institution of Glasgow (see below), an opportunity arose for this to be carried out by a public establishment, something that had never been previously undertaken in Britain.

In October 1836, Nichol wrote to the Astronomer Royal, Sir George Airy, and to J. D. Forbes (Professor of Natural Philosophy at Edinburgh University), commenting that the Glasgow Observatory may have been scientifically useless in the past, but it had the potential to be expanded and might act both as a teaching centre and as a popular establishment. Murray[5] has noted that Nichol, on his appointment, formed two classes, one 'Popular' and the other 'Scientific'. The former was intended for those who desired to receive a general view of astronomy, without having to use mathematics, and met twice a week. The Scientific class met four days a week; it was intended for students who desired to study astronomy for professional purposes, particularly navigation and engineering. Nichol promoted his ideas at the time of his inauguration on 1 October 1836, by saying in the introduction:[6]

> As the public cannot be aware of the nature of the course of introduction purporting to be given in the Astronomical Class, I have thought proper to publish the following details of my plan, as well of considerations which induced me to adopt it. It is my earnest hope, that the arrangements about to be specified will enable me to render the Chair I have the honour to occupy, promotive to a considerable extent of the important objects of the Foundation.

Later, he commented:

> The great and increasing attention now paid throughout Europe to the improvement of the means of internal intercourse, demands a supply of young men, trained in the theory of engineering, and practically conversant with the processes of surveying in all its departments; but as this demand became urgent only in recent years, our Institutions have not hitherto fully supplied the means

of communicating the necessary instruction. The Class here referred to is the only one of the kind in Scotland; – it may be termed the school of engineers. The Students are exercised in the different Astronomical calculations, and made acquainted with the construction and use of the tables founded upon them. The theory and adjustment of the various instruments are all minutely described. But perhaps the most important and valuable feature of the Class is this, – the Student has ready access to the Observatory, and is obliged to perform numerous computations from actual observations.

Some years later, Murray[7] wrote:

What success Professor Nichol had with these classes I do not know. They had ceased to exist before I became a student; but the scheme is notable as being an anticipation 90 years ago to the present doctrine that it is part of the function of a University to train students in science for professional life.

The skill and detail with which Nichol prepared his lectures can be seen from his notebook[8] for the session of 1837–8. His beautiful handwritten script declares that the 'First Section of Practical Astronomy' would deal with the 'Nature and use of those instruments by the aid of which the Practical Part of this science is carried on'. The instruments to be described were the transit instrument, the mural circle and the equatorial. The second section related to the application of these instruments, with detailed examples given of how arithmetic is applied to reduce any recorded measurements. Within the presented algebra is a section related to the fitting of data by the method of least squares. The third section related to problems regarding the planetary systems, and the course terminated with a discussion of the investigation with respect to the figure of the Earth. The cover page of his notes and an example of the presentation style of the algebra associated with the problem of atmospheric refraction are reproduced in Fig. 6.3.

Nichol was most generous with his teaching abilities, and gave courses well outside the remit of pure astronomy. During his latter years, Meikleham, the Professor of Natural Philosophy, was unable to teach on account of illness, and Nichol took his place. Dr Meikleham, according to Lord Kelvin,

taught his students reverence for the great French mathematicians Legendre, Lagrange, Laplace. Dr Nichol added Fresnel and Fourier to this list of scientific nobles; and by his own inspiring enthusiasm for the great French school of mathematical physics, continually manifested in his own experimental and theoretical teaching of the Wave Theory of Light, and of practical Astronomy, he largely promoted scientific study and fairer appreciation of science in the University of Glasgow.

In 1856, Nichol undertook the work of the Chair of Natural History during Professor William Couper's illness, lecturing for him on geology and crystallography, with an Allen Thomson adding a few lectures on fossil zoology.

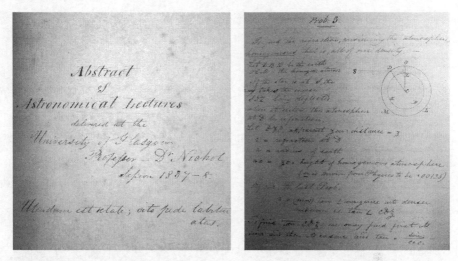

Figure 6.3   The cover page and a page of notes made by Professor Nichol for his lecture course of 1837–8. Problem 3, given as an example of his style, refers to the apparent positions of stars affected by refraction of the Earth's atmosphere.

His interest in and knowledge of geology was apparent as he contributed the introductory article[9] in the form of a preface to Johnston's *The Physical Atlas of Natural Phenomena*, published in 1856. He also offered to conduct the moral philosophy class in September 1837 when Mylne's energies were failing, but this was assigned to a Dr Fleming who later became professor of that department. This latter gentleman appeared to be quite reactionary. In 1843, he dissented from purchasing the Observatory at Horselethill; in 1846, he disapproved of the actions of the Faculty in reference to the sale of the old College (see Chapter 1, p. 10). Around ten years later, he objected to the appropriation of a room by the Professor of Natural Philosophy, William Thomson (who later took the title of Lord Kelvin), for expansion of laboratory space.

The students had no reason to regret attending courses on any subject presented by Nichol. His influence was not limited to the classes he taught, but extended to the whole University student population as a body. He imparted to them something of his own bright spirit, rousing them to aspire to higher things. Nichol was a man with a liberal mind and marked characteristics, and he exercised his initiative in a way that was sometimes frowned upon by his colleagues. Early and later portraits of him are displayed in Fig. 6.4, and an image of his bust is presented in Fig. 6.5.

Professor Nichol was also very keen to offer his teaching talents to the local citizens of Glasgow, and his inspiring influence on them was equally great. Towards the end of 1836, he delivered a course of popular lectures in the Assembly Rooms in Ingram Street on the then recent discourses

Figure 6.4 An early and a later portrait of the charismatic Professor John Pringle Nichol. The round chair may be that described by his son on p. 161.

Figure 6.5 An image of a bust of Professor John Pringle Nichol.

regarding nebulae, attracting such crowds of listeners that the event had to be repeated in the evening. What he had to say had been first written up in a note[10] extolling the discoveries made by Sir William Herschel on the structure of the Milky Way, on double stars and on nebulae. This early popular presentation offered an insight into the possibility of stars being born out of diffuse material as a result of rotating and spiralling motions. A few years later, it was followed by further speculation[11] that nebulae had evolutionary phases and that they might be birthplaces of the stars.

This latter commentary also dealt with some philosophical and theological aspects of the subject. The style of these two papers clearly conveys the literary skills of the author, with the expectation that they would be used further in his various books. On this particular topic, Nichol's later books became a source for the concept of an evolutionary 'nebular hypothesis', whereby, in commenting on Herschel's theories, he proposed[12] that there was a close relation between astronomical results and those of contemporary geology.

Such was the success of his public educational enterprise that it was proposed that the professors of botany, chemistry and natural history, as well as those of arts subjects, might give similar popular courses of lectures to the citizens of Glasgow, but nothing came of the suggestion (see Murray).[13]

As a lecturer to an audience of laypersons, Nichol was unrivalled. His statements were clear and precise, arranged in logical sequence and easy to follow, but beyond this he had extraordinary rhetorical power, and carried his audience from point to point, and from argument to argument, in language that swept along to a magnificent climactic summary. Murray[14] has provided anecdotes as to the power Nichol had by relating that Professor Story, one time Principal of the University, on unveiling the memorial stained glass window[15] in the Bute Hall, said:

> One of my own early reminiscences as a young student in Edinburgh was going to hear the Professor of Astronomy in Glasgow lecturing on the subject of his science one evening at the Philosophical Institution of that City. I was struck with the flow of lucid exposition and brilliant description which he gave us, as far as I remember, without note or any assistance, except the brightness and fertility and readiness of his own mind. I remember the tall, handsome figure, the imposing looks of this man from Glasgow, with which we young students in Edinburgh were very much impressed. I never, I think, saw him again, but I remember him as if I had seen him yesterday.

In addition, a Dr Donald Macleod described his experience as follows:

> The Professor of Astronomy was certainly the most eloquent lecturer I ever listened to. Well do I remember the thrills of awe and admiration in which he held us spell-bound, so that, like St Paul, we sometimes could not tell whether it was 'in the body or out of the body' that we were raised to magnificent visions of the universe, and he unfolded the Nebular Hypothesis, then less familiar than it is now.

At the time of Professor Couper's death, the inventory of instruments was much the same as it had been in Patrick Wilson's time, and the collection had deteriorated. The list has been collated by Gavine[16] and reads as follows:

> Large, medium (broken tube) and small instruments, 10-foot Herschel reflector. Dolland achromatic and stand. Small refractor, Astronomical clock with gridiron compensation. Clock with mercury compensation. Equal-altitude instrument. Troughton reflecting circle. Hadley sextant. Azimuth compass. Micrometer

and sun glasses. Two old globes. Two Short reflectors. Also in the observatory there were: 3 ft mural quadrant by Bird. Small mural quadrant. Zenith sector. Journeyman clock. Astronomical circle for transits, azimuths and amplitudes. Rain gauge.

In support of Nichol's appointment, the Faculty recommended the immediate repair of the instruments, preferably by Adie of Edinburgh. They also recommended that there should be regular meteorological observations, and that a committee, or Board of Visitors, be formed to inspect the instruments and accounts annually. By 1 May 1837, £120 had been paid out of the Wilson Fund on repairs, and £58 7s 3d on a new alt-azimuth instrument.

Nichol continued to keep an eye on politics, particularly with regard to connections with France. In 1840, he published a work[17] in the form of a letter to Sir Archibald Alison, a lawyer and historian who is described on the title page as 'The Historian of Modern Europe'. Written a quarter of a century after the end of the Napoleonic Wars, the letter explained that tensions between Britain and France had again been raised to breaking point, and warned that relations must be improved to avoid another war. Nichol stated that he addressed the letter to Alison, as opposed to an MP or the Prime Minister himself, because of the historian's 'acquaintance with the influences guiding European politics' and his 'generous sympathy with the welfare of all mankind'.

6.2 THE HORSELETHILL OBSERVATORY

Largely owing to the euphoric enthusiasm following Nichol's lectures, a Grand Dinner[18] for the Friends of Astronomical Science in Glasgow was held in the old Town Hall at the Cross on 16 December 1836. Two hundred gentlemen of Glasgow and the neighbourhood attended, and the occasion involved many speeches and what seemed to be a riot of toasts. Key to the outcome of the occasion were the comments made on the interplay of astronomy with commerce, and the importance thereof. After the first speech, the toast was to the 'Success in the proposed Astronomical Institution of Glasgow, and Mr Nichol, Professor of Astronomy'. Professor Nichol responded in customary flamboyant manner and later proposed a toast to the 'Ladies of Glasgow'; another toast by Dr Laurie was to the 'Memory of James Watt'. There was great intent that the venture should succeed, bearing in mind the previous failure of their predecessors of 1808. It was resolved to establish a body that would be called the Astronomical Institution of Glasgow, and to raise money by subscription for the erection of a new observatory with good instruments, for which £2,000 was deemed adequate. Nichol was carried along by this vision, and was encouraged by Sir David Brewster. The commercial distress of the depression of 1837 slowed down the collection of funds, but, by the new year, £1,195 had been raised.

The Institution published a minute of 30 May 1838, resolving to establish a new observatory as soon as possible, hopefully to coincide with the 1840 British Association Meeting in Glasgow. It would be a joint venture of the Institution and the College, the former supplying the building, the latter procuring the instruments which Nichol was already beginning to order. Negotiations for a transit telescope by Ertel were already underway. The subscribers drew up a Memorial to the Lords Commissioners of the Treasury, applying for a grant of £1,500 to purchase about two acres of land for the building; the document referred to the increase of shipping in the Clyde and how the project would help by increasing revenue from customs duty. Copies of it were passed to the Royal Astronomical Society, with a note from Nichol that a transit circle was on order and that £1,000 was available for an equatorial. The Astronomer Royal, Airy, and the Royal Astronomical Society Council agreed to support the plan on condition that the government should receive some kind of security, such as published observations, and a satisfactory arrangement for supplying time to ships, which in Britain was generally poor. The work at Glasgow was therefore to include meridian and extra-meridional observations, chronometer rating and instruction. Nichol went to London to present the application to the government, and a grant of £1,500 was successfully obtained on the condition that there should be extra-meridional and magnetic observations, as well as regular meridional work, the latter providing the generation of correct time for relaying to the Clyde. The results of the observations were to be dispatched annually to the Admiralty and to the Royal Observatory at Greenwich. A site was acquired, and building commenced immediately and was completed by early 1841. The architect was G. Murray, who also participated in the development of the City Hall and Candleriggs Bazaar. Professor Nichol and his family moved[19] into the attached house that was provided as accommodation in the spring of 1841.

The site selected was on Horselethill (see Figs 6.6 and 6.7) in the parish of Govan, located several miles west of Glasgow, north of the river and in open countryside. When, in 1861, John Nichol (Jnr) described[20] the location he noted that:

> Twenty years ago the Observatory stood alone on the hill over which it still presides: a site selected as the highest point between Glasgow and the Kilpatrick range and, as it then appeared the least likely to be interfered with by other buildings.

Horselethill was an old thirteen shilling and fourpenny land of old extent, which then formed part of the estate of Kelvinside, but it was necessary also to acquire a small portion of the adjoining land of Dowanhill. The site and buildings are sometimes referred to as the Horselethill Observatory, sometimes as the Dowanhill Observatory. The purchase was made by a consortium that included Professor Nichol and other Trustees of the Institution. Remembering the previous episode of a newly built church compromising

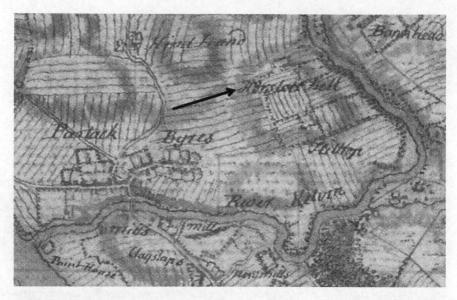

Figure 6.6 The upper map is a small section of General Roy's 1747 military map, showing the agricultural nature of Horslethill about 100 years prior to the building of the Observatory. A section of William Forrest's map of 1816 shows the rural location of Horselethill on which the Observatory of 1841 was built. At the bottom edge 'Observatory' can be seen, indicating the location of Garnethill Observatory which would have been operating at the time when the map was prepared. Gilmorehill, located just south of Horselethill, and now the present site of Glasgow University, can also be seen to be free of buildings. Other panels from both maps, not included above, show the College in High Street with the Macfarlane Observatory in its grounds.

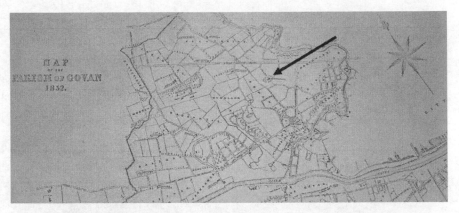

Figure 6.7 A 1852 map of the Parish of Govan shows Horselethill Observatory ten years after it had been built, isolated from the urban development that would eventually negate its operation fifty years later.

the viability of observational work at the Macfarlane Observatory, to guard against obstruction in the line of the transit instrument as well as against smoke, the purchasers obtained an obligation from the Dowanhill Trustees[21]

> not to erect nor allow to be erected on a space of ground measuring ten feet in width from east to west and two hundred feet from the southern boundary of the ground now feued in a line directly south of the position of the transit or meridian instrument already erected or to be erected by the said second party any building the highest point of which shall be above the level of the ground on which the said instrument shall be placed or to erect any furnace from which smoke may issue so as to interfere with the use of said instrument, such servitude to cease and determine forever whenever the buildings on said ground shall fail to be used for astronomical purposes.

In 1841, before the Tudor-style building was complete, the College and Institution came to an agreement whereby the professors were to be honorary shareholders. The College would decide upon the opening hours on condition that it maintained the fabric and requisite instruments, and that shareholders and their families and friends were to be freely admitted; the Professor of Practical Astronomy was to carry the title of Observer.

The plans of the Observatory were altered several times. It was to be of an elliptical shape inside, with portraits of Kepler, Tycho and other notables. The Professor's house was to the right of the entrance gate, the Observatory to the left. A large room for teaching and computing, which also held the portable instruments, led off to the transit room containing the meridian circle telescope. Access to the dome, which awaited the equatorial, was by a side stair; one of the Ramage reflectors (see p. 161) obtained by Nichol was

Figure 6.8   A sketch of the Observatory c. 1843. The framework of the 15-inch Ramage telescope can be seen to the west of the building.

mounted on a Herschel-style arrangement outside and the wooden observatory for magnetic studies was on the extreme west. An 1843 sketch (see Fig. 6.8) shows the openness of the site, with the 15-inch Ramage Telescope frame exposed to the elements. Simpson's colour wash picture of 1848 seen in Plate 8 reveals that the telescope had by then been removed. The artist produced many sketches and paintings related to the Crimean War and he became affectionately known as 'Crimea' Simpson.

An early photograph showing the solid nature of the building is presented in Fig. 6.9. Its development was reported[22] in the *Topographical, Statistical, and Historical Gazetteer of Scotland*, with the information gathered around 1840.

Figure 6.9 An early photograph of the Horselethill Observatory with rough ground surrounding the building.

The situation is one of the finest near Glasgow. With a perfectly uninterrupted horizon towards the south, it commands Arran and all the Cowal Hills on the west, and its view towards the north reaches to the Trossachs. The building now erected, and nearly finished, is very interesting. It is divided into two parts – the dwelling-house for the professor lying on the right of the entrance gate, and the observatory on the left. The observatory consists of the following apartments:-First, a large room destined for the custody of all the minor instrument when not in use, and the conducting of computations, and which is also fitted to serve the purposes of a class or lecture-room. From this room we enter the great transit room, where a very fine instrument from Munich is about to be placed on the two pillars now erected in its centre. Ascending by a side stair, we reach the top of the circular tower, in which a dome will soon be placed, and which is set apart for a large equatorial. If a commanding view of the heavens can at all be got near Glasgow, it must be from this room. The large reflectors will be placed outside in the grounds; and the magnetic observatory, for which the preparations are completed, will be towards the extreme west of the space within which the other erections are placed.

In May 1842, Nichol prepared a report[23] for the Committee of the Institution commenting that, although the project of establishing the Observatory was not complete, it was time for a formal communication to be made on its progress. In the preamble he noted that:

An effective Observatory, which shall be capable of joining in the discussion of the delicate and vast problems with which the modern Astronomy is engaged, requires for its equipment instruments of two different classes – *viz.* such as are fitted for discerning and fixing the place of objects, by noticing their transit across the meridian, and others that enable the observer to examine an object closely and with continued attention in any part of the sky.

He then commented on the reasons for purchasing Ertel's design of transit telescope, noting that the 'continental type' had circles on the extremities of the rotation axis to indicate the zenith distances of the observed stars, providing a combination of a standard British transit instrument and a mural circle. With such an instrument both elements that fix a star's position can be obtained simultaneously. The telescope, delivered from Munich, had a diameter of 6 inches, far beyond the customary range. He also noted that a heliometer was too expensive. He suggested that the recently purchased 22-inch Ramage reflector might have its focal length of 55 feet reduced to 14 feet, making it more manageable for the dome. On the performance of the 15-inch (25 feet in focal length) Ramage Reflector brought up from Greenwich, he echoed Pond in extolling the brilliancy of its metal:

The brilliancy of the metal – an attribute for which all Ramage's specula were distinguished – enables it to yield a light superior to anything I had before seen from a reflector. The blaze of its image of the larger stars is absolutely painful, and some other fainter nebulae appear as in relief on the dark sky, with remarkable splendour.

He neglected to mention the poor figure, though, which had not escaped the notice of Airy, James South and John Herschel. It has been noted by Learner[24] that a reasonable image was only possible if the aperture was reduced to 10 inches, and even then, with half of its area covered, it was still far from perfect. The telescope tube was mounted on a turntable and A-frame support (similar to that used by Herschel), but this too had its problems. While being tested at Greenwich, one of the assessors fell off the high platform from which observations were made.

Nichol wrote that work would be undertaken on astronomy, meteorology and terrestrial magnetism; an exquisite thermometer by Greiner of Berlin was available, and a wooden pavilion in the west portion of the grounds had been erected to house the equipment for magnetic studies. The demise of this structure, caused by a 'hurricane' in 1845, has been dramatically portrayed[25] by Nichol's son, and the instruments were subsequently transferred to the Department of Natural Philosophy.

Shortly after Nichol's Report of 1842, the whole project began to falter. Nichol had committed the University to various expenditures in the way of equipment (see below), but a more serious demand was yet to be made on the University's finances. Again, as with the previous venture of 1808, pecuniary difficulties soon arose, inducing the directors to offer to sell the buildings and grounds to the College. On 5 May 1843, the Faculty, notwithstanding dissent from Fleming, noted earlier as being generally obstructive, resolved to pay off the Institution's debts. By early 1844, the problems were dire and Coutts[26] notes that the Faculty and the subscribers to the Astronomical Institution had forwarded a joint memorial and petition to the Lords of the Treasury, narrating the history and misfortunes of the Institution and asking the Treasury to agree to the buildings and grounds being conveyed to the College. Some miscalculations had been found with regard to the property; there was a second feu duty of £17 10s in addition to one for £30, which had hitherto been the only one taken into account, and the revenue of the Institution, instead of being £12, was only £2 12s and this was already set aside to cover an outstanding bill of £1 16s. After some correspondence and explanations, in March 1845, the Treasury gave their consent, on the condition that the University accept the responsibilities the Institution had incurred when the original grant of £1,500 was given, and that the Astronomical Institution should be relieved from responsibility. This is clearly set forth in a letter[27] addressed by the Secretary of the Treasury to the University authorities, and reproduced in Fig. 6.10. With the acceptance of the terms of this letter, the Observatory buildings became the property of the University of Glasgow.

There was a further delay which interfered with the routine meteoro-

Treasury Chambers, 18 Mar, 1845

GENTLEMEN,

The Lords Commissioners of Her Majesty's Treasury having had under their consideration the several memorials transmitted to them, relating to the proposed sale to you of the property of the Glasgow Astronomical Institution, I am commanded by their Lordships to convey to you their consent to the purchase in question, upon the understanding that you take upon yourselves all the obligations and conditions to which the properties of the Institution are liable to the Crown, thereby relieving the Institution of its responsibility.

I am, GENTLEMEN,

Your obedient Servant,
C.E. TREVELYAN.

The Principal and the
Faculty of the College of Glasgow.

Figure 6.10 The letter from Her Majesty's Treasury transferring the Horselethill Observatory to the University of Glasgow.

logical observations made at the government's request, and the Faculty were concerned about the state of the abandoned building. Tradesmen began to clamour for payment since the College had taken over the Institution's debts of £1,849, plus £258 4s 5d interest, and the annual feu duty of £47 10s 6d. The creditors of the Institution agreed to accept fifteen shillings in the pound, and, after further protest from Fleming, the transaction was completed on 11 November 1845. The Wilson Fund (see Chapter 4, p. 102) had been subjected to some strain in providing the equipment, and, in April 1845, directions were given for the investment of a full capital sum, and for the gradual extinction of debt by applying half its annual proceeds. The transfer was finalised in 1846 with the Observatory becoming University property, free of incumbrances excepting the feu duty. The University instruments of the old Macfarlane Observatory were then moved to Horselethill.

It was a condition of the takeover by the University that members of the Astronomical Institution should be entitled during their lifetime to visit the Observatory. Murray[28] mentions that his father was an original subscriber, and that he was taken by him to the Observatory in 1847.

John Nichol, Professor Nichol's son, later appointed Professor of English Literature at Glasgow University, contributed many interesting recollections and touching stories of life in the Observatory during his father's lifetime. They are collected in his memoir[29] in the form of letters written to his wife. His first visit to the Horselethill Observatory is vividly recalled[30] as follows:

One afternoon in the autumn of 1840, I and my father and uncle William went to take what seemed to me a very long walk into the country. We passed the outskirts of the town – then about Sauchiehall Street; left behind a few detached houses, which remain to recall the old days of St George's Road; crossed over the fields where Queen's Crescent was afterwards erected; and went on by an ill-made narrow road that skirted the great black quarry; and reached a farm-house on Horslet Hill, and plunged through ploughed fields to the top, where a crowd of masons were planting the foundations of the future Dome. 'Miratur molem Aeneas'; it seemed to me as if they were preparing to build a city. I only remember my wonder, and my weariness on drawing near home, and that I was dragged along between my uncle's and my father's arms.

Young John Nichol's uncle William, mentioned above, was his father's brother, a lithographer who contributed figures to be included in *Views of the Architecture of the Heavens*, one of J. P. Nichol's classic books on astronomy, and who is still remembered in his own right. He had businesses in Edinburgh and produced many artistic prints, but lived in an unsettled way, moving to a new address almost every year. According to Schenck,[31] it seems probable that he was adversely influenced by the distinguished and high-profile career of his more illustrious brother, for, from a phrase in one of his letters, it appears he hankered after a life with closer academic involvement. In 1848, he abandoned lithography, resigned as Secretary to the Edinburgh Philosophical Society and applied for the post of Secretary to the Mechanics Institution in Liverpool. A number of his influential friends provided generous testimonials, and his application was successful; he moved to Liverpool in 1849.

Another of young John's early recollections of the time when he lived in the precinct of the Old College provides descriptions of his nursery and toys. He recalls:

My father's study and my mother's room – the red room – were specially sacred. Well I recollect one afternoon stealing into the latter, and looking in amazement at her watch, standing in its case on the table with the mirror in it. If I could get into that old house, all covered over with cobwebs though it is likely to be, I could lay my hand on the very corner where my father used to be sitting in his round study chair, when I was called to account for my misconduct and sentenced to an hour's imprisonment. Ah, what tears were shed on those occasions!

He enjoyed looking through the telescope of the Macfarlane Observatory with his father, but noted his 'fear' of some of the other demonstrations of physics that were conjured for him. He writes:[32]

I used to mount up the rising ground on the College Green to the old Observatory, which you may remember, and see my father with his fur cap on winter nights peering through the tube of the old telescope. I was alive to the wonders of the old tube. I remember seeing Orion through it for the first time, and Jupiter, and the moon. The experiments in the Natural Philosophy class which my father taught before we went to Germany, used to rivet my attention. I will not forget coming into his study – that dread study with the round chair in it – and seeing him hold a mass of flame in a spoon, and throw it into his hand, and toss it into another, as I fancied, by the help of the Devil himself. When I came back from the Rhine, my own researches with the electrical machine removed some of those wonders from the land of magic into that of discovery; but my first shock was a terror to me for a long time to come!

Both quotations immediately above refer to Professor J. P. Nichol's round chair, which can be seen in Fig. 6.4. At the time of the move to Horselethill in the spring of 1841, son John was eight, and he remembered the occasion and the new life out in the country with great fondness. He again recalls:

I remember peeping into the dining-room while it was getting its magnificent coating of oak, and better still the evening after all was over, and the furniture had got settled into its place, and the doors were thrown open. How I danced through the room, gazing at the roof that seemed so lofty, and marching with pride through the spacious halls that were to give me shelter and inspiration for so many years. The new house, now the old house, on the hill, how grandly it stood there all alone – nothing near but the little farm in the valley, nothing to break the view from the white battlements, with the long sweep of upland to the south, the snowy mountains in the north, the gleaming strip of the river beneath in the twilight, and the great city bathed in the morning sun or glittering with a thousand lamps at night. About a mile off was the village – a village then, and not a dingy suburb where our old gardeners used to lodge, and whither I used to walk in the morning to bring the letters from the post, or occasionally in the forenoons, with Agnes and our house-maid, to fetch a basket full of fruit during the first summer, before our own garden began to pour forth its abundance. Our garden, what a delight was that! Here, over broad green sward and winding walk, I was free to run about alone, or in a world peopled only with my own fancies.

The 'snowy mountains' referred to above would have included Ben Lomond, which can still be seen from the location between the houses and the trees.

6.3 EQUIPMENT EXPANSION

All through the period prior to the University taking over the Observatory, Professor Nichol accrued many pieces of equipment, sometimes with no apparent plan for their use. With more or less complete access to Institution and College funds, he contemplated the purchase of a large transit circle from Ertel of Munich, a zenith sector and a great equatorial telescope at least equal to the Great Dorpat Refractor, perhaps even a heliometer. He obtained a 'curious and delicate' Steinheil prism circle and for £100, 'almost by accident', he obtained two reflecting telescopes by John Ramage of Aberdeen,

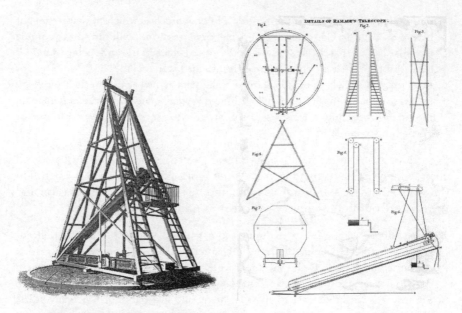

Figure 6.11   A sketch of the 15-inch Ramage telescope at Greenwich, with details of the mechanical arrangement, prior to its relocation at the Horselethill Observatory.

mentioned earlier. One of them, a Herschel front-view type with a 15-inch aperture and a focal length of 25 feet (see Fig. 6.11), had been erected by Ramage at Greenwich in 1825, but it was never used and Airy finally ordered its removal. For a number of years it was located at Horselethill on a small plot to the west of the main building. It can be seen clearly in the drawing of Fig. 6.8, with its design provided in Fig. 6.11. A dramatic silhouette of its structure against the twilight sky can be seen in Fig. 6.12, which also shows the comet of 1843, referred to in Chapter 4 in respect of observations made in South Africa using a previous Glasgow telescope.

The second telescope was a prodigious reflector with an aperture of 22 inches and a focal length of 50 feet, which had been stored away at Aberdeen before Ramage's death in 1835. Nichol's idea was to have the focal length reduced to a more manageable proportion; in the meantime, he had it erected in a temporary mounting made for £16 by a wright, so that it would sweep in a restricted arc: 'I fear from its unwieldiness that I shall not be able to give it an azimuthal motion but even in the Meridian it may serve as a portent pioneer'.

John Nichol[33] recalled that, when he was a child, the outdoor telescope provided him and his playmates with a ready-made gymnasium:

On the great stone circle there rose then the framework of the immense fifty-feet telescope, rising with its double ladders like the 'mast of some great Ammiral,' to a height perilous on that hill of the winds; for now and then in March, when the terrible storms came, it fell with a crash, and lay on the ground covering more roods than Milton allots to the prince of the fallen angels. Between the ladders there was a box swung, something like a cage, in which the astronomer – the bird of prey for the time – was lifted up to a giddy height, while looking in search for his moons or stars through the gigantic tube which was our boast, until Lord Breadalbane's bounty and the discovery of a less cumbrous method of attaining a higher power came to supersede it. Around this circle and about this framework, I, and my cousin, and some young friends used to disport ourselves, and as we swung in the cage and clambered to supreme heights of the ladder, we were admirals, commanders, conquerors in some great naval fight, with our battle flag, in the shape of some audacious handkerchiefs, flaunting from the peak of the great wooden cone. Those were the days of the telescope's glory: in after years, when the scaffolding was all taken to pieces, it used a lie, a neglected hulk, along the ground beneath the windows of the Transit Room. *Sic transit!*

It seems likely that young John's reference to the 'fifty-feet telescope' is a confusion with the 'Lord Breadalbane' Telescope (see p. 171) to which he also refers, and that it was the smaller Ramage telescope that was initially erected to provide his playground. The 'March' referred to above was probably 1845, when the same storm destroyed the magnetic observatory. The speculum appeared to have survived the destruction of the telescope frame as it was referenced under 'Instruments and Parts of Historical Interest' within an inventory[34] dated 1906. The larger mirror spent most of its early

THE GREAT COMET OF 1843
Glasgow Observatory 25th March.

Figure 6.12   A sketch made at the Horselethill Observatory of the Comet of 1843.

years in the hands of Grubb in Dublin with the aim of reducing its focal length, so that it could be made more manageable for location within the dome to the rear of the Observatory building.

The young John Nichol also describes his father's lecturing skills and lays bare his own shortcomings with delightful honesty:[35]

> I was familiar with most of those facts and theories concerning the earth, and moon, and stars, which a large audience came to learn, for twenty successive weeks, in the Merchants' Hall. Without that power which could at will arrest and make wonder the eyes of his hearers; without that grace which gave to his speaking the charm of a perfect actor; without that fire of eloquence which raised me into another world every time I heard it, I could have delivered most, as I was once constrained to deliver one, of my father's lectures to a disappointed audience. I could follow the minute explanation of the planetary perturbation which one winter drove most of his students away, while Jack and Smith and I sat taking notes. But I never was, and it may be questioned if I was capable of becoming, an astronomer. I only knew enough of the science to know its difficulty, and had attained but the first step of wisdom in arriving at a due sense of my ignorance. Yet many a glorious night we spent together in that dome watching the shadows creep up the sides of Tycho, straining our sight to see the fields under the Apennines green as with waving corn; marking the changes in an aery comet, or the steadfast glories of the Lyre; watching the Pleiades, 'like a swarm of fire-flies tangled in a silver braid,' the snows gathering about the poles of the red planet Mars, or the winds blowing across the belt of Jupiter; or gazing till our eyes grew dim on the fathomless abysses of Orion blazing from a past eternity on the world. Many a night when the clouds came chasing over the sky have I walked with him on the roof, while the furnace fires were looming over heaven, and the lights of the city seemed to dazzle the distance, while pacing to and fro till Hesperus brightened through the rifts of rain, we have felt our hearts drawn together and our souls raised upward by the stars.

On several occasions, Professor Nichol seems to have upset his Faculty colleagues. His first reprimand resulted from his spendthrift ways, as can be seen by the number of his acquisitions. Within about a year from the date of his induction, a sum of £204 was spent on repairs and improvements to the Observatory, and by January 1843, more than £700 more had been spent on procuring instruments, a large portion of the amount being paid to Ertel & Son of Munich for equipment (see below) made to order.

The Faculty continued to buy. In 1843, £399 9s was paid out mainly to Merz, Troughton and Simms, to R. Bryson for a sidereal clock, to J. Gordon (London), to J. Dunn (Glasgow), to R. Lockhart & Co. and to J. R. Adie (Liverpool), who was forced to bring an action against the College. The total collection was now insured for £1,500, and the professors were alarmed at the drain of capital. Coutts[36] has noted that in 1843, as a result of Nichol's unilateral way of effecting purchases, the Faculty laid down a rule that no single member should in future order instruments on his own responsibility. Nevertheless, they were soon forced to embark on an even greater expenditure on Nichol's behalf, as related above. By 1843, it was becoming clear that the Astronomical Institution was falling into financial distress; it was rescued by the university's purchase of the Observatory, the monies representing a considerable but visionary investment so providing 'the substitution of an efficient for an inefficient body in fulfilment of conditions in which the Treasury sum was advanced'.

On 1 May 1845, a new catalogue of the instruments was laid before the Faculty, which Gavine[37] lists as comprising:

> The great 'Munich' telescopes with 4 eyepieces, great level and 4 supplementary levels; reversing instrument with 2 collimators, 2 levels and 2 supplementary levels; Bryson transit clock, Ramage 25-foot reflector, box of eyepieces and finder; Herschel 10-foot reflector, 4 eyepieces and Troughton eyepiece micrometer; 79 mm Dolland achromatic with ring micrometer; 6 m reflector in College; 533 mm Ramage speculum; Cassegrain reflector (Short?); with divided objective micrometer; and Gregorian reflector (Short?); old transit telescope in College; great universal instrument; great theodolite; Craig altazimuth circle; Steinhill prism circle; 2 telescopes and case; 2 reflecting circles with stand, one by Troughton, Troughton sextant, levels; artificial horizon; portable reflecting circle; heliotropes; gridiron pendulum clock; 3 journeyman clocks; meteorological and magnetic instruments (full list); and miscellaneous old or broken instruments, books and catalogues.

Later, in December 1846, the Board of Visitors inspected the Observatory, reporting that the instruments were in good order, but that some of those marked in the catalogue could not be produced for examination as they were noted as being in Ireland, some loaned to Lord Rosse, with the speculum of the larger Ramage reflector in the hands of Grubb in Dublin for refiguring. Maconochie, the Professor of Law, strongly disapproved of the state of affairs, and shortly afterwards the Faculty issued imperative orders that no officer of the College, having the care of its property, should remove or lend any instruments from the place appointed for its custody.

6.4 THE ERTEL TRANSIT TELESCOPE

Having visited Germany in 1839, Nichol ordered various optical pieces from George Merz of Munich, for the construction of a heliostat (see Fig. 6.13). He was also adamant, against all advice, that the main instrument should be a great transit circle by Ertel of Munich, for which a deposit of £203 was paid out of the Wilson Fund, and customs remission sought from the Exchequer. The total cost eventually ran to £970 5s 3d. A letter from Ertel in the form of an invoice is presented in Fig. 6.14, and a translation of the list of contents of the consignment is given in Fig. 6.15. Although Forbes, who later became

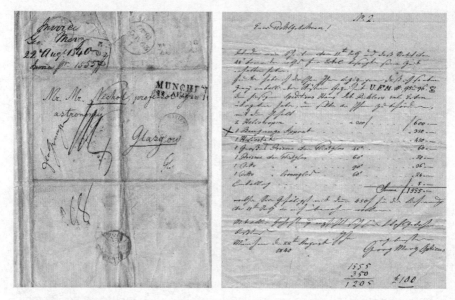

Figure 6.13 Invoice No. 1550, dated 22 August 1840, for a heliostat and various optical pieces, from George Merz in Munich to Professor Nichol.

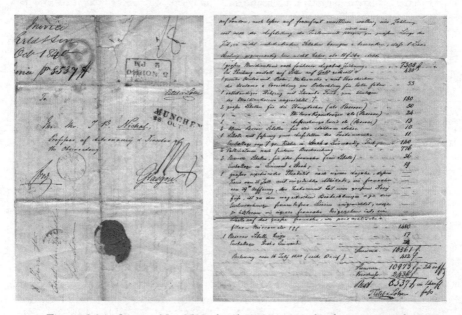

Figure 6.14 Invoice No. 8537, dated 28 Oct 1840, for the transit circle, from George Ertel to Professor Nichol. Above the list of equipment was a letter which has been translated for inclusion as Figure 6.15.

To Professor Nichol in Glasgow   Munich 28 October 1840

Most Highly Honoured Sir,

I have duly received your most esteemed letter of 10th inst. and learned therefrom that an earlier letter of mine indeed went missing, but this has fortunately not been so important, as both I and the instruments are still here. The meridian circle with all accessories, the collimators, the magnetic instrument together with all apparatus are completely ready and already excellently packed. However, I do not venture to dispatch the instruments before I receive, following on this letter, your own instruction to do so, as I think that the instruments are best stored here, where inspection can be made from time to time. If on the other hand we send off the instruments now, they might be stored in a damp place and thereby be damaged by the spring. Will you therefore be so kind as to write me as soon as possible, as to whether I am to dispatch the instruments at once, which can be done at any moment. I have in mind to leave here by March, insofar as that will be agreeable to you. Of the handling and setting up, and of the type of packing, I make no mention here, since I can explain anything in person at the opening of the chests and also at the installation.

That the fitting up should not be undertaken until next spring, my father considers – as also I do myself – to be a very prudent arrangement.

Herewith I take the liberty of submitting a detailed account for all the instruments, which amount you will most kindly remit to me in several small bills of exchange drawn on London, or better still on Frankfurt. Our payment made only after the installation of the instrument, on account of the great length of time, would bring us a not inconsiderable loss, and we note that £1 Sterling at present stands no higher than 11f24d.

1	large Meridian Circle, according to earlier specification and drawing	7308f
	the divisions on gold instead of silver	420
1	second eyepiece with tube, micrometer, and with displacement of the	
	eyepiece, and arrangement for illumination of bright wires	55
1	complete levelling gear, with screw and drive, arranged for the rotation	
	of the meridian circle	180
2	large spirit-levels for the main axis (as spares)	50
1	" " " " " microscope bearing (as spare)	24
1	" " " " " finding circle (as spare)	13
2	small spare levels for the steel axis	10
	Packing of 7 large chests in straw and canvas	140
2	collimators in accordance with earlier description	776
2	spare levels (for each telescope one level)	36
	Packing in canvas and straw	19
1	large repeating theodolite to our own specification, with circle of 25 inches	
	with alidade fitted in, one telescope of 39" aperture, the instrument	
	has a large tripod , and is fitted for magnetic observations and for	
	the examination of Fraunhofer lines, for which finally one of our	
	own telescopes is included, a level for the large telescope, a filar-	
	micrometer, etc.	1480
1	spare level for this	17
	Packing, straw and canvas	22
	Total	10561f
	Account of 14 July 1840 (see letter)	412
	Total	10973f
	Deposit	2436
	Net	8537f

T Ertel & Son

I think that it was better to prepare the lifting apparatus here, since without this the instrument probably cannot be set up. In future we cannot prepare such a meridional instrument, I shall certainly bring with me and pains for the fulfilment of the contract.

My father, mother and brothers/sisters send their regards, and while I look forward to a speedy reply

I remain
etc. etc.
p & p T Ertel & Sohn
George Ertel

Figure 6.15 A translation of the letter and invoice (No. 8537) of 28 October 1840, from George Ertel in Munich to Professor Nichol, relating to the shipment of the transit telescope.

Professor of Natural Philosophy at Anderson's University, encouraged Nichol with regard to obtaining a large telescope, he was less enthusiastic about the transit circle. A smaller one, he suggested, would suffice to regulate the clocks, and he advised Nichol to 'keep your subscribers in good humour'.[38]

The instrument and the room designed for housing it have been described by Grant.[39] The telescope was 90 inches in length, and the object glass had a clear diameter of 6 French inches, Nichol[40] describing it simply as having a diameter of 6 inches. According to the specification of the letter (see Figs 6.14 and 6.15), the scales on the circles were engraved in gold, rather than the usual silver, adding 420 francs, or about £100, to the cost. Nichol appeared to be going to extremes with his extravagances, which were of no real benefit in terms of the instrument's function. At the time of delivery, the new observatory was only at foundation stage, and Professor Thomas Henderson at Edinburgh took charge of the incoming assignment. Gavine[41] has noted that, through this involvement, Henderson was intrigued by the universal circle of Uitschneider and Fraunhofer, with circles of 25 and 30cm and a prism telescope with a focal length of 50cm, which was to be erected first on the transit pillar of the Macfarlane Observatory. Apparently the Germans praised its precision. Henderson vainly tried to divert Nichol away from his lectures and writings of 'glowing hyperbolic language' to attend to his instruments. A silhouette drawing of the transit telescope included by Nichol[42] in his *Cyclopaedia of Physical Sciences* to depict the design of such instruments is presented in Fig. 6.16.

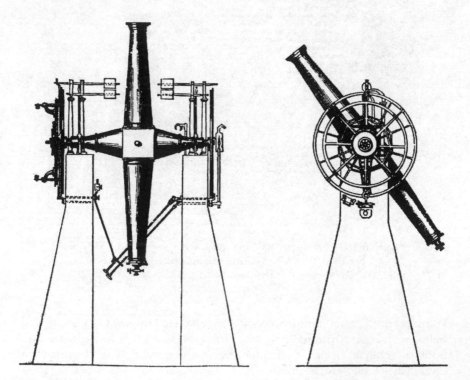

Figure 6.16   Sketches of the Ertel transit circle at the Horselethill Observatory.

The Ertel transit telescope was in place by the end of 1841, but Nichol had difficulty using it, as it was apparently 'less solid' than English instruments. Correspondence passed between him and Airy in which it is clear that Nichol had taken on instrumentation beyond his technical capabilities. Airy assured him that he was in a more fortunate position than his colleagues in established observatories in that he had freedom to try out his instruments in order to find the best plan of work. Under Nichol's direction, the transit telescope never operated successfully and he was unable to establish an absolute time reference for transmission to the Clyde. It was reported[43] that he was unable to make much use of the meridian circle on account of the optical inferiority of its object glass. In around 1858, he informed the Royal Astronomical Society that he had ordered, at his own expense, a new one from Thomas Cooke in York.

At around the same time, there was activity within the city to establish a time-ball for the benefit of shipping in the Clyde. Commencing with the recommendations of Sir Thomas MakDougall Brisbane for the establishment of a time-ball at both Glasgow and Greenock, the story of this venture has been documented by Kinns,[44] and further comments on its operation will be made in Chapters 7 and 8. Nichol exchanged correspondence with Sir George Airy over the matter, and was concerned about the accuracy with which the original signal was being obtained and then presented to mariners by the dropping of the time-ball. He suggested that the Astronomer Royal for Scotland, Professor Piazzi Smyth, might be asked to become involved and to pay a visit to Glasgow to assess the situation. There appears to be no evidence that Nichol proposed that the University Observatory could provide direct support, a little surprising as the development of a time service for the city was a major element in the conditions for obtaining monies to build the facilities at Horselethill. After fifteen years of experimentation with the Ertel transit telescope, it looks as though Nichol was not in a position to offer a professional facility of accurate time.

As it turned out later, it was the Ertel telescope that became the mainstay of observational work in Glasgow following Nichol's death; whether the proposed change of objective lens was made remains obscure. The realisation of the true potential of transit telescope observations, in a poor climate, had to wait for the arrival of Professor R. Grant, the next incumbent of the Chair.

6.5 METEOROLOGY

With the establishment of the Observatory, Nichol was able to undertake regular weather recordings. His first report covered the period of 1843 to 1844 and the data were presented in the form of beautifully drawn diagrams. An example of the presentation[45] is depicted in Fig. 6.17 showing the frequency of the wind directions and their speeds over the period.

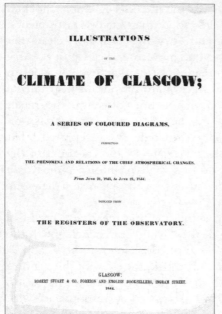
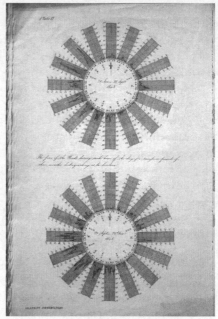

Figure 6.17   The title page of *Illustrations of the Climate of Glasgow 1843–4* and one of the beautiful diagrams depicting the wind directions and their speeds during most of 1843.

Such preliminary observations set a scheme which was maintained at the Dowanhill site for 100 years.

6.6 HORSELETHILL UNDER NICHOL

The commentaries[46] by John Nichol (Jnr) of the period give the impression that while his father lectured, taught, wrote and socialised, the routine observations were carried out by a succession of assistants who earned little more than bed and board. Noted personnel included a Mr Mackenzie who became a vicar, William Leitch, later DD and Principal of King's College, Halifax, Nova Scotia, and John Rollo, who remained for eighteen years. Affectionate details[47] of their characters and traits are provided by John Nichol (Jnr), and he notes John Rollo's loyalty, saying that he 'stayed with us to the last; he seemed to become part of the house, as regular as the transit-room clock'. Nichol had envisaged offering salaries to two assistants for rating the chronometers and undertaking the meteorological and magnetic readings, but he was allowed a clerk at no more than £5 a year.

In 1847, Nichol requested leave of absence to visit America, ostensibly for a lecture tour, but partly to escape from 'financial difficulties'. There was some objection, but, provided he 'behaved' on his return, the Faculty agreed, and John Rollo was left in charge. On 1 November, before Nichol's departure, the Board of Visitors inspected the Observatory and found all in order, apart from some rain ingress and the absence both of the 22-inch Ramage speculum and of the eyepieces of the Herschel telescope. The speculum was discovered to be in the hands of Grubb in Dublin for its focal length to be reduced to ~14 feet. It was also found that Nichol had privately hired Dunn, the Edinburgh instrument maker, at a cost of £10 a year to look after the instruments. The Faculty hurriedly enforced an earlier suggestion by the Professor of Law that no individual might remove property. Dunn was told to do nothing until he received official instructions.

By 1851, the Faculty were worried about Grubb's retention of the speculum. Nichol was requested to do something about it, or the matter would be taken out of his hands, but Grubb and Piazzi Smyth gave assurances that the work was well advanced. Nichol requested that a mounting be made ready. The following April, he submitted a four-volume report including meteorological observations as well as latitude and longitude values for the Observatory. The speculum was back from Grubb, but the optical figure, or shape of the mirror's surface, was unsatisfactory, and Nichol had to use all his powers of persuasion on his colleagues for it to be sent back again, and, at the same time, to agree to a plan for mounting it by Randolph and Elliot, who had built the dome. In addition, he was allowed to buy a micrometer by Grubb, similar to the one made for Lord Rosse, costing about twenty guineas. The cost of figuring the speculum, however, was borne by the Rector of the University, John Campbell, Marquess of Breadalbane, an amateur astronomer. When the speculum returned from Grubb, the Board of Visitors were satisfied, and the reflector, complete with clock drive, became known as the Breadalbane Telescope. It was apparently used by Nichol and the Marquess around 1855 for lunar photography and was said to be remarkable for the time, but their images were not published and none can be found in the archives.

During the late 1840s, Nichol's health declined and, as a result of his physician's prescription, he became addicted to opiates. An account of his drug addiction and its cure by hydrotherapy at the Ben Rhydding Hydro in Wharfedale is recorded in his book *Memorials from Ben Rhydding*.[48] This is a remarkable monograph with a hotchpotch of content. The first chapter deals with his illness, probably some form of mental breakdown; laudanum gave him great relief, but with it came the realisation of being trapped by its spell. His writing on the episode gives an intense insight into the various ways by which he tried to break the addiction and how his hydrotherapy treatment succeeded.

My illness was of long standing, and its secondary forms or consequences had become far more disastrous and menacing than itself. Five years ago I was stricken by a low nervous fever, the issue of profound and protracted anxieties, involving more than prosperity, or even life. As usual in such cases, sleep fled; and my physician, than whom a kinder never existed, fondled me with opiates. Intense irritation of the stomach – a frequent concomitant or result of such fevers – was likewise a sequel in my case; and it seemed necessary that the use of opiates and anodynes should be prolonged. At length was I reduced to reckon, as part of my daily food, a detestable mixture of deadliest poison – a mixture of composition I am most unwilling to tell, for never did demon put on more deftly the garments of an angel of light. Its early insidious soothings – who could forget them! How ethereal and soft those billows of luxurious ease flowing round and round the vexed body, and wooing pain to rest by the gentlest lullaby! Sleep under that anodyne is, at first, the sleep of a sinless infant, neither dreamful nor yet wholly unconscious – mingling with it that delicious feeling of pleasure in mere *being*, which the late metaphysician, Dr Thomas Brown, considered a separate and distinctive *sense*.

The later chapters give descriptions of the beauty and delights of Wharfedale, including such idyllic places as Bolton Abbey. It is also striking that Nichol was aware of the fashions and developments of society, and he made several comments on the development of 'Bloomerism', which began to appear soon after 1849. The term within quotes refers to the ladies' fashion of soft, long and baggy trousers with a narrow ankle cuff worn underneath a skirt. This, and its later forms, are sometimes referred to as marking the beginning of the emancipation of women. At the time, Bloomerism was vigorously ridiculed by cartoons in *Punch* magazine.

6.7 NICHOL'S BOOKS

Murray[49] has commented on the wide circulation that Nichol's books found and has mentioned that one or more of them might have been seen on the bookshelves of most Glasgow houses seventy or eighty years before, now one hundred and sixty years ago. In his first discourse, written with the general public in mind, he presented an overview of the then known Universe. This was followed immediately by a second tome providing a complete description of the Solar System. Both books were extremely successful, running to several editions, but with slight changes to their titles. His writings invariably had a flamboyant style and displayed his penchant for purple prose. It has been noted[50] that a reviewer of the *System of the World*,[51] published in 1846, thought his style was over the top, and said that 'his very liberal and very unconsidered use of words such as inconceivable, infinite, eternal . . . we find them applied to things which are not only quite measurable, but which have been measured'. It has also been noted that his prolific writing, which no doubt contributed to his ill health, was stimulated by his need to recover from the debts he had incurred through his extravagances in equipping the Horselethill Observatory.

George Eliot has provided testimony on the effect that Nichol's books had 'on minds equally inspired by a zeal for truth and a love for beauty'. In 1841, she wrote:

> I have been revelling in Nichol's 'Architecture of the Heavens' and 'Phenomena of the Solar System', and have been in imagination winging my flight from system to system, and from universe to universe, trying to conceive myself in such a position and with such a visual faculty as to enable me to enjoy what Young enumerates among the novelties of the stranger man when he bursts the shell to:
> Behold an infinite of floating worlds,
> Divide the crystal waves of ether pure,
> In endless voyage without port.

The same two books were also an inspiration to James Nasmyth, the inventor of a telescope design that provided a fixed focal position along its declination axis, a scheme adopted by some of the large instruments currently in use. In January 1840, Nasmyth wrote a letter[52] to Nichol, in which he commented on the pleasure the books had given him, and he also offered a sketch for possible inclusion as a frontispiece for later editions of his discourses on the Solar System. Whether or not Nichol responded is unknown, and he appeared to continue to provide his own local material for illustrations, as in Fig. 6.12.

Among J. P. Nichol's many readers was the writer Edgar Allen Poe. In his story *Murders in the Rue Morgue*, published in 1841, he introduced a fictional detective called Dupin who mentioned 'Dr Nichols' and the 'nebular cosmogony' (note the additional 's' in the name). While Nichol was on a tour of New York in the winter of 1847–8, Poe attended one of his lectures. This may have been the trigger for him to write a prose poem entitled *Eureka: An Essay on the Material and Spiritual Universe*, published in 1848. He quoted several phrases and ideas from Nichol's book *Views of the Architecture of the Heavens*,[53] and enlarged upon them.

His writings also inspired the thinking and work of Thomas De Quincey, perhaps best known for his *Confessions of an English Opium-Eater*. As noted in an article by Jonathan Smith,[54] De Quincey was given succour for several weeks in 1841 under the Nichol family's roof at the Observatory to avoid his Edinburgh creditors. From the friendship that developed, De Quincey wrote an article on the nebular hypothesis. Later, as a result of Lord Rosse's observations with his large telescope of the Orion Nebula and of its resolution with embedded stars, general thinking within scientific circles related to the hypothesis began to change. In 1846, De Quincey published a revised review article on the subject, in which he commented particularly on the contributions made by Lord Rosse and Nichol, despite Nichol advising him not to do so. Recently, Smith[55] has explored the episode, noting that some of the points of discussion taken from Nichol's books had been misconstrued and interpreted erroneously, the conflicts revealing what can be considered as a gulf between poetry and science. The episode of De Quincey's sojourn in the Glasgow Observatory inspired Douglas Dunn to write a play[56] entitled *The Telescope Garden*, with music by David Dorward.

The two mainstays of Nichol's popular books had the themes of the Solar System and the more general Universe, with titles sometimes changing from one edition to the next. For example, the second edition of *Phenomena of the Solar System* had the new title of *Contemplations on the Solar System*.[57] An American edition[58] of 1842 carried the title *Phenomena and Order of the Solar System*. In the first edition[59] of this book (see the left-hand panel of Fig. 6.18), the orbital behaviours of the planets are given with their visible features presented. In the section on the Sun, a copy of Alexander Wilson's drawing of the apparent depressions associated with sunspots is included. The Moon features prominently with discussions on its lack of water and the 'fact' that it had a thin atmosphere that could be seen at the time of solar eclipses. This latter interpretation is wrong, of course, the phenomenon being related to the solar corona; the correction was promoted by Professor Grant (see Chapter 7, p. 191). Lunar formations are described with staggering

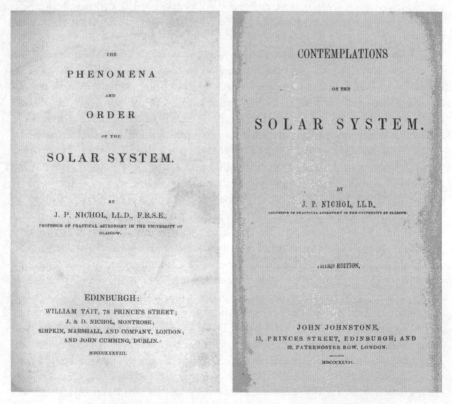

Figure 6.18   The title pages of the first and third editions of Nichol's books on the Solar System.

hyperbole, with copious lithographs prepared by a short-lived partnership of the author's brother, William, and a William Forrester. One of the plates is a copy of the famous lunar map of Beer and Mädler of Berlin. Images of the Apennines and of the crater Plato are provided in Fig. 6.19. Shortly before his death, Nichol was preparing an essay[60] on the physical features of the Moon, and particularly on the chronological relations of the long streaks or rays that diverge from many craters, intersect one another and are often dislocated along their path.

An additional chapter appears at the beginning of the third edition[61] of Nichol's discourses on the Solar System, discussing the very recent discovery of the planet Neptune. It includes a copy of the 'Special Report' by Professor James Challis of the Observatory at Cambridge on the unfortunate way in which the British astronomical establishment 'missed out' on the discovery. This whole topic was later presented as an individual book[62] based on a lecture given to the Edinburgh Philosophical Institution. In this tome, Nichol eloquently describes how the perfect elliptical orbital figure of a planet, solely under the gravitational influence of the Sun, is modified or 'perturbed' by the presence of the other planets in the Solar System. The major part of the book then deals with Bouvard's work on the detailed behaviour of the orbit of Uranus and on the difficult task of determining its perturbations and using them to predict the position of a 'missing' planet. Its final section deals with the trans-channel controversies related to assigning the claim of discovery to Leverrier, whose name was originally given to the new planet, and to the 'apparent' slowness of the Astronomer Royal, Sir George Airy, and of Professor Challis in acting on the first predictions of its location, made by John Couch Adams in England. He was not complimentary about the French claim that a formal publication of an observational discovery should take precedence over unpublished labours undertaken slightly earlier by Adams. He made reference to Gallic intolerance, quoting[63] François Arago (in very rough French) as saying:

M. Adams n'a le droit de figurer, dans l'histoire de la decouverte de la planete LEVERRIER, ni par citation detaile, ni meme par le plus legre allusion . . . Aux yeux de tout homme impartiel: cette decouverte restera un des plus magnifiques triumphes des theories astronomiques, une des gloires de l'Academie, un des plus beaux titres de notre pays a le reconnaissance et a l'admiration de la posterite!

Nichol dealt with these expressed sentiments and answered all the arguments with dexterity, but in a footnote[64] says: 'How often French public acts and disquisitions remind one of the ancient – We and the barbarians'. The roles that Airy and Challis played (or didn't play) are also reasoned out. Finally, in the last pages, he appears charitable to all, proclaiming that the discovery of Neptune, and the way in which two people took on the grand chore of the mathematical groundwork to predict the location of the new planet, was far

Figure 6.19 Plate XVI from *Phenomena of the Solar System* comprises dramatic lithographs of the Apennines and the crater Plato by Nichol's brother, William, in partnership with Forrester.

Figure 6.20 Plate XI from *Contemplations of the Solar System* illustrates the somewhat fanciful appearance of a lunar crater.

more important than the ensuing petty squabbles. More is presented on this topic in Chapter 7.

Another edition of the Solar System work, entitled *The Planetary System: Its Order, And Physical Structure*,[65] appeared in 1851. Although not the best example of the extremes of Nichol's flamboyant style, the following quotation[66] illustrates his mastery of language. It relates to viewing the Moon through a telescope, a pleasure that every living person deserves and should experience:

> It is scarcely possible to conceive a more remarkable contrast than that between the appearance of the Moon to the naked eye, and that which she presents to the telescope, whether in quadrature, or when she is full. Instead of the plain bright surface, sending from all its parts an illumination not far from equable, we discern a body of most strange character, broken by irregularities which, in extent and form, present few analogies with the mountainous regions of our own globe. The reality of these, as well as the singularity of their contours, the briefest glance at the crescent luminary is sufficient to establish. The incomplete edge is, in that case, under the influence of the morning or evening light; and all phenomena of lightened peaks, dark valleys, and long shadows, which occur in a broken district of the Earth in such circumstances, are there distinctly visible, but on a scale far more grand.

A second popular and classic book,[67] first published in 1837, was *Views of the Architecture of the Heavens – In a Series of Letters to a Lady*. Its style and content are considered to be of such importance that it was reprinted in 2009 (see Reference 53). In 1840[68] and 1842,[69] editions were published in the United States. Again, Nichol's brother, William, provided the lithography for the earlier Edinburgh productions, and the images in the 1843 edition,[70] under partnership with Taylor, represent his best illustrative work. The frontispiece provides an image of Enke's Comet and the drawings include the following plates:

Plate 1 Constellation Hercules Insc. Nichol Lithog. Edin
Plate 2 Rationale of Herschel's Method
Plate 3 An object near the outermost limit of Telescopic Observation
Plate 4 Group of 11 stars Insc. Nichol & Co., Lith Edin.
Plate 5 Nebulosity Insc. Nichol, lithog.
Plate 6 Nuclei Insc. Nichol, lithog.

A fifth edition[71] appeared in 1845, followed, in 1850, by a new edition,[72] dedicated to Lord Rosse. It included drawings by David Scott (which, after his death, were completed by his younger brother William Scott) that had a marble-like texture and revealed an amazing wealth of thought. The lithographs portraying the celestial objects were signed 'R. Dale Sc. Edin.'.

Thoughts on Important Points Relating to the System of the World appeared[73] in 1846. The second edition,[74] published in 1848, contained (opposite p. 73) a magnificent lithograph of the Whirlpool Nebula by John Le Conte. In the same year, an American edition was published[75] following Nichol's visit there. Later the same year, a revised version appeared,[76] this time with the title *The Stellar Universe; Views on its Arrangements, Motions, and Evolutions*. A copy of the frontispiece of this edition is shown in Fig. 6.21 and is an inferior version of the larger lithograph of the Whirlpool Nebula mentioned above.

In 1844, a book entitled *Vestiges of the Natural History of Creation* was published anonymously; it also took the 'nebular hypothesis' as the back-drop for a theory of cosmic evolution. Its content was highly controversial as it suggested that 'all' of nature is the unfolding of deep natural laws that prescribe the evolution of the heavens, the creation of Earth and the evolution of life. It envisaged a progression from swirling clouds of dust that first formed the Solar System to the final development of organic life. These ideas countered the Biblical teaching of the whole of creation being established in seven days, and they were attacked by the Church. Several different names were suggested as to who the author might be, including the moral philosopher Alexander Bain, the novelist Catherine Crowe and the phrenologist George Combe. An inspection of the writing style and traces of dialect within it indicated a Scottish voice, so that, based on Edinburgh or Glasgow gossip, the finger of suspicion was pointed at Nichol. Eventually, the author was declared to be Robert Chambers, the co-founder of the large mass-circulation publishing house.

According to Secord,[77] Nichol disliked the book even before reading it.

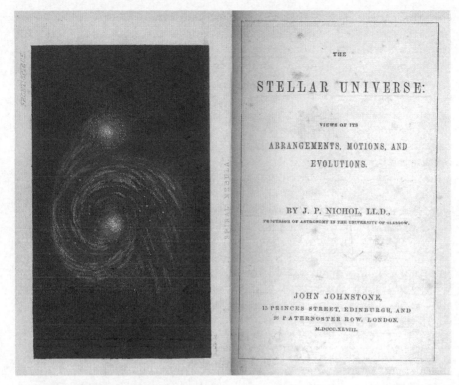

Figure 6.21 The frontispiece and title page of Nichol's *The Stellar Universe*, published in 1848. The engraving is of the Whirlpool Nebula, M 51.

In a letter to William Thomson, he dismissed it as 'a foolish book' that had attracted 'still foolisher reviews' so raising the spectre of 'Atheism, etc.' In respect of the growing 'science versus religion' debate, the tone of *Vestiges* was blunt. It was very different from the delicate compromises Nichol had engineered in his own *Architecture of the Heavens* and was written in such a way that the nebular hypotheses had become linked with challenges to the nature of the human soul.

In addition, the publication of *Vestiges* appears to have scuppered Nichol's plans to produce a multi-volume work on cosmology, dealing with the heavens, Earth and Man. Nichol had told Chambers about the scheme while on a tour of Ireland in 1837 and believed that his erstwhile friend had plagiarised his ideas. Long quotations he had originally written for *Chambers's Journal* had been blatantly copied. In a letter to Combe in 1848 he complained: 'I think I can promise you as curious a history of the unblushing & wholesale appropriations of another mans [sic] plans & thoughts, as probably ever occurred'. The issue needed 'to be decided by some code of Morals'. Nichol's suspicions of the authorship of *Vestiges* had been confirmed when, two years earlier, he had heard the secret from Chambers himself, who asked for advice on some of the astronomical content. All subsequent editions of his *Architecture* were sharply distinguished from the themes of *Vestiges*, but none contained mention of the latter work.

Again, at about this time, Secord[78] notes that the emphasis of Nichol's writings changed with the discovery by Lord Rosse that the core of the Orion Nebula was resolvable and contained stars. This was certainly notable in his comments and criticisms related to an article written by De Quincey, advising the author that it should not be published. He began to focus on the broader and more conjectural questions of the structure, size and fate of the entire Universe. As part of this development, a new view of nebular condensation emerged in which gravity brought individual stars together. Nebulae were the origins of stellar clusters and not the remnants of a primordial 'Firemist', a flamboyant term used by Nichol.

During his lecture tour of the United States in 1847–8, several of Nichol's books were again published there, and some of the references listed at the end of the chapter have notes to that effect. His *Cyclopaedia of the Physical Sciences*,[79] published in 1857, was his last complete major work. It was a most comprehensive and admirable work in its day. According to one commentator, it was said to be 'almost unparalleled for the extent and accuracy of the information that it contains in a small bulk'.

John Pringle Nichol was an advocate[80] for the establishment of national and absolutely non-sectarian schools, and his voice was significant in promoting legislation to achieve this. Later in his life he presented his ideas on school education, and reports of his talks were circulated in pamphlets;[81] the latter reference noted that he had extensive experience, having been head of various common schools over an eighteen-year period prior to his University post.

6.8 CONCLUSION

Nichol spent his last failing years in teaching and writing. He had been full of bright ideas which had borne little fruit. Dismissed by Airy and Forbes as a mere showman, he had been a thorn in the flesh of his professional University colleagues, whom he had persuaded to spend enormous sums on prestigious pieces of equipment. These appeared to have been wasted through misuse and, in the inferior climate, had provided no permanent results, not even an accurate measurement for the new Observatory's position. Yet this chaotic man, through his travels and writings, spread astronomical knowledge perhaps further than any fellow-countryman of his generation, and as a teacher had few equals.

According to Coutts,[82] Nichol was 'the bright particular star' of his time at Glasgow. He greatly enlarged the stock of instruments, began afresh the conducting of observations, revived the class for students and guided the

Faculty in the purchase of the new Observatory at Horselethill. He wrote and edited numerous works on science, and crowds gathered to his popular lectures, an audience of more than a thousand flocking to hear him discourse on astronomy. While he made science popular, it was not at the expense of making it worthless, for he threw out ideas that guided and stimulated men who later rose to distinction, and inspired others to have an enthusiasm akin to his own. He had a comprehensive and accurate knowledge of a vast range of subjects, astronomy and geology, mathematics and physical science, philosophy and political economy, and was deeply interested in education with its associated social matters. He was as popular a writer as he was a speaker. His printed output was prolific and included papers on literary and social subjects, many of them written under an anonymous pen-name. He made contributions to Griffin's *Cyclopaedia of Biography*, including so many names that it served as a brief summary of the history of modern science. At the time of his death he was engaged as one of the editors of Mackenzie's *Biographical Dictionary*. In 1847–8 he had spent the winter in the United States where he had delivered several courses of lectures. On his return he gave graphic accounts of his experiences and his impressions of the new civilisation of the West. The task of collecting these and condensing them into a substantial volume had begun, but it remained unfinished.

His son, John, paid a final tribute[83] by writing:

> Generous beyond the point of prudence, and sanguine of impossible things, he had too little perseverance to lay the foundation of even the most practicable of his distant dreams. Easily depressed by failure, he was as easily elated by success . . . everything he thought was noble, and everything he said was inspiring, and everything he did was majestic.

John Pringle Nichol died in Rothesay on 19 September 1859, leaving an estate[84] of £1,935 15s 6d. A brief obituary[85] appeared in *Monthly Notices of the Royal Astronomical Society*. On 17 October 1903, a beautiful window, presented by Professor Jack, was unveiled[86] in the University Bute Hall, commemorating John Pringle Nichol, his son John and his daughter Agnes, whom Professor Jack had married. The design of 1893 was by Sir Edward Coley Burne-Jones, the English painter, designer and illustrator.

At the top of the window appears Urania, the heavenly Muse. The figures in the upper division of the window (see Plate 9) represent Copernicus, Galileo, Kepler and Newton, the four great founders of astronomy. The four in the second division represent Bacon, Burns, Byron and Carlyle; the younger Nichol had produced memorable commentaries on the work of each of the four. The four figures on the bottom row represent motherhood, charity, patience and faith. At the bottom of the window, almost hidden by the bench, are the inscriptions for John Pringle Nichol and his two children (see Plate 10).

On its unveiling, Lord Kelvin[87] made some touching remarks about

Nichol, who, as the stand-in for Dr Meikleham, was his first teacher of natural philosophy. Later, he was able to see the manuscript of the originally prepared lectures and commented:

> I was greatly struck with it, and much interested to see in black and white the preparations he made for the splendid course of natural philosophy that he put us through during the session 1839–40. In his lectures the creative imagination of the poet impressed youthful minds in a way that no amount of learning, no amount of mathematical skill alone, no amount of knowledge in science, could possibly have produced . . . I remember the enthusiastic and glowing terms in which our professor and teacher spoke of Fourier, the great French creative mathematician who founded the mathematical theory of the conduction of heat . . . The knowledge of Fourier was my start in the signalling through submarine cables, which occupied a large part of my after life.

Again, at the unveiling[88] of the Bute Hall Memorial Window, Professor Jack said that 'Nichol's books presented the latest and best results of the science of his time, and the figures and facts of his science were illuminated by a matchless eloquence warm with the creative imagination of a poet'. With respect to his public lectures, Gilfillan[89] described Nichol as 'the prose laureate of the stars'. Not only was he a teacher of his beloved astronomy, he was also a true professor of eloquence.

A quotation of Napoleon Bonaparte seems a very appropriate way to summarise Nichol's life. It proclaims:

> Men of genius are meteors
> whose destiny is to be consumed
> in lighting up their century.

References

1. MacLehose, J. (1886), *Memories and portraits of one hundred Glasgow men who died during the last thirty years and in their lives did much to make the city what it now is. No: 71 – John Pringle Nichol*, Glasgow: James Maclehose & Sons, vol. 2, pp. 249–52 [GUL History qDX203 MEM].
2. Burnett, J. (2004), 'Nichol, John Pringle (1804–1859), astronomer and political economist', *Oxford Dictionary of National Biography*, Oxford: Oxford University Press.
3. Eyre-Todd, George (1905), *Official Guide to Glasgow and its Neighbourhood by the Tramway Car Routes*, Glasgow: Robert MacLehose & Co., Ltd, p. 37 [GUL Sp Coll Mu26-a.12].
4. 'Certificates in Favour of Mr J P Nichol, candidate for the Professorship of Practical Astronomy in the University of Glasgow', University College London Library. Available at: http://www.jstor.org/stable/60207853
5. Murray, David (1927), *Memories of the Old College of Glasgow – Some Chapters in the History of the University*, Glasgow: Jackson, Wylie and Co., p. 266.

Plate 1 A modern megalithic observatory in Glasgow's Sighthill Park at the spring equinox. Light from the setting Sun is passing through the open structure of the University Tower on the horizon, slightly to the left of the central stone. The enlargement shows this to excellent effect.

Plate 2 Smith's map of 1828 depicts the extent of the College Grounds along the High Street, which runs up to Glasgow Cathedral. Since its establishment in 1757, the southern skyline of the Observatory had been affected by the building of St John's Church at the north end of McFarlane Street. The University had also established buildings to the north.

Plate 3   The memorial and statue of James Watt in the south-west corner of George Square, Glasgow.

Plate 4   A portrait in oils of Alexander Wilson (1714–86), Professor of Astronomy at Glasgow University from 1760 to 1783.

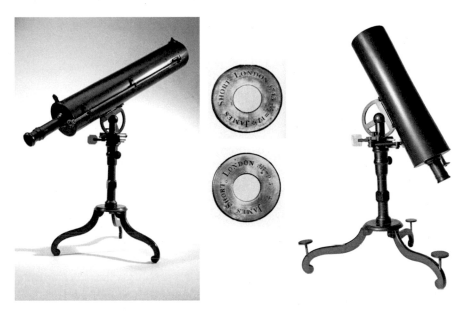

Plate 5 The left-hand picture shows the James Short telescope, with details of its eyepiece ring in the upper centre (see GLAHM 105684 in the Appendix of Chapter 1), probably used by Patrick Wilson to observe the Transit of Venus of 3 June 1769 at the Macfarlane Observatory of the Glasgow College. With less certainty, the right-hand image may be the telescope, with identification by its eyepiece ring in the lower centre (see GLAHM 113782 and 105681 in the Appendix to Chapter 1), used by Professor Alexander Wilson and referred to in his table of timings as given in Fig. 4.5.

Plate 6 The High Possil Meteorite is on display at the Hunterian Museum. Its landing site was marked 200 years later with a commemorative stone inscribed: 'THE HIGH POSSIL METEORITE – Near this site on the 5th April 1804 fell the first recorded meteorite in Scotland. It can be seen in the Hunterian Museum University of Glasgow'.

Plate 7 Fulton's Orrery as displayed at the Kelvingrove Museum, Glasgow, is remarkable in the detail of planet and satellite movements that it mimics. A poster advertising an exhibition of Fulton's Grand Orrery in the Argyll Arcade Saloon is shown to the left.

Plate 8 A colour-wash drawing of the Observatory in 1847 by 'Crimea' Simpson. Note that, relative to Fig. 6.8 (1843), the telescope on the west side of the building no longer stands.

Plate 9 Part of the main stained glass memorial western window in the Bute Hall, Glasgow University, depicting Copernicus, Galileo, Kepler and Newton.

Plate 10 The stained glass memorials to John Pringle Nichol and his son, both at the foot of the large western window in the Bute Hall.

Plate 11 In preparation for the Transit of Venus of 8 December 1874, Professor Grant produced a 'Companion' containing background material related to the event. With a beautiful royal blue cover, the title and cartoon were embossed in gold.

Plate 12 A picture by David Small (1888) showing a colourful scene of the activity on the Clyde, with the time-ball tower to the distant right.

Plate 13 From comparison with the images of Fig. 8.2, the modern easterly panorama of the Clyde from the BT building shows the dramatic changes of 150 years. St Andrew's Church remains in the distance, towards the left, but gone is the smog and grime of the factory chimneys.

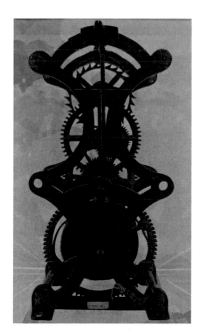

St. George's-Tron
Parish Church

Plate 14 The mechanism with pendulum lever of the Old College Tower Clock, the first clock put under telegraphic control from the Observatory at Horselethill. The image on the right shows the clock tower of St George's Tron Parish Church, the first city clock to be linked by telegraphic wire.

Plate 15 The Muirhead mean time clock, key to providing accurate time for the city and port of Glasgow, remains in the possession of the University within the Melville Room. The bottom-right section shows the terminals on the side of the case used to dispatch the beating of regular seconds to the clocks around the city.

Plate 16 The building block containing 38 Clyde Place, with the doorway reflecting the new developments on the Broomielaw where the time-ball once operated. The emblem with the ensign operating a sextant is from an envelope of Duncan McGregor & Co., posted in 1887.

6. Nichol, J. P. (1836), *Outline of the Plan of Instruction proposed to be Followed in the Class of Practical Astronomy, in the University of Glasgow*, Glasgow: Glasgow University Press [GUL Sp Coll Mu22-a.11].

7. Murray, *Memories of the Old College of Glasgow*, p. 267.

8. Nichol, John Pringle (1837–8), 'Abstract of Astronomical Lectures delivered at the University of Glasgow, Session 1837–8' [GUL Sp Coll MS Gen 879].

9. Nichol, J. P. (1856), 'Notes on some general principles in geology and their applications', in Johnston, Alexander Keith, *The Physical Atlas of Natural Phenomena*, Edinburgh & London: William Blackwood, new and enlarged edition, pp. 1–8 [GUL LRA – Store HF00175].

10. Nichol, J. P. (1833), 'On some recent discoveries in astronomy', *Tait's Edinburgh Magazine*, 4: 19, 57–64.

11. Nichol, J. P. (1836), 'State of discovery and speculation concerning the nebulae', *London and Westminster Review*, 3: 2, 390–409.

12. Schaffer, S. (1980), 'Herschel in Bedlam', *The British Journal for the History of Science*, 13, 230.

13. Murray, *Memories of the Old College of Glasgow*, p. 269.

14. Ibid., p. 268.

15. (1903) 'Unveiling of a stained glass window designed by Henry Holiday, in the Bute Hall, University of Glasgow, on 17th October 1903: In memory of John Pringle Nichol, LL.D., Professor of Astronomy, 1836–1859; his son John Nichol, LL.D., Professor of English Language and Literature, 1862–1889; and his daughter Agnes Jane Nichol or Jack, born 1839; died 1901', Glasgow: James MacLehose [GUL Sp Coll MacLehose 787; GUL Sp Coll Mu22-f.7].

16. Gavine, David M. (1981), 'Astronomy in Scotland 1745–1900', unpublished PhD thesis, Open University, 66.

17. Nichol, J. P. (1840), 'Our Relations with France, and the Prospects of a General War, Briefly Considered, in a Letter Addressed to Archibald Alison, Esq.', Glasgow: John Smith & Son [GUL Sp Coll Mu22-a.9; GUL Sp Coll RB 4802].

18. Glasgow University Manuscript 117/27402. A Report of the Proceeding at the Dinner of Friends of Astronomical Science – 16 Dec 1836 [Glasgow University Archives].

19. Knight, William Angus (1896), *Memoir of John Nichol, Professor of English Literature in the University of Glasgow*, Glasgow: James MacLehose & Sons, p. 32 [GUL English MN71 KNI; GUL Sp Coll MacLehose 106; GUL Sp Coll Mu7-c.15].

20. Ibid., p. 31.

21. Murray, *Memories of the Old College of Glasgow*, p. 270.

22. *The Topographical, Statistical, and Historical Gazetteer of Scotland*, 1842. Vol 1, A–H, Glasgow: A. Fullarton and Co., pp. 656–7.

23. Nichol, J. P. (1842), *The Present State Of The New Observatory: Addressed To The Committee Of The Institution*, Glasgow: John Clark [GUL Sp Coll Bf72-k.19]. The frontispiece of this copy carries the inscription 'The Very Revd. Principal Macfarlane – With Dr Nichol's Comps. – P.'

24. Learner, Richard (1982), *Astronomy Through the Telescope: The 500 Year Story of the Instruments, the Inventors, and Their Discoveries*, London: Evans Bros, Ltd, p. 74.

25. Knight, *Memoir of John Nichol*, p. 34.
26. Coutts, J. (1909), *A History of the University of Glasgow (From its Foundation in 1451 to 1909)*, Glasgow: James MacLehose & Sons, p. 389 [GUL Education S271 1909-C – two copies; GUL Sp Coll Mu21-a.2; Sp Coll Bh12-e.14; GUL Sp Coll MacLehose 824.
27. Grant, Robert (1883), *Catalogue of 6415 Stars for the Epoch 1870*, Glasgow: James MacLehose & Sons; London: Macmillan & Co., p. vi.
28. Murray, *Memories of the Old College of Glasgow*, p. 271.
29. Knight, *Memoir of John Nichol*.
30. Ibid., p. 32.
31. Schenck, D. H. J. (1997), personal communication to the author containing a mini-biography of William Nichol.
32. Knight, *Memoir of John Nichol*, p. 31.
33. Ibid., p. 33.
34. Glasgow University Manuscript 117/27409. Catalogue of Apparatus in the Department of Astronomy – 1947 [Glasgow University Archives].
35. Knight, *Memoir of John Nichol*, p. 70.
36. Coutts, *A History of the University of Glasgow*, p. 388.
37. Gavine, 'Astronomy in Scotland 1745–1900', 77.
38. Ibid., 73.
39. Grant, *Catalogue of 6415 Stars for the Epoch 1870*, p. xi.
40. Nichol, *The Present State Of The New Observatory*.
41. Gavine, 'Astronomy in Scotland 1745–1900', 73.
42. Nichol, J. P. (1857), *A Cyclopaedia Of The Physical Sciences, Comprising Acoustics, Astronomy, Dynamics, Electricity, Heat*, London and Glasgow: Griffin [GUL LRA Store HA02493].
43. Royal Astronomical Society (1859), 'Report of the Council to the Thirty-ninth Annual General Meeting of the Society', *Monthly Notices of the Royal Astronomical Society* [hereafter MNRAS], 19, 141.
44. Kinns, Roger (2010), 'Time balls, time guns and Glasgow's quest for a reliable local time service', *Journal of Astronomical History and Heritage*, 13, 194–206.
45. (1844) *Illustrations of the Climate of Glasgow 1843–4*, Glasgow: Robert Stuart & Co. [GU Observatory].
46. Knight, *Memoir of John Nichol*.
47. Ibid., p. 36.
48. Nichol, J. P. (1852), *Memorials From Ben Rhydding: Containing The Place, Its People, Its Cures*, London: Charles Gilpin; Edinburgh: James Nichol [GUL Sp Coll bG59-i.5]. A reproduction of this text is available from BiblioLife Publishers, India.
49. Murray, *Memories of the Old College of Glasgow*, p. 269.
50. Burnett, 'Nichol, John Pringle (1804–1859)'.
51. Nichol, J. P. (1846), *Thoughts on Important Points Relating to the System of the World*, Edinburgh and London: John Johnstone.
52. Broughton, L. N. (1931), 'Some Unpublished Astronomical Correspondence', *Popular Astronomy*, 39, 171–3.

53. Nichol, J. P. (1837), *Views of the Architecture of the Heavens – In a Series of Letters to a Lady*, Edinburgh: William Tait. This book was reproduced in 2009 as part of *Cambridge Library Collection – Life Sciences*.

54. Smith, J. (1993), 'DeQuincey's revisions to "The System of the Heavens"', *Victorian Periodicals Review*, 26: 4, 203–12.

55. Ibid., 203–12.

56. Dunn, Douglas (1986), *The Telescope Garden*. Broadcast on BBC Radio 3 on 11 July 1986 at 7.30 p.m.

57. Nichol, J. P. (1844), *Contemplations on the Solar System* (a 2nd edition of *Phenomena and Order of the Solar System*), Edinburgh: William Tait [GUL LRA I7-f.6].

58. Nichol, J. P. (1842), *Phenomena and Order of the Solar System*, New York: Dayton & Newman.

59. Nichol, J. P. (1838), *Phenomena and Order of the Solar System*, Edinburgh: William Tait.

60. Royal Astronomical Society, 'Report of the Council to the Thirty-ninth Annual General Meeting of the Society', 141.

61. Nichol, J. P. (1847), *Contemplations on the Solar System*, 3rd edition, Edinburgh and London: John Johnstone.

62. Nichol, J. P. (1848), *The Planet Neptune: An Exposition and History*, Edinburgh and London: J. Johnstone [GUL LRA I7-f.8]. This book is now available in the form of a reproduction as a Kessinger Publishing's Legacy Reprint.

63. Ibid., p. 115.

64. Ibid., p. 118.

65. Nichol, J. P. (1851), *The Planetary System: Its Order, And Physical Structure*, Edinburgh: James Nichol; London: H. Balliere; Glasgow: Griffin & Co. [GUL I7-g.16].

66. Nichol, *Contemplations on the Solar System*, 3rd edition, p. 141.

67. Nichol, *Views of the Architecture of the Heavens*.

68. Nichol, J. P. (1840), *Views of the Architecture of the Heavens – In a Series of Letters to a Lady*, New York: H. A. Chapin & Co.

69. Nichol, J. P. (1842), *Views of the Architecture of the Heavens – In a Series of Letters to a Lady*, 2nd edition, New York: Dayton & Newman.

70. Nichol, J. P. (1843), *Views Of The Architecture Of The Heavens – In a Series of Letters to a Lady*, 4th edition, Edinburgh: William Tait [GUL LRA I7-f.7].

71. Nichol, J. P. (1845), *Views Of The Architecture Of The Heavens – In a Series of Letters to a Lady*, 5th edition, Edinburgh: William Tait.

72. Nichol, J. P. (1850), *The Architecture of the Heavens*, London: John W. Parker [GUL LRA I7-b.3].

73. Nichol, *Thoughts on Important Points*.

74. Nichol, J. P. (1848), *Thoughts on Important Points Relating to the System of the World*, 2nd edition, Edinburgh and London: John Johnstone [GUL LRA I8-c.22].

75. Nichol, J. P. (1848), *Thoughts on Important Points Relating to the System of the World*, Boston, MA and Cambridge, MA: J. Munroe & Co. This was the first American edition, revised and enlarged.

76. Nichol, J. P. (1848), *The Stellar Universe : Views on its Arrangements, Motions, and Evolutions*, Edinburgh and London: John Johnstone [GUL LRA Store HA09054].

77. Secord, J. A. (2000), *Victorian Sensation: The Extraordinary Publication, Reception, and Secret Authorship of Vestiges of the Natural History of Creation*, Chicago, IL and London: University of Chicago Press, p. 466 [GUL History DM300 SEC].

78. Ibid., p. 467.

79. Nichol, *A Cyclopaedia Of The Physical Sciences*.

80. Willm, J. (1847), *Education of the People: A practical treatise on the means of extending its sphere & improving its character; With a preliminary dissertation on some points connected with the present position of education in this country by J. P. Nichol*, Glasgow: William Lang; London: Simpkin, Marshall & Co.; Edinburgh: A. & C. Black; Dublin: J. M'Glashan [GUL E-Book].

81. Nichol, J. P. (1851), *On the Existing Obstructions to the Institution of a National System of Education*. A pamphlet published by David Robertson, Trongate and John Robertson, 5 Maxwell Street and produced from an address given by the author at a meeting of the Glasgow Public School Association, held in the Merchants' Hall, Glasgow, on 11 November 1851. The material came from a report in the *North British Daily Mail* and was completed by Nichol's rough notes; Nichol, J. P. (1858), *Moral Training In Our Common-Schools: Suggestions Of Certain Practical Methods Of Increasing Its Efficiency: Being the Substance of a Discourse read before the Glasgow Branch of the Scottish Educational Institute, on 20th March, 1858*, London and Glasgow: Richard Griffin & Co. [GUL Sp Coll Mu Add. q94].

82. Coutts, *A History of the University of Glasgow*, p. 388.

83. Knight, *Memoir of John Nichol*, p. 58.

84. Burnett, 'Nichol, John Pringle (1804–1859)'.

85. Royal Astronomical Society (1859), 'Report of the Council to the Fortieth Annual General Meeting of the Society', *MNRAS*, 20, 109–49.

86. (1903) 'Unveiling of a stained glass window'.

87. (1903) 'Lord Kelvin and his first teacher in natural philosophy', *Nature*, 68, 623–4.

88. (1903) 'Unveiling of a stained glass window'.

89. Gilfillan, G. (1848), 'Popular Lecturers – No. 1 – Prof. Nichol', *Tait's Edinburgh Magazine*, 149, 152.

Chapter 7

The Glory of Glasgow

7.1 PROFESSOR ROBERT GRANT

Following Nichol's death, the University appointed Robert Grant[1] to the Regius Chair of Astronomy in 1859. He was a native of Grantown in Strathspey, a tradesman's son, born in 1814. The local endowed school provided him with the usual educational subjects, including Latin and Greek, to prepare successful pupils with the prerequisites for entry to the Scottish universities. He commenced Latin at the age of nine and made rapid progress in this as well as in all other subjects. At the age of thirteen, he was overtaken by an illness that confined him to the care of his relatives for the next six years, putting on hold his educational advancement. From then on he became his own instructor and, with the help of books, he continued assiduously to study Greek, Latin, French, Italian and elementary mathematics. The return of Halley's Comet in 1835 appears to have given an impulse to his astronomical studies. After a short spell of study at King's College, Aberdeen, he gravitated to London, where he was employed as book-keeper within a counting-house run by an elder brother. During the next four years, he spent his leisure time reading all the standard works of astronomy and mathematics. Around this time, he first conceived the notion of writing a brief history of physical astronomy, from the earliest ages to the current time. He had already collected a wealth of material suitable for this venture but delayed the production until he had taken the opportunity of consulting the works on astronomy contained in the great public libraries of Paris. In 1845, he took up residence in Paris and remained there for two years, supporting himself by giving English lessons. During the greater part of this time, he had access to the library of the Académie des Sciences in the Institut de France, a privilege he received through an introduction by one of his pupils. He regularly attended courses of astronomical lectures delivered by Arago at the Observatory and by Leverrier at the Sorbonne. Later, it is noted that one of his favourite books was Sir John F. W. Herschel's *Treatise on Astronomy*, published in 1851.

On his return to London in 1847, Grant devoted most of his time to the preparation of his classic work, *History of Physical Astronomy*.[2] At this time, he was still unknown to the leading astronomers of the day, but fortunately had been introduced to a Mr Robert Baldwin, an enterprising publisher of an excellent series of works entitled *The Library of Useful Knowledge*, which received popular attention. Grant proposed to contribute to the series with a concise history of the principal investigations and discoveries in astronomy which helped to build up the current knowledge of the physical movements and aspects of the heavenly bodies, or, in his own words:

> to exhibit a view of labours of successive enquirers in establishing a knowledge of the mechanical principles which regulate the movements of the celestial bodies, and in explaining the various phenomena relative to their physical constitution which observation with the telescope has disclosed.

His proposal was immediately accepted, and the work was published in separate parts, or numbers, the first appearing in September 1848. After the ninth issue, this publication form was discontinued. The complete work of the *History* was eventually published in March 1852, by which time Grant had been elected as a Fellow of the Royal Astronomical Society and had served as editor of *Monthly Notices* from 1853 to 1860. Some of the personal disadvantages under which Grant had prepared his work were described in the Annual Report of the Council of the Society in February 1853:

> During the past year Mr Grant has completed his *History of Physical Astronomy*, a work the first numbers of which formed his introduction to the Society. It is very seldom that the Council are called upon to notice any work of so peculiar a merit. If Mr Grant had possessed perpetual access to extensive libraries, and peculiar inducements to consult them; if he had passed many years in historical research, and matured his judgment by long habits of discussion with others of like pursuits, and had learned by small attempts how to venture on larger efforts, the production of a sustained history, of extent and matter such as he has published, would have excited admiration, and would have called forth some remark on the true secret of success in such undertakings. But when it is considered that the author had no more connection with libraries than anyone may command for the asking, and no peculiar calling to tempt him into research; that he has not reached middle life, and that his opportunities were furnished, as is understood, by the leisure of ill-health or convalescence; that he had, before the publication of this work, little or no sustained opportunity of communication with those who could have helped to direct his reading or form his judgment, and that he never published anything whatever previously to the appearance of this well-conceived, well-executed, and greatly-wanted history, the Council, while they state that the work has taken a place among standard books of history from its appearance, under circumstances which greatly enhance the merit of its author, congratulate the Society upon the prospect of numbering Mr Grant among its executive members.

His writing was clear, very much to the point and in strong contrast to Nichol's verbose 'purple prose'. An excellent example of his succinct style can be appreciated by reading his *Elementary Survey of the Heavens*, in which he provides definitions of various astronomical terms to form part of Johnston's *Atlas of Astronomy*.[3] At the time of its publication, it was noted that any reader of Grant's *History of Physical Astronomy*[4] could not fail to be struck by the rare skill and power displayed in the discernment of the salient points of some of the most difficult researches in astronomy, many of which, by their nature, were controversial, as well as of an abstruse character. It was commented that, in most cases, the author had most scrupulously endeavoured to assign to each of the great men who had been working independently on the same line of research their proper share in the common labour. On the award to Grant of the Gold Medal of the Royal Astronomical Society on 8 February 1856, the President, Manuel J. Johnson, remarked that

> nowhere is this more conspicuous than in the discussion relative to the discovery of the planet *Neptune*. By a simple narration of facts he has placed the history of the great event in so clear and so true a light that I believe I am not wrong in saying he has gained an author's highest praise under such circumstances – the approval of both the eminent persons concerned.

As described in Chapter 6, ten years previously, there had been much controversy relating to Sir George Airy, the Astronomer Royal, and his 'slowness' in taking up the mathematical predictions of the autumn of 1845 made by John Couch Adams in respect of the presence of a planet beyond Uranus that affected the latter's orbit. In France a few months later, Leverrier also provided a celestial location for the disturbing planet; the information was dispatched to the Berlin Observatory, and there it was Galle who recorded its position and movement within a few weeks, the planet being discovered on 23 September 1846. It was immediately referred to as Leverrier's Planet, much to the chagrin of many English astronomers when they realised that Adams's work had not been appreciated fully and the 'discovery opportunity' of finding this new planet beyond Uranus had been missed. When it was claimed that it was an Englishman who had first made the prediction for Neptune's position, the French became aggrieved. The celestial coordinate position predictions made by Adams and Leverrier were within 1° of each other. In his *History of Physical Astronomy*,[5] published shortly after the discovery controversy, Grant, very conscious of the attitude of the French, presented a measured and balanced account of the events from the viewpoints of both sides of the Channel. The descriptions of the roles played by Airy and Challis, the Professor of Astronomy at Cambridge, in not taking up the work of Adams with the urgency that it deserved, were somewhat tempered relative to the 'bad press' they had received at the time. The detailed account of the discovery of Neptune that Grant provided must have placed him in good stead in the eyes of the Astronomer Royal.

Professor Grant's time spent in France and his intimate knowledge of the French language served him well in the translation of François Arago's

Astronomie Populaire[6] which he edited with Admiral W. H. Smyth; the two-volume work was published in 1855 and 1858. A part of this was reprinted in 1861 under the title of *A Popular Treatise on Comets*.[7] In 1858, he became a protégé of Airy, who gave him an unprecedented year-long training course in observational astronomy at Greenwich to fit him for a post. This opportunity, and his appointment to the Regius Chair in Glasgow, may not have been unconnected with his favourable treatment of the episode concerning the discovery of Neptune in his *History of Practical Astronomy*, which was especially noted in the Royal Astronomical Society's President's Address as having gained the approval 'of both the eminent persons concerned'. According to Hutchins,[8] the appointment of Nichol's successor lay technically with the Home Secretary, although it would actually have been made on the Lord Advocate's recommendation. The post was not directly within Airy's patronage, unless he was approached for advice by the Minister. However, as with Adams's election at Cambridge, an example of local versus disciplinary networking now occurred in which Airy's role proved decisive. When the University parochial group tried to 'push an old Glasgow student named Jack' (Nichol's former student and young son-in-law, William Jack), Professor J. D. Forbes of Edinburgh University took a special 'and unsolicited interest' in Robert Grant's success. He commended Grant's 'singleness of purpose', like that of Thomas Henderson in Edinburgh, and managed to get Airy to write a private letter of recommendation, with permission for him to use it. Another candidate for the Chair was an Arthur Gayley, and William Thomson recommended George Gabriel Stokes, the illustrious physicist. Grant, however, backed by Forbes and Adams, was successful and embarked on a dedicated and fruitful tenure lasting thirty-three years, with observational work of international significance. Portraits of him soon after his Glasgow appointment and in older age are presented in Fig. 7.1.

Grant's appointment to the vacant professorship took place in November 1859, but he did not take up residence at the Observatory until May 1860. The stipend was £270, with accommodation provided in the attached house. Very soon after his arrival, he was asked to help out with the running of the natural philosophy class after Lord Kelvin met with an unfortunate accident in 1859. This he did in combination with Professor Darwin Rogers (see Murray).[9]

Within weeks, he journeyed to Spain to observe from the summit of the Sierra de Toloño the total eclipse of the Sun of 18 July 1860. One of the debated questions of the mid-nineteenth century was the location and origin of the red, scarlet or vermillion sierra – the choice of colour varying from one document to another – observed on the lunar limb during solar eclipses. In his *History of Physical Astronomy*, Grant had discussed the reports of the observations of all previously recorded eclipses. 'While engaged in this enquiry,' he remarked,

Figure 7.1 A portrait of Professor Grant in middle age is on the left, while the right-hand panel shows him in later life; a question might be posed as to whether the watch chain fob is his RAS gold medal.

I endeavoured to show that the phenomena observed during solar eclipses, whether corona, streaks or patches of light, or prominences great or small, are all solar phenomena. While the results of observation were either positively in favour of the solar aspect of the case, or adverse to the lunar aspect, there did not appear to exist a shadow of a proof in support of the lunar origin of the phenomenon.

It was a source of great satisfaction to Professor Grant that, having been a member of the HMS *Himalaya* eclipse expedition to Spain in 1860, he had the opportunity of witnessing the event of the 'scarlet sierra' of which he had given so plausible an interpretation ten years before. He took with him a Dolland refractor belonging to the University Observatory, the instrument having been fitted with measuring apparatus of some kind. At the time he asked William Thomson to propose useful experiments that might be undertaken.[10]

Professor Grant's observations of the 1860 eclipse were simply visual, and he reported on several aspects such as the colour of the Moon prior to the eclipse, the colour of the sky and his failure to see the passage of shadow bands, probably because of his location on the southern slope of Sierra de Toloño. All his comments are interspersed with those of other contributors within a tome[11] of 1879 which is dedicated to reports and measurements of several different solar eclipses. A distillation of one of his observations is presented in Fig. 7.2.

562   *Drawings of the Corona of* 1860, *July* 18*th.*

Prof. Grant.   42° 38' N. } SIERRA DE TOLONO,
                                    2° 40' W. }   18th July, 1860.

MS. Reports of the Himalaya Expedition.

The breadth of the corona appeared to be equal to about a fourth
of the moon's diameter. From its exterior margin there issued forth
diverging beams of light of unequal length. These beams were not of
uniform brightness throughout their whole
extent. At their junction with the corona
they appeared to be of equal brightness
with the latter, but from thence they gradually faded off until they were finally lost
in the dark ground of the heavens; their
average length was about 30'.

Long curved ray   One very long beam, a little to the
to the right of
the vertex.   right of the vertex, exhibited a decided cur-
vature. The exterior margin of the corona
was very imperfectly defined. A similar remark applied to the terminating sides of the
diverging beams of light.

The light, both of the corona and of
the beams issuing from it, appeared to me
to be perfectly motionless. The above-

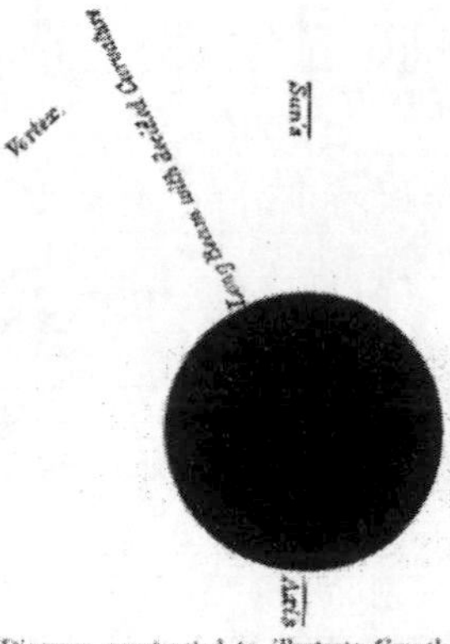

Diagram constructed to illustrate Grant's
description of the corona of 1860, July
18th.

mentioned features of the corona were observed with my naked eye, and
not in the inverting telescope.

Figure 7.2   A detail from Professor Grant's report of the solar eclipse of 18 July
1860, observed from Sierra de Toloño. The original manuscript formed part of the
Reports of the Himalaya Expedition.

Professor Grant's grounding in the classics proved an advantage when he
responded to a note by a Dr Julius Schmidt concerning the description of
a solar eclipse in AD 968. Grant[12] redirected the discussion to the works of
Philostratus and then to Plutarch, quoting several passages in ancient Greek.
He noted that Plutarch made comment on solar eclipses in general, using
graphic language, stating that:

> but even if the moon should at any time conceal the whole of the sun, still the
> eclipse is deficient in duration as well as in amplitude, for a certain effulgence
> is seen around the circumference which does not allow a deep and very intense
> shadow.

Grant suggested that it was probable that the total eclipse mentioned by
Plutarch was the same one that Philostratus had described, and it may be
presumed that the writings of both these commentators contain the earliest
recorded allusions to the 'corona'.

Grant's knowledge of the history of science was also important when it came to debunking the 'Chasles' episode. Between 1867 and 1869, Michel Chasles presented a series of manuscripts to the Académie des Sciences, which suggested that some of Isaac Newton's claims to original discovery were unfounded, and that they should be ascribed to Pascal. Professor Grant made incisive contributions to the discussion by writing two letters[13] to Monsieur U. J. Leverrier in Paris, the first dated 12 September 1867, and the second 31 October of the same year, dismantling Chasles's propositions as forgeries. In the summary of his first letter he states: 'There is only one possible solution to the difficulties which I have proposed, and it is this: the entire mass of the documents communicated to the Academy of Sciences by M. Chasles are pure forgeries.' Additional responses to the affair came from three other British men of science: David Brewster, Augustus De Morgan, who became Professor of Mathematics at London University at the early age of twenty-two, and Thomas Archer Hirst, Professor of Physics, also at London University, who later succeeded De Morgan to the Chair of Mathematics.

In an article printed in *The Times* of 20 September 1867, Grant exposed the fallacies of Chasles by showing that certain numerical results contained in the so-called Pascal papers corresponded to numbers in the third edition of Newton's *Principia*, published in 1726; Pascal had died in 1662.

David Murray,[14] who had the privilege of being shown around the Observatory by Professor Grant on several occasions, described him as an accomplished astronomer, but without the gifts that had rendered his predecessor so popular. He had, however, other qualities. When the movement for the higher education of women began in Glasgow, Professor Grant undertook to lecture on astronomy; he proved to be an excellent presenter, and his courses were popular and attracted a good student following. He had no graces of oratory, no flowers of rhetoric and he spoke with a strong Aberdonian accent, but he spoke easily and his statements were clear and concise and always to the point. His listeners were able to grasp just what he intended them to know, and they left the lecture room with a precise and definite picture in their minds. He presented courses of astronomical lectures on several occasions, the most important of which were delivered in 1854 and 1855 at the London Institution, Finsbury Circus, and in 1868, 1869 and 1870 at the Royal Institution. Shortly afterwards, he delivered three similar courses of lectures at the University of Glasgow, inviting both ladies and gentlemen to attend (see Fig. 7.3), and had the distinction of lecturing to women before they were formally admitted to the University.

Although there was no formal degree course established in astronomy, Professor Grant believed strongly in the importance of the subject as part of education, and also in its use for general practicalities. In his Report of 1865[15] to the University Senate, it was recorded that the usual course of lectures on astronomy was delivered in the College during the session, with attendance about the same as on former occasions. It was suggested that with

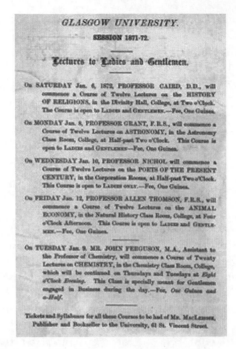

Figure 7.3 For the University session 1871–2, a course of twelve lectures on astronomy was advertised for ladies and gentlemen at a fee of one guinea. It is also noteworthy that Professor Nichol, son of James Pringle Nichol, was offering a course entitled 'Poets of the Present Century' for ladies only.

the move from the Old College and the erection of the new University buildings in the vicinity of the Observatory, this might exercise a favourable influence on attendance numbers. A well-attended course of popular lectures on astronomy was also delivered at the Observatory. Similarly, in the Report of 1867,[16] he noted that the small transit instrument recently fitted in the building for the Ochtertyre Telescope was available for the students attending the lectures on astronomy which he delivered at the College. Arrangements were being made for its use in the present session by a few young gentlemen who were engaged in the preparatory course of study connected with the engineering profession.

His optional classes usually attracted about ten students, with one lecture given per week for twenty weeks from December to March. This system continued throughout Grant's career. Astronomy was not a major part of the physics or mathematics syllabus, but it was present in the University examinations. The papers for the Ordinary MA degree in mathematics and natural philosophy contained sections on geometry, algebra and trigonometry, dynamics, astronomy and physics. As an example of the standards of knowledge required, the gist of the astronomy questions in the paper for 1882–3 appears below.

ASTRONOMY

1. Explain what measurements are necessary for specifying the position of a celestial object, the plane of the ecliptic being taken as the plane of reference.
2. Describe shortly the instruments used for obtaining the zenith distance of a celestial body and the time of its meridian passage.
3. Explain the principle for finding Greenwich time by means of lunar observations. What other methods may be followed?
4. What is meant by the diurnal libration of the Moon? What is its cause?
5. What is the lowest latitude at which the phenomenon of a non-setting sun may be observed?
6. How is it known that the Moon turns on her axis in the same time as she revolves round the Earth?
7. Show how the distance of Mercury from the sun may be obtained from observation of his greatest and least apparent diameters.
8. Give a brief explanation of the theory of periodic shooting stars.

One of Grant's concerns was the distribution of accurate time to the city. As part of the commissioning of the Horselethill Observatory, the University had a 'contract' with Her Majesty's Government to render the resources of the Observatory for the transmission of correct time to the mariners on the Clyde so that they could regulate their chronometers. Although Professor Nichol had invested in the astronomical equipment required to execute the obligation, after fifteen years nothing had come to fruition on this front. Having been appointed University Observer, and recognising the conditions of the Observatory licence, Grant immediately took up a scheme of protracted observations using Ertel's meridian instrument to provide time; this also provided a basis for investigations of stellar proper motions. In his Report of 1865 to the University Senate[17] he reiterates that:

> It is of importance to bear in mind that the office of Astronomical Observer in the University is not a mere accidental appendage to the Chair of Astronomy. On the contrary, it was instituted by the Crown as an Office co-ordinate with, and distinct from, the Professorship, and on this ground is equally entitled to the watchful consideration of the authorities under whose control it is placed. I may be permitted, therefore, to express a hope that the University will not relax its efforts to secure for the Observatory such a measure of public support as is consistent with its excellent instrumental equipment, its general activity, and its well founded claims as one of the most ancient University Observatories in the British Isles.

His persistence in persuading the city authorities and his success in getting help from the telegraph companies is clearly apparent in all of his Reports to Senate and in other documents and papers.

By 1850, Greenwich Mean Time (GMT) had been accepted as the

standard for running the country's railways, and, using chronometers, 'time' was carried from station to station. To determine GMT locally (remembering that the 'G' is an abbreviation for 'Greenwich' and not 'Glasgow') and to provide it both for shipping on the Clyde and for the operations of city commerce, a person of vision with strong resolve was required. Although other cities used astronomical observations to provide time in order to operate a one o'clock time-ball, or a single clock face, the Glasgow scheme was unique in terms of its scale with widespread functionality over a small area. More than fifteen clocks were controlled around the city and the system operated for over fifty years. The operations were not surpassed by any other city around the world. The story of its development is so special that Grant's endeavours deserve a more expansive presentation, and this is reserved to form Chapter 8.

7.2 OBSERVING AT HORSELETHILL

At the commencement of Grant's command in Glasgow, the Observatory on Horselethill, just to the north-east of Dowanhill, remained fairly isolated with clear vistas all around. The picture in Fig. 7.4 made in around 1864 clearly shows the observatory dome on the skyline on the far right. Beneath it is the downward flow of Victoria Crescent; the building at the centre is Crown Circus which remains in its grandeur to this day. The sweep up to it shows the absence of any other properties as it ran from Byres Road which, at the time, was but a rural track. Within a few years, the urban spread picked up at a pace that would erode the practicability of the Observatory site, but for twenty-five years under Professor Grant's directorship, the observations made there proved their worth beyond what might have been imagined.

Included in his Report of 1861 to the University Senate,[18] Grant summarised the role of any Observatory. He suggested that such an institution should contribute by:

Figure 7.4 In 1864, the Observatory to the side of Dowanhill on the far right commanded a clear skyline in all directions. The site had a rural surround, except for Crown Circus in the centre of the figure, and was approached by a clear driveway up from Byres Road marked by the line of trees across the foreground of the panorama.

1. Advancing science through the course of Astronomical Observation,
2. Disseminating accurate time to one or more stations more central to any adjoining conurbation,
3. The diffusing of a practical knowledge of Astronomy, throughout the Country, by connecting the Observatory closely with the course of Scientific Instruction in the University.

In Grant's statement in the preface of his Glasgow Star Catalogue,[19] the instruments at the Observatory in 1860 were: a transit circle by Ertel & Sons of Munich, acquired by Nichol; a Dolland refractor, 3 inches in aperture; a sidereal clock by Bryson of Edinburgh; a mean time clock, probably that by Muirhead of Glasgow; and several other instruments of minor importance. The Bryson clock had a dead beat escapement and a mercury compensation pendulum. Although another sidereal clock by Frodsham of London was purchased in 1861, it was the Bryson that appeared to be the mainstay for the observations made with the transit telescope. In Fig. 7.5 can be seen photographs of the piece as it stands in the Turnbull Room, named after Bishop Turnbull, the founder of the University, as noted in Chapter 1 (see p. 4). According to an inventory[20] of Observatory apparatus, the Bryson clock ran at sidereal rate until 1894 and was then readjusted in 1905 to keep mean time. It was removed from the Observatory and taken to the University in 1938.

The establishment also contained the Breadalbane Telescope, noted as being of 20 inches in aperture, mounted equatorially. The instrument was unsuitable for observations in its existing condition, and considerable expenditure would have been required to have it upgraded for any practical use. After being in post for a year, in his first Report to the Senate[21] of 1861 Grant mentions that an arrangement had been made for illuminating the face of the sidereal clock, probably the Bryson, and that the mean-time clock, presented by Mr Muirhead, had been recently repaired and cleaned. In the same report he describes the unsatisfactory nature of the large reflector, stating that 'it can only serve as a gazing instrument' and was unsuitable for measuring purposes. This inadequacy was rectified shortly afterwards with the installation of a magnificent 9-inch refractor, as mentioned in Professor Grant's Report of 1864.[22]

Professor Grant's vision of the potential and practical application of the existing Ertel meridian telescope, purchased by Professor Nichol, was key to the immediate future of the Glasgow Observatory. The telescope was 6 inches in diameter with a graduated circle 3.5 feet in diameter. Rather than simply limiting the instrument's use to generating a time service, its practicality was extended by a modification that allowed the zenith distances of stars to be measured at the time of meridional passage, so providing accurate positional measurements of stars and minor planets to about ninth magnitude.

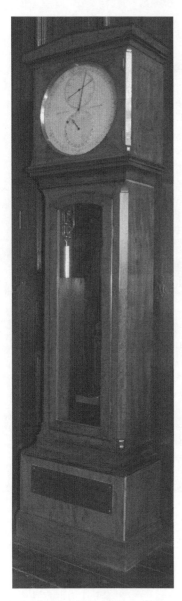

Figure 7.5 The Bryson sidereal clock, dated AD 1835, was the mainstay for the stellar positional observations made by Professor Grant over a twenty-year period commencing 1860. It currently resides in the Turnbull Room within the Main Building of the University of Glasgow.

Professor Grant commented that the weather on the whole during the period 1866 to 1867 was unusually unfavourable for astronomical observations, even given the poor climate of Glasgow. And yet, by taking advantage of the unclouded state of the sky on every possible occasion, considerable numbers of star-places were determined. The table in Fig. 7.6 shows that it was possible to take measurements on 155 nights in the interval covering 29 March 1866 to 29 March 1867, and that 2,042 Right Ascension (RA) values

Month.	No. of Observations in R. A.	No. of Observations in N. P. D.	No. of Observing Nights.
1866, March,	20	16	2
" April,	172	142	15
" May,	158	110	17
" June,	118	62	12
" July,	124	95	12
" August,	160	135	12
" September,	113	82	14
" October,	135	120	9
" November,	121	94	11
" December,	126	118	11
1867, January,	243	210	14
" February,	222	224	11
" March,	330	328	15
	2042	1736	155

MONTH.	Clock Stars.		Bessel or B. A.C. Stars.		Sums.	
	R. A.	N. P. D.	R. A.	N. P. D.	R. A.	N. P. D.
April,	4	3	3	3	7	6
May,	44	27	51	53	95	80
June,	18	0	0	0	18	0
July,	36	8	19	19	55	27
August,	101	71	302	324	403	395
September,	67	44	198	207	265	251
October,	116	93	401	405	517	498
November,	149	124	516	519	665	643
December,	79	67	375	377	454	444
January,	61	39	104	104	165	143
February,	23	13	32	32	55	45
March,	82	62	265	265	347	327
April,	30	20	57	57	87	77
Sums	810	571	2323	2365	3133	2936

Figure **7.6** The left-hand panel summarises the meridional observations from 29 March 1866 to 29 March 1867, taken from Professor Grant's 1867 Report to the Senate, while the right-hand panel is for 25 April 1870 to 20 April 1871, taken from the 1871 Report.

($\equiv$ to a determination of time) for stars were obtained. North Polar Distance (NPD) measurements numbered 1,736. It was noted too that the reduction of these observations, which forms by far the most onerous part of the operations connected with the meridional instruments, was well advanced.

Such an accumulation of data bears testament not only to the kinds of observation being made, that is the time measurement of a star crossing the meridian, each observation requiring about 30 seconds of clear vision, but also, and more importantly, to the dedication and perseverance of the staff in being vigilant and taking advantage of those brief opportunities during the tedious nights. In his Report to the Senate of 1868,[23] Grant stated that the weather had been very unfavourable for astronomical observations, especially since the commencement of the year. This can be readily confirmed by noting that during the three months of January, February and March, the days on which rain fell numbered sixty-nine. As for the present-day climate of Glasgow, it might be said that '*Plu(ie)s ça change, plu(ie)s c'est la même chose*'.

The Report of 1871[24] continued to show the great drive in undertaking the transit telescope observations; the statistics for the period 25 April 1870 to 20 April 1871, revealing the significant accumulation of data, are contained in the right-hand panel of Fig. 7.6. This interval was a year that exceeded any previous one in terms of the number of observations made. Determinations of 3,133 RA values and 2,936 of NPD were obtained. A key element of Grant's drive was that the nightly measurements were continually reduced and not allowed to accumulate without attention. In the Report, he noted that 'the reduction of the measurements was very much in hand'. He also noted that the star places determined from year to year were steadily accumulating, and that they constituted a mass of materials, which, it was to be hoped, would ultimately be employed in the construction of a substantial

Figure 7.7 The Observatory at Horselethill remained in a rural location until the expansion of the city, which included the College's move in 1871 to its present site at Gilmorehill.

Catalogue of Stars. As will be seen from Section 7.5 below, this was indeed achieved.

All through this period, the site at Horselethill still remained relatively rural, as the map in Fig. 7.7 suggests. This same figure illustrates the impending relocation of the College from the city to what is now called the West End. In 1871, the University moved from the Old College site in the High Street to its present location at Gilmorehill. To mark the occasion, the professors posed for an exodus picture on the 'Lion and Unicorn' staircase which later became incorporated on the west side of the imposing main building on the new campus (see Fig. 4.12). Although Professor Grant had his home, office and operations at the Horselethill Observatory, he took part in the occasion, and he is clearly identifiable in the photograph depicted in Fig. 7.8.

7.3 METEOROLOGICAL OBSERVATIONS

As seen in the previous chapter, some meteorological work had been conducted by Nichol, and this was continued and developed by Professor Grant. In his 1866 Report to the Senate,[25] it was noted that meteorological observations had continued as usual. The Robinson's Cup Anemometer, fitted at the beginning of 1865, had been in constant use, with the exception of a few days when it was rendered unserviceable after being blown down during the

Figure 7.8 The exodus of the professors from the Old College by the 'Lion and Unicorn' staircase in 1871 shows Professor Robert Grant in the middle of the group.

storm of 31 December 1865. At the time of the Report, the Meteorological Department of the Board of Trade, in association with the Royal Society, was considering setting up a small number of meteorological stations equipped with self-recording instruments. Glasgow Observatory was one of the stations that had been recommended for this purpose, and it was expected that the result of the deliberations would be shortly communicated to the public.

Soon afterwards, Glasgow Observatory became one of the seven stations in the British Isles selected to run a system of meteorological observations using self-recording instruments, with the operations being executed, and the results published, at the government's expense. This scheme of observations commenced in January 1868. The instruments included the Robinson Anemometer, a barograph and a thermograph.

After three months of operation, Grant's Report to the Senate of 1868[26] noted that all astronomical and meteorological instruments were in excellent working order. The expediency of reducing the dimensions of the pressure plate of Osler's anemometer, a step resorted to in 1865, had been signally illustrated on the occasion of the great storm that occurred on 24 January 1868. Previously, on similar occasions, the instrument almost invariably broke down from the fury of the wind at the very time when its indications would have been most desirable. In this instance, the instrument rolled through the entire storm without sustaining the slightest damage, and was the means of furnishing an interesting record of the direction and force of the wind on that memorable occasion. The appendix to the same Report to the Senate provided a lithographic representation of the barograph and anemometer tracings.

In Grant's Report to the Senate of 1871,[27] it was noted that meteorological observations, established under the Committee of the Royal Society, were in full operation, with the results transmitted weekly to Kew. This work continued without interruption until December 1883, when the Meteorological Council reduced the number of participating stations to three, these being Valencia, Kew and Aberdeen; connections with the stations of Armagh, Stonyhurst and Falmouth were also dropped in addition to that of Glasgow. Through his strenuous exertions, Professor Grant obtained support[28] from the Town Council and the Clyde Trust to provide the pecuniary assistance necessary to continue the observations as before. In addition, he was supportive[29] to the Scottish Meteorological Society for setting up an observatory on Ben Nevis.

Details of many of the instruments in use, including some remarkable automated measuring devices, can be found in an analysis by Grant[30] of recordings made at the Observatory in 1887. On 31 August that year, 1.772 inches of rain fell in two hours. The average annual rainfall over the years 1878–87 was 36.046 inches.

7.4 SOME GENERAL ASTRONOMY

7.4.1 The Observatory

When Professor Grant was appointed, the Observatory had been in the possession of the University for about fifteen years. Although many pieces of equipment had been purchased by Nichol, in no way could it be described as a fully functional institution. The structure itself was substantially built and, at the time when Grant took charge, it was relatively isolated from houses and the urban sprawl that was inevitably to appear later.

Partly through its advertisement of providing GMT to the city, the Observatory was highlighted as a feature to be seen on any visit to the city. Tweed's 1872 *Guide to Glasgow*[31] provides a description in the following manner:

> Leaving the Botanic Gardens, let us proceed still westward, having the fences of Glasgow on our right, and the magnificent range of Grosvenor Terrace on our left. Behind this range is the *Royal Observatory* on Downhill. It is in charge of the Professor of Astronomy of Glasgow University, and is finely situated on the

Figure 7.9 The Observatory at Horselethill at the time of Professor Grant. The left-hand panel shows open cultivated fields but with large stone townhouses beginning to appear. The main building with the dome behind is depicted in the photograph on the right-hand side, behind the fence and entrance gates, this latter image appearing in King's 1896 publication.

highest ground in the neighbourhood. Further west the houses get more scattered, but buildings are still going up on every side, there apparently being no end to the extension of Glasgow in this direction. On surmounting the rise in the Great Western Road, we obtain a wide view westward, which includes the great mass of *Gartnavel Royal Lunatic Asylum*, one of the most extensive in Scotland.

It should be pointed out that the Observatory never acceded to being designated as 'Royal', and, of course, such a reference to Gartnavel is no longer politically correct. In a similar vein, a 1905 guide[32] also notes: 'Gartnavel Asylum keeps its own sorrowful chapter of wrecked lives; and the Observatory on Dowanhill has its share in making history'. The buildings at Gartnavel with the tall chimneys remain, although a view of them, from the noted site in the Great Western Road, has gone as a result of the development of the substantial stone houses along its length. A pair of images of the Observatory on Horselethill are displayed in Fig. 7.9.

7.4.2 The Ochtertyre Refractor

Professor Grant's appointment to the Regius Chair of Astronomy at Glasgow involved the running and development of the Observatory. In his Report to the Senate of 1864,[33] he reiterated what he thought was the purpose of any astronomical observatory. In this discourse he described the status of the Greenwich Observatory and the role that the Astronomer Royal played in respect of duties associated with the production of the *Nautical Almanac*, the examination and repair of chronometers for the use of the Admiralty and the inspection of lighthouses. Grant was very conscious of the changes affecting the pursuit of astronomical research, with telescope activity grappling more and more with the understanding both of the nature of objects and of the physical processes controlling their behaviour. It is interesting to note that in the debates of the time as to whether observatories should be established in order to pursue physical astronomy beyond the remit of Greenwich, Grant did not think[34] that a new state observatory was the answer, instead believing that developments might best be made by private endeavour. It is an open question as to whether this fairly substantial discussion contained in his Report is simply in respect of the politics affecting the prime importance of Greenwich as the National Observatory relative to the developments of the Royal Observatory at Edinburgh. It may also have been related to a more personal concern over the local issues and dealings with Professor Piazzi Smyth, and the standing of the Edinburgh Observatory with respect to that of Glasgow University.

In the same Report of 1864,[35] he listed the three leading objectives of the Observatory, much in the same way as he had done in 1861, these being:

1. The advancement of astronomy by a course of accurate observations,
2. The transmission of correct time from the Observatory to the City of Glasgow and
3. The diffusion of a knowledge of the principles of Observing Astronomy by connecting the Observatory closely with the course of scientific instruction in the University.

Certainly, since taking the Chair, Grant had been very positive about the achievements made in his first three years. As for the second objective, the meridional observations of the stars, commenced the previous year, had been pursued without interruption. Despite the slowness[36] of the Town Council, persuasive progress had been made with them and with other agencies, and the beginnings of an electric distribution network for disseminating accurate time was well underway. With respect to the third objective, instruction was being given to 'a small but intelligent class at the College', with the Observatory occasionally thrown open for inspection and use by students.

With reference to the first objective, Grant had inherited the large, although ineffective, Breadalbane Reflector (20-inch), and commented that:

> No Observatory, in the present day, can be considered as complete, or capable of entering into honourable competition with other establishments of a similar kind, which is not furnished with an equatorially mounted refractor of at least eight or nine inches aperture.

Professor Grant corresponded several times with Airy and, according to Hutchins,[37] he wrote that his 'great desire was to obtain an equatorial, for I conceive it to be the most injudicious for a little Observatory to extend its resources upon meridional observations'. He also needed a first-class sidereal clock because it was 'such a horrid climate that without a [perfect] clock' he could not catch enough clock stars to be sure of accurate time for his observations.

OCHTERTYRE OBSERVATORY.

Published by G. M°Culloch, Bookseller, Crieff, 1858.

Figure 7.10 The Observatory at Ochtertyre, near Crieff, Perthshire, where the 9-inch refractor was originally housed. The social activity on the observing platform can be clearly seen.

In 1862, entirely by his own effort, his aspirations were answered through the purchase of the 9-inch Cooke refractor of 1852 from the estate of Sir William Keith Murray at Ochtertyre, near Crieff, in Perthshire. His private observatory is depicted in Fig. 7.10, and other related photographs can be found in Mayall's *Crieff & Strathearn Through Time*.[38] The main telescope was the largest refractor in Scotland. Through personal enterprise, Grant obtained a sum of just over £1,000 from businessmen and benefactors in the city. A list of all the contributors can be found in the document[39] prepared for the inauguration of the telescope in a specially constructed building attached to the Observatory. The transportation of the telescope and circular roof beam by rail from Crieff, an observatory building by John Honeyman and the restoration of the grounds totalled just over £1,160. The price of the telescope was £350, and a Frodsham clock was purchased for £80, with the total expenditure leaving a small surplus from the accumulated fund.

The telescope was inaugurated at a ceremony in May 1863, at which the University Principal, professors and subscribers were invited to a tour of inspection of the octagonal building, with its adjoining small transit room, north-east of the main observatory block. Grant, in his address, spoke of the lack of observational results coming from Scotland, and took the opportunity to relate that Edinburgh Observatory functioned only as an academic institution that had failed to produce results since its establishment in 1834. His apparent angst here no doubt relates to the growing animosity between him and the Astronomer Royal for Scotland, Professor Piazzi Smyth, son of Admiral Smyth, with whom Grant had collaborated earlier. More on this east–west confrontation, and the concern that Edinburgh was claiming powers which it was not entitled to, will be presented in the following chapter. Grant extolled the importance that astronomy had within society by saying of the *Nautical Almanac*:[40]

> I hold in my hand a volume which may be considered as combining within its moderate dimensions, the principal Astronomical Theories of modern times, and the results of all the Observations made at Greenwich since the middle of the last century. It consists almost entirely of figures; but in regard to the scientific value of its contents, and their bearing on the practical purposes of life, it is unquestionably one of the most wonderful volumes in existence. It is the 'Nautical Almanac,' a work which may be purchased for half-a-crown at any marine bookseller's shop. To the Astronomer and the Mariner this volume is alike indispensable . . . There are, doubtless, several gentlemen present on this occasion who have a direct interest in the shipping of the Clyde. Are you aware how absolutely dependent your richly laden barques are upon the numerical indications of this unpretending volume? . . . How many thousands of our countrymen and country-women are annually wafted to the opposite shores of the globe, little demeaning to what extent their safety is dependent on this book of figures. Such then is the important part which Greenwich Observatory plays in upholding the interests of this great country . . . I may add, that the Astronomer Royal, the Superintendent of the National Almanac, and the Astronomer at the Cape, are placed by Her Majesty's Government under the direct control of Lords Commissioners of the Admiralty. They are, in fact, recognised as occupying the official positions in virtue of which they are actively engaged in upholding the interests of the country as a great maritime power.
>
> I hope that the youth, not merely of the City of Glasgow but of the West of Scotland generally, will avail themselves of the valuable instrumental means which the Observatory now possesses for instruction in practical Astronomy. I would remark also, that to gentlemen who have a little leisure time, and a taste for scientific pursuits, this establishment now possesses instruments admirably adapted as models for the prosecution of Astronomical Observations. Many gentleman in England have done good service to Astronomy, while prosecuting the science as mere amateur observers.

The instrument itself (see further discussion in Chapter 11) is seen in operation in Fig. 7.11.

Figure **7.11** The 9-inch Ochtertyre Telescope in excellent condition, after relocation to University Gardens in the late 1930s.

7.4.3 Telescopic Observations

In his 1864 Report to the Senate,[41] Grant delivered a very positive account of the functioning of the Observatory through recordings with the transit telescope and observations with the Ochtertyre Refractor. With the 9-inch refractor installed, a programme was commenced that was devoted to observing double stars. By 1866,[42] measurements of 708 objects selected from Struve's Catalogue had been made. According to the Report of 1867,[43] the Ochtertyre Refractor provided 466 measurements of double stars. Although some basic statistics of the numbers of measurements that were made each year are available, detailed results of this programme do not seem to have been published.

At this time, Mr Mondeford R. Dolman was the Assistant at the Observatory and Mr A. M'Gregor the Junior Assistant. Positions of minor planets were also recorded with the Ochtertyre Telescope, and the results were presented by Dolman[44] himself in 1866.

In 1878, the Ochtertyre Telescope was used for recording[45] a Transit of Mercury and an occultation of Mars. The transit event was observed by Grant's Principal Assistant, Mr Arthur Bowden, who noted the appearance of a bright spot within the disc of the planet at about the time of ingress. This was not seen by Professor Grant himself when the planet had wholly entered the Sun's disc. The occultation event was duly timed.

In preparation for the Transit of Venus on 8 December 1874, Professor

Grant produced[46] a handsome small tome (see Plate 11) which sold well and went to a second edition. Eight years later, on 6 December 1882, he was fortunate to record the second of the pair of such events in Glasgow. Visual observations[47] were made with the Ochtertyre Telescope reduced in aperture to 5 inches and using a magnifying power of 120. Details were not supplied whereby the apparent solar brightness was further reduced. His description of the internal contact suggested the appearance of a dark ligament, which became more elongated and more attenuated towards the solar limb as time progressed. He reported that he saw no trace of an atmosphere about the planet. It may also be noted that Lord Kelvin paid a visit to the Observatory and was offered the small Dollond 3-inch telescope to view the event, and no dark ligament was perceived by him.

The total eclipse of the Moon on 28 January 1888 provided timings[48] of occultations of faint stars, with six disappearances and eight reappearances noted. The observer was a Mr John McKinnell, who used the 9-inch refractor. The exercise was undertaken in connection with Döllen's method for correcting the apparent diameter and parallax of the Moon.

In addition to the core work of stellar measurements, the Ertel meridian telescope was also used to record the positions of minor planets, and the results of these exercises were communicated to the journal *Astronomische Nachrichten*, for example the determined positions in 1860–1 of Eunomia, Euterpe, Pallas, Amphitrite, Flora and Massalia.[49] In 1871,[50] the records of positional measurements were reported for twenty-six asteroids and Neptune.

7.4.4 Meteor Showers

In addition to telescope observations, Professor Grant made visual recordings of meteor showers, reporting his descriptions to the Royal Astronomical Society. In 1866 and 1868, he observed the Leonids. In this era, the Leonid shower was notoriously active, and Grant's descriptions are detailed and graphic. On the night of 13–14 November 1866, he noted[51] that multitudes of meteors were as bright as first magnitude, many were as bright as Jupiter at opposition, while some rivalled Venus at her greatest brilliancy. The trajectories presented arcs of 50°, 60° or 70°, with durations usually not exceeding three seconds. He also described the colours of the 'flashes' and the dissipation of the after-trails. It may be noted that Grant refers to times expressed in old GMT notation, later referred to as Greenwich Mean Astronomical Time (GMAT), with 00h 00m occurring at midday. Using this notation, at 13h 15m, an hour and a quarter after midnight, the recording rate was fifty-seven per minute. At 14h 41m, he noted a 'blaze of light' in Ursa Major; although he saw it only out of the corner of his eye, preoccupied as he was with note-taking, a residual, slightly curved band of light indicated a meteor of remarkable brightness, with a lustre far in excess of any other of that night. He noted:[52] 'I could only compare its appearance in this respect to that presented in a dark night by the blazing furnace of one of the great iron-works in the neighbourhood of Glasgow'. (For obvious reasons, this Glasgow standard unit for surface brightness, based on the effect of an iron smelter operating at a distance of 10 miles, was not accepted for regular use.) Over some twenty minutes, the trail dispersed with the arc becoming more elongated and pointed, 'suggesting its resemblance to a merry-thought or the outline of a heart'. By noting the meteor vectors on a star map, he calculated the radiant point as: RA = 147° 35′ and Dec = 22° 53′ N. Descriptive details by John Plummer, the Observatory Assistant, of the very bright meteors are appended to the same paper.[53]

Two years later, the Leonid shower was again observed,[54] after being hidden by cloud in the early part, with an occurrence closely matching the previous description. Several of the brighter meteors maintained trains that persisted for two to three minutes. With his Junior Assistant, John McKinnel, Grant counted 256 meteors in the interval between 16h 56m and 18h 06m, that is in 1 hour 10 minutes. Grant suggested that the number 'really' visible was probably three times this. His final comment relates to the fact that the shower of 1868 was much inferior to that of 1866, giving reason to believe that its source was characterised by a more uniform structure. The event of 1866 is sometimes referred to as the 'Leonid storm'.

In 1885, Grant[55] observed a meteor shower on 27 November, with a radiant at RA = ~24° and Dec = ~45° 30′ N, in Andromeda. According to the tabulated records of Mr H. Urquart, one of the Observatory Assistants, for over an hour the rate was just greater than sixty per minute. In a concluding note, Grant says that he had the good fortune to see the great shower of 13 November 1866, and that of 27 November 1872, from the Glasgow Observatory. He commented that:

> The shower of the *Leonids* appeared to me to be beyond comparison the grandest of the three apparitions both in respect to the abundance and the magnitude of the meteors. On the other hand, the recent apparition of the *Andromedes* offered a most striking resemblance in every respect to its brilliant predecessor of 1872.

7.5 THE GLASGOW STAR CATALOGUE

From 1860 to 1881, Grant's energies were directed to meridian work. His original intentions[56] were to observe a limited number of the brighter reference stars of magnitude six to eight, selected from the British Association Catalogue. In late 1866, he found it necessary to enlarge the field, and included stars to magnitude nine; he then added about 6,000 stars from the first volume of Weisse's Bessel Catalogue of 1820–1.

The explanation for this major change of priority may reside in a remark made by T. R. Robinson of Armagh, who noted that many small stars

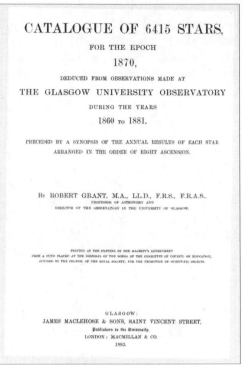

CATALOGUE OF 6415 STARS,

FOR THE EPOCH

1870,

DEDUCED FROM OBSERVATIONS MADE AT

THE GLASGOW UNIVERSITY OBSERVATORY

DURING THE YEARS

1860 TO 1881.

PRECEDED BY A SYNOPSIS OF THE ANNUAL RESULTS OF EACH STAR
ARRANGED IN THE ORDER OF RIGHT ASCENSION.

By ROBERT GRANT, M.A., LL.D., F.R.S., F.R.A.S.
PROFESSOR OF ASTRONOMY AND
DIRECTOR OF THE OBSERVATORY IN THE UNIVERSITY OF GLASGOW.

PRINTED AT THE EXPENSE OF HER MAJESTY'S GOVERNMENT
FROM A FUND PLACED AT THE DISPOSAL OF THE LORDS OF THE COMMITTEE OF COUNCIL ON EDUCATION,
ADVISED BY THE COUNCIL OF THE ROYAL SOCIETY, FOR THE PROMOTION OF SCIENTIFIC OBJECTS.

GLASGOW:
JAMES MACLEHOSE & SONS, SAINT VINCENT STREET,
Publishers to the University.
LONDON ; MACMILLAN & CO.
1883.

Figure 7.12.   The spine and the title page of the *Glasgow Catalogue of 6415 Stars.*

have considerable proper motions caused by their velocities, this attribute revealed by secular positional drifts relative to the more distant stars. Grant's[57] imagination seems to have been fired by the opportunity to play a part in solving a big question, with the need to 'throw new light on the great cosmical problem . . . the motion of the solar system in space'.

From a continuous devotion to this exhausting work over some twenty years, his first *Catalogue of 6415 Stars*[58] was published on 22 May 1883 (see Fig. 7.12). The reductions of the measurements were meticulous in allowing for all the instrumental characteristics and for atmospheric refraction. Not only was it found useful in supplying accurate mean places of a large number of stars of the eighth and ninth magnitudes, suitable for employment as reference stars in connection with the observations of minor planets and comets, but the places were sufficiently reliable for the determination of proper motions of ninety-nine stars, which, with only a few exceptions, had previously escaped detection. In the introduction to his *Catalogue*, Grant[59] noted that stars with Glasgow numbers 2206 and 2207 (Lalande 17050, 17053) had a common proper motion, confirming that they have a physical connection. The same conclusion also applied to the stars 3367/8. The first page of the catalogue is presented in Fig. 7.13.

Date.	No. of Obs.	Mean R.A., 1870. h. m. s.	Date.	No. of Obs.	Mean N.P.D., 1870. ° ′ ″	Date.	No. of Obs.	Mean R.A., 1870. h. m. s.	Date.	No. of Obs.	Mean N.P.D., 1870. ° ′ ″
Means and Sums.			Means and Sums.			Means and Sums.			Means and Sums.		

1.   W. B. (1) XXIII. 1222.

Date.	No. of Obs.	Mean R.A.	Date.	No. of Obs.	Mean N.P.D.
1867.73	1	0 . 0.24.56	1867.73	1	83 . 50 . 51.16
1870.80	1	24.56	1870.80	1	49.58
1872.65	1	24.51	1872.65	1	52.19
			1880.83	1	51.90
1870.39	3	0 . 0.24.54	1873.01	4	83 . 50 . 51.21

2.   W. B. (1) XXIII. 1226.

1867.75	1	0 . 0.38.53	1867.75	1	83 . 1.13.05
1868.79	1	38.38	1868.79	1	13.57
1870.70	1	38.39	1870.70	1	13.74
1869.08	3	0 . 0.38.43	1869.08	3	83 . 1.13.45

3.   W. B. (1) XXIII. 1233.

1872.94	1	0 . 0.56.30	1872.94	1	82 . 20.49.88
1877.91	1	56.38	1877.91	1	47.96
			1878.79	1	48.69
1875 43	2	0 . 0.56.34	1876.55	3	82 . 20.48.84

4.   W. B. (1) XXIII. 1235.

1874.86	1	0 . 1 . 0.30	1877.75	1	89 . 37 . 45.43
1875.94	1	0.49	1878.74	2	44.17
1876.61	1	0.32			
1875.80	3	0 . 1 . 0.37	1878.41	3	89 . 37 . 44.59

5.   4 CETI.

1862.83	1	0 . 1 . 4.47	1862.83	1	93 . 16 . 23.13
1865.78	2	4.37	1865.78	2	21.65
1869 83	2	4 55	1869.83	2	22.43
			1872.90	2	21.69
1866.81	5	0 . 1 . 4.46	1868.55	7	93 . 16 . 22.09

6.   5 CETI.

1876.88	1	0 . 1 . 32.57	1878.81	3	93 . 10 . 17.60
1877.85	1	32.64			
1880.91	3	32.60			
1879.49	5	0 . 1 . 32.60	1878.81	3	93 . 10 . 17.60

7.   21 ANDROMEDÆ *...a.

1860.91	8	0 . 1.40.29	1860.97	1	61 . 37 . 40.01
1861.01	4	40.16	1861.39	2	39.35
1862.87	5	40.29	1866.81	4	39.67
1863.62	8	40.33	1867.77	2	39.84
1864.04	3	40.41	1870.90	5	39.45
1865.83	1	40.32	1871.87	1	39.40
1866.84	4	40.34	1872.89	2	39.17
1867.77	2	40.37	1877.77	3	39.56
1868.79	1	40.44	1878.88	15	39.48
1869.73	2	40.29	1879.78	2	39.48
1870.89	4	40.33	1880.88	10	39.80
1871.32	2	40.28			
1872.85	4	40.33			
1873.14	1	40.21			
1875.88	2	40.29			
1878.88	12	40.28			
1879.42	4	40.38			
1880.69	11	40.28			
1881.01	1	40.32			
1870.87	79	0 . 1.40.31	1875.39	47	61 . 37 . 39.57

8.   W. B. (1) XXIII. 1251.

1867.76	1	0 . 1.44.88	1867.76	1	84 . 6 . 27.10
1871.70	1	45.08	1871.70	1	26.02
1880.02	1	44.92	1880.02	1	26.24
1873.16	3	0 . 1.44.96	1873.16	3	84 . 6 . 26.45

9.   W. B. (1) XXIII. 1253.

1879.90	1	0 . 1.51.15	1871.77	1	79 . 33 . 9.80
1880.73	2	51.17	1879.90	1	9.48
			1880.77	2	8.89
1880.45	3	0 . 1.51.16	1878.30	4	79 . 33 . 9.27

10.   PIAZZI XXIII. 282.

1865.88	3	0 . 2 . 3.39	1865.88	3	92 . 56 . 47.09
1877.86	1	3.36			
1868.87	4	0 . 2 . 3.38	1865.88	3	92 . 56 . 47.09

11.   W. B. (1) XXIII. 1256.

1871.86	1	0 . 2 . 5.07	1871.86	1	76 . 34 . 52.14
1872.94	1	5.23	1872.94	1	53.76
1880.75	2	5.27	1878.70	1	51.38
			1879.88	1	52.13
1876.57	4	0 . 2 . 5.21	1875.85	4	76 . 34 . 52.35

An Asterisk appended to a Star indicates that the proper motion has been applied to the annual place of the Star. The values of proper motion in R. A. and N. P. D. so applied are inserted in their proper places in the Catalogue.

Figure 7.13 The first page of the *Glasgow Catalogue of 6415 Stars* providing, for each star, the details of the dates of measurements of the Right Ascension (RA) and North Polar Distance (NPD).

In effecting his study of the possible proper motions, Grant compared the new Glasgow positional measurements with those in previous catalogues, such as Weisse's Bessel compilation and Lalande's *Histoire Céleste*, with appropriate corrections for precession, for an epoch of 1870. As a rule of thumb, because of the date differences associated with the Bessel/Glasgow and Lalande/Glasgow comparisons, he noted that the displacement effects of proper motion should be in the proportion of two to three. He was extremely careful to consider whether any positional discordance for stars, which drew attention to themselves in this manner, resulted from their proper motion or could be attributable to errors of observation and reduction. The details of measurements with discussion were presented for forty-three stars in a paper[60] presented to the Royal Astronomical Society. More locally, on 5 November 1884, at the Philosophical Society of Glasgow, he lectured[61] on the history of stellar proper motion studies prior to presenting the work undertaken at his Observatory. At this talk, he discussed a table listing forty-four proper motion measurements, as given in Fig. 7.14. He commented on how the resolved stellar motions of stars might be ascertained by combining proper motions with radial velocity measurements to provide the true 3D motions.

For undertaking observations, Grant required support in the form of night assistants and computers, an old-fashioned term for people with mathematical skills. Mention has already been made of Professor Grant's plea in his Memorial,[62] in which he sought financial help directly from Her Majesty's Government. According to Hutchins,[63] in December 1863, he asked Airy to send a 'trained assistant' to work the circle and equatorial; he offered £100 per year and the opportunity to attend the University. Airy persuaded one of his computers, M. R. 'Augustus' Dolman, to go north, and he proved very satisfactory. But when the government made its typical response and refused the University's plea for help towards the cost of an assistant, Grant had no choice but to let Dolman leave to become the Durham Observer in late 1865. In April 1866, the Glasgow Senate granted an annual subsidy of £100 for all purposes; this was continued over Professor Grant's reign. John I. Plummer worked for him for two years before also moving on to Durham.

In the Acknowledgements to this first catalogue, Professor Grant named some of the people who had contributed to the great work, writing:

I wish it were in my power to congratulate all the young gentleman whose services I have here recorded, on the final termination of our common labours. That, alas! five of them are now sleeping their last sleep, Dr Seeling, Mr Dolman, Mr M'Gregor, Mr Fisher, and Mr Stuart. Two of these, Mr M'Gregor and Mr Stuart, may be said to have retained their connection with the Observatory to the last. Mr M'Gregor entered the Observatory in the year 1862, and continued to act as an assistant to the end of 1871, when he was compelled to resign in consequence of ill-health, and died at his home in the country about eighteen months

Reference No.	No. of Star in Glasgow Catalogue.	Magnitude of Star.	Annual Proper Motion.	Direction of Proper Motion.
1	125	8	0″·60	180° 0′
2	407	7	0 ·41	208 26
3	696.	8	1 ·13	97 7
4	748	8	0 ·60	131 57
5	1341	8	0 ·91	200 18
6	1667	6-7	0 ·33	125 8
7	1760	8	0 ·30	180 0
8	1763	8	0 ·33	156 35
9	2120	8	0 ·30	239 6
10	2205	8	0 ·53	200 53
11	2206	8	0 ·31	163 24
12	2207	9	0 ·31	163 24
13	2235	7	0 ·40	186 24
14	2272	6-7	0 ·33	144 28
15	2326	8-9	0 ·40	184 17
16	2650	7-8	0 ·45	154 37
17	2693	7-8	0 ·40	186 17
18	2920	7-8	0 ·33	244 59
19	3040	7-8	0 ·30	285 11
20	3122	8	0 ·40	160 46
21	3123	7	0 ·60	165 47
22	3126	7-8	0 ·31	171 53
23	3308	8	0 ·30	223 35
24	3438	8	0 ·46	257 23
25	3524	8-9	0″·39	215 5
26	3537	9	0 ·32	139 20
27	3562	6	0 ·49	171 7
28	3572	8	0 ·38	245 23
29	3613	7-8	0 ·36	138 50
30	3648	8	0 ·60	174 23
31	3703	8	0 ·39	202 22
32	3717	8	0 ·32	185 19
33	3752	6-7	1 ·39	246 40
34	3845	8	0 ·45	161 5
35	3978	6	0 ·76	161 41
36	4066	7	1 ·49	197 32
37	4112	7	0 ·48	189 3
38	4162	6-7	1 ·62	204 40
39	4203	6	0 ·41	180 0
40	4273	7	0 ·34	381 12
41	4294	7	1 ·38	207 51
42	4330	8-9	0 ·59	312 46
43	4616	7-8	0 ·54	160 53
44	4843	7	0 ·44	199 39

Figure 7.14 Proper motions of forty-four stars determined at the Glasgow Observatory as presented by Professor Grant. The 'direction of proper motion' is described as an angle that increases north through east. After compiling this list, Grant noted that the proper motions of Nos. 1, 17, 24, 32, 33, 35, 36, 38 and 41 had already been recognised by Argelander, and that No. 1 was also contained in Ellery's Melbourne Catalogue.

afterwards. His intelligence and industry, his scrupulous attention to accuracy as an observer and computer, and his honourable conduct throughout could always be relied upon. Mr Stuart was connected during the period of six years with the Observatory. He too was the most trustworthy observer and computer, and a young man of exemplary conduct in every respect. He left the Observatory for his holidays in the autumn of 1880 and was unfortunately drowned a few days afterwards while bathing in the river Spey.

Five years after the publication of the *Glasgow Catalogue*, Grant[64] noted that those stars that exhibited large discordances in their position relative to Bessel's places had been remeasured and commented that the results would be published in a Supplementary Catalogue.

The *Second Glasgow Catalogue of 2156 Stars*, in which Professor Grant[65] was involved at every stage through to completion, was published only a few weeks after his death in 1892. Its title page is reproduced as Fig. 7.15. This catalogue contains the mean places of all stars observed between 1886 and 1892, with the object of explaining certain inconsistencies in the first *Glasgow Catalogue*, or where there was any suspicion of proper motion

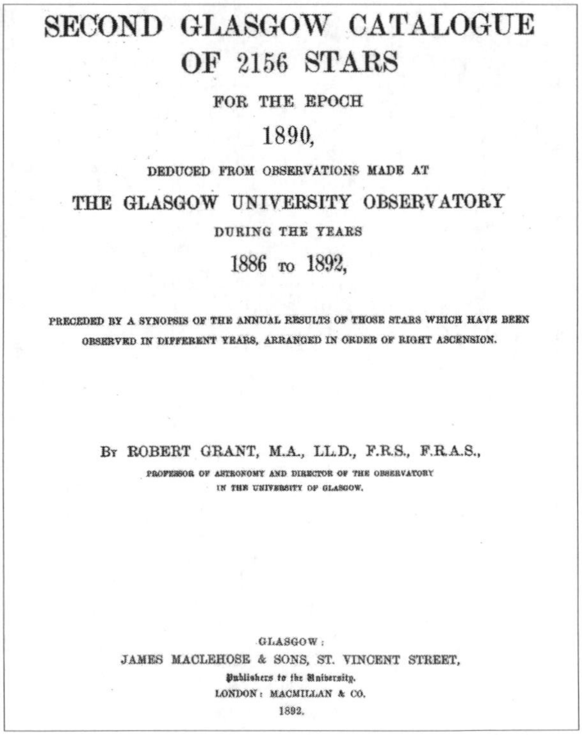

SECOND GLASGOW CATALOGUE
OF 2156 STARS

FOR THE EPOCH

1890,

DEDUCED FROM OBSERVATIONS MADE AT

THE GLASGOW UNIVERSITY OBSERVATORY

DURING THE YEARS

1886 TO 1892,

PRECEDED BY A SYNOPSIS OF THE ANNUAL RESULTS OF THOSE STARS WHICH HAVE BEEN
OBSERVED IN DIFFERENT YEARS, ARRANGED IN ORDER OF RIGHT ASCENSION.

BY ROBERT GRANT, M.A., LL.D., F.R.S., F.R.A.S.,

PROFESSOR OF ASTRONOMY AND DIRECTOR OF THE OBSERVATORY
IN THE UNIVERSITY OF GLASGOW.

GLASGOW:
JAMES MACLEHOSE & SONS, ST. VINCENT STREET,
Publishers to the University.
LONDON: MACMILLAN & CO.
1892.

Figure 7.15 The title page of the *Second Glasgow Catalogue*, published in 1892.

> the number of revolutions of the micrometer, a contingency of not unfrequent occurrence in the former system. It may be stated further that the gas-light, used for the illumination of the microscopes, was superseded by a paraffin oil hand-lamp. The instrument, although reversible, was not used in reversed positions. In consequence of the peculiar form of

Figure 7.16 The Introduction of the *Second Catalogue* reveals an interesting development in the operation of taking readings from the circle scales.

shown by earlier comparison with the corresponding places of Weisse Bessel stars and of Lalande. In the introduction, he gave the details of 192 new determinations of proper motion.

During its operation under Professor Grant, the Ertel telescope was cleaned regularly and occasionally modified to improve its working. A note in the Introduction to the *Second Catalogue* mentions that the method for illuminating the microscope, in order to take readings from the circle, was changed from a gas light to a paraffin lamp (see Fig. 7.16). Such was the pace of technological progress of this era.

In presenting the work at Glasgow, Grant noted that the observations were chiefly made by Mr James Connell, Mr John Dey and Mr John M'Kinnel. He personally revised the greater portion of this work in its final stages prior to publication, but his health must have been failing fast, as in the early autumn he wrote of his assistant:

> I am glad to acknowledge my special obligations to Mr Connell, who was engaged in the work of the Catalogue almost from its beginning, and on whose intelligence, attention, and sense of responsibility I could always rely. He was especially helpful to me in the final stage of the Catalogue passing through the press, when, in consequence of severe illness by which I was overtaken at the time, the aid of an assistant familiar with the whole details of the work was indispensable.

This acknowledgement, taken directly from the *Second Catalogue*, is presented in Fig. 7.17. Though still an invalid, Professor Grant had the gratification of completing the Introduction, and he continued to examine and correct all the proof-sheets of the catalogue, even down to the last sheet, which he looked over on the morning of his death.

> The observations for the Catalogue, both in R.A. and N.P.D., were chiefly made by Mr. James Connell, Mr. John Dey, and Mr. John M'Kinnell. I am glad here to acknowledge my special obligations to Mr. Connell, who was engaged in the work of the Catalogue almost from the beginning, and on whose intelligence, attention, and sense of responsibility I could always rely. He was especially helpful to me in the final stage of the Catalogue passing through the press, when, in consequence of severe illness by which I was overtaken at the time, the aid of an assistant familiar with the whole details of the work was indispensable.

Figure 7.17 Professor Grant pays a sincere acknowledgement to one of his key assistants in the production of the *Second Catalogue*.

7.6 PROFESSOR GRANT'S LEGACY

According to Hutchins,[66] the Horselethill Observatory under Grant represents an extraordinary contrast to Nichol's years, explained by the very different aspirations and skills of the incumbent. Grant obtained a privately subscribed new refractor, and was granted a commitment by the University to a small but essential annual subsidy; his time service and meteorology gave the Observatory civic utility. Grant's meridian work was overtaken by grander schemes abroad, such as the *Carte du Ciel* and the Harvard photographic surveys, but his was more than a personal achievement. He built a tradition of utility for the Observatory, and also trained young men in practical skills.

When Professor Jack[67] presented a Memoir to the Philosophical Society of Glasgow, he commented that the *Glasgow Catalogue* was a gigantic piece of work, which would have been accepted with gratitude from any fully equipped Observatory, having a complete and well-paid staff. The preparation of a work of such magnitude must necessarily have involved an immense amount of observation and calculation at a well-equipped site, but when it is considered that this important catalogue was the outcome of an observatory provided with only a meagre allowance for one assistant, it is difficult to form an adequate conception of the heavy responsibility which must have fallen on Professor Grant during the progress of the calculations and the arrangement of the work for publication.

In an obituary, Professor Copeland[68] comments that one point connected with the *Catalogue* deserved special mention, namely that, although the observations from which it was derived extend over twenty-one years, the work appeared within two years of the close of the series. This promptitude excites the greater admiration when we learn that, as well as Professor Grant's personal share in the work, no fewer than thirteen young assistants at various times took part in the observations, and two others in the computations. Many of these changes in personnel, each of which brought its quota of extra work to Grant, were no doubt in some measure due to the smallness of the allowance provided for assistance: £100 per annum. Professor Grant, however, was the last man to waste his energies in useless complaint, and dismissed this point with the remark that

> in recent years the work of scrutinizing, reducing to a common epoch, and combining together the vast mass of the observations of the catalogue, extending over a period of more than twenty-one years, has pressed very heavily upon slender resources of the observatory.

Professor Grant and his assistants achieved an unbelievable body of work in conditions of increasing difficulty. Given an already damp, cloudy and windy climate, the site at Dowanhill had deteriorated greatly since it was established among the ploughed fields of Nichol's day. At the end of this era, the Observatory was surrounded by large stone houses with smoke emanating from their copious chimneys. Industry, too, was spreading along the river to the south. With his achievements to hand, Grant was always able to contribute to the ever-present west–east rivalries, commenting on the comparative inactivity of the Edinburgh station under the clearer skies of the east.

As related earlier, Professor Grant was awarded the Gold Medal of the Royal Astronomical Society in 1856 and, on 1 June 1865, he was elected Fellow of the Royal Society. An honorary degree of MA was conferred on him in 1855 by the University of Aberdeen and that of LLD in 1865. He also filled the office of President of the Philosophical Society of Glasgow for three years.

References

1. Much of the biographical material related to Robert Grant can be found in an obituary by 'E. D.'. See Dunkin, E. D. (1893), *Monthly Notices of the Royal Astronomical Society* [hereafter *MNRAS*], 53, 210–18.
2. Grant, Robert (1852), *History of Physical Astronomy*, London: Henry G. Bohn.
3. Johnston, Alexander Keith (1869), *Atlas of Astronomy*, Edinburgh: Blackwood. New and enlarged edition published with *Elementary Survey of the Heavens* by Robert Grant [GUL Astronomy qH890 1869-J].
4. Grant, *History of Physical Astronomy*.
5. Ibid., pp. 168–201.
6. Smyth, W. H. (Admiral) and Grant, Robert (1855, 1888), *Popular Astronomy. A translated and edited version of* Astronomie Populaire *by François Arago*, 2 vols, London: Longman, Brown, Green, and Longmans.
7. Smyth, W. H. (Admiral) and Grant, Robert (1861), *A Popular Treatise on Comets*, London: Longman, Green, Longman, and Roberts.
8. Hutchins, Roger (2008), *British University Observatories 1772–1939*, Aldershot: Ashgate Publishing Ltd, p. 141.
9. Murray, David (1927), *Memories of the Old College of Glasgow – Some Chapters in the History of the University*, Glasgow: Jackson, Wylie and Co., p. 272.
10. Thomson, Sir William (1860), 'On the Importance of making Observations on the Thermal Radiation during the coming Eclipse of the Sun', *MNRAS*, 20, 317–18.
11. Ranyard, A. C. (ed.) (1879), 'Observations Made During Total Solar Eclipses', *Memoirs of the Royal Astronomical Society*, XLI. See pp. 82 and 85; pp. 112 and 114; pp. 123 and 129; pp. 139 and 146; pp. 171 and 175; p. 194; pp. 526 and 562.
12. Grant, Dr. R. (1871), 'Schreiben des Herrn Dr. R. Grant, Directors der Sternwarte in Glasgow, an den Herausgeber', *Astronomische Nachrichten* [hereafter *AN*], 77, 223–4.
13. Grant, Robert (1867), 'Two Letters on the Question of the Authenticity of the Documents respecting Newton, which have been communicated by M. Chasles

to the Academy of Sciences of the French Institute, addressed to M. U. J. Leverrier, Director of the Imperial Museum, Paris' [GUL Res Ann Y4-e.27].

14. Murray, *Memories of the Old College of Glasgow*, p. 271.

15. 'Report of The Professor of Astronomy in The University of Glasgow, Addressed to the Committee of Senate Appointed to visit the Observatory and Read at the Visitation of the Committee on the 24th of April, 1865' [GUL Res Ann I15-d.5].

16. 'Report of The Professor of Astronomy in The University of Glasgow, Addressed to the Committee of Senate Appointed to visit the Observatory, and Read at the Visitation of the Committee on the 29th of March, 1867' [GUL Res Ann I15-d.5].

17. 'Report of The Professor of Astronomy, 24th of April, 1865'.

18. 'Report of The Professor of Practical Astronomy in The University of Glasgow, Addressed to the Committee, and read at the Visitation of the Committee, on the 17th of April 1861' [GUL Res Ann I15-d.5].

19. Grant, Robert (1883), *Catalogue of 6415 Stars for the Epoch 1870*, Glasgow: James MacLehose & Sons; London: Macmillan & Co, p. vii. The cover page carries the note: 'Printed at the expense of Her Majesty's government from a fund placed at the disposal of the Lords of the Committee of Council on Education, advised by the Council of the Royal Society, for the Promotion of Scientific Objects'.

20. Glasgow University Manuscript 117/27409. Catalogue of Apparatus in the Department of Astronomy, 1947 [Glasgow University Archives].

21. 'Report of The Professor of Practical Astronomy, 17th of April 1861'.

22. 'Report of The Professor of Astronomy in The University of Glasgow, Addressed to the Committee of the Senate Appointed to Visit the Observatory, April 28, 1864' [GUL Res Ann I15-d.5].

23. 'Report of The Professor of Astronomy in The University of Glasgow, Addressed to the Committee of Senate Appointed to visit the Observatory, and Read at the Visitation of the Committee on the 17th of March, 1868' [GUL Res Ann I15-d.5].

24. 'Report of The Professor of Astronomy in The University of Glasgow, Addressed to the Committee of Senate Appointed to visit the Observatory, and Read on the Day Visitation, 20th of April, 1871' [GUL Res Ann I15-d.5].

25. 'Report of The Professor of Astronomy in The University of Glasgow, Addressed to the Committee of Senate Appointed to visit the Observatory and Read at the Visitation of the Committee on the 29th of March, 1866' [GUL Res Ann I15-d.5].

26. 'Report of The Professor of Astronomy, 17th of March, 1868'.

27. 'Report of The Professor of Astronomy, 20th of April, 1871'.

28. 'Glasgow Observatory – Report for 1885' (1886), MNRAS, 46, 212.

29. Home, David Milne (1883), 'Ben Nevis Observatory', *Nature*, 27, 411.

30. Grant, R. (1888), *Analysis of the Results of Meteorological Observations made with Self-Recording Instruments at Glasgow University Observatory in the Year 1887*, Glasgow: Robert Maclehose [GUL LRA Ann Ea2 - f.9].

31. Tweed, J. (1872), *Tweed's Guide to Glasgow and the Clyde, with tours of Oban, Glencoe, and Inverness, and a finely executed map of the City of Glasgow*, Glasgow: John Tweed. Reprinted in 1972 by the Molendinar Press.

32. Eyre-Todd, George (1905), *Official Guide to Glasgow and its Neighbourhood*, Glasgow: Robert MacLehose & Co., Ltd, p. 37 [GUL Sp Coll Mu26-a.12].

33. 'Report of The Professor of Astronomy, April 28, 1864'.

34. Williams, M. E. W. (1987), 'Astronomy in London: 1866–1900', *Quarterly Journal of the Royal Astronomical Society*, 28, 10–26.

35. 'Report of The Professor of Astronomy, April 28, 1864'.

36. The *Glasgow Herald*, 18 May 1876, 'Transmission of Correct Time from the Observatory to the City and Port of Glasgow'.

37. Hutchins, *British University Observatories 1772–1939*, p. 142.

38. Mayall, Colin (2010), *Crieff & Strathearn Through Time*, Stroud: Amberley Publishing.

39. Grant, R. (1863), 'Inauguration of the Ochtertyre Equatorially Mounted Refracting Telescope' [GUL Sp Coll BG57-c.12].

40. 'The Glasgow Observatory' (1863), *Astronomical Register*, 1, 88–9.

41. 'Report of The Professor of Astronomy, April 28, 1864'.

42. 'Report of The Professor of Astronomy, 29th of March, 1866'.

43. 'Report of The Professor of Astronomy, 29th of March, 1867'.

44. Dolman, M. R. (1866), 'Observations of the Minor Planets made with an Equatorial of nine inches aperture at the Observatory of Glasgow University (Communicated by Prof Grant)', *AN*, 64, 71–2.

45. Grant, Professor R. (1879), 'Observations of the Transit of Mercury, May 6, 1878, and of the Occultation of Mars by the Moon, June 30, 1878, Made at the Glasgow Observatory', *MNRAS*, 39, 167–8.

46. Grant, Robert (1874), *The Transit of Venus in 1874*, Glasgow: J. MacLehose [GUL Res Ann, I6-i.6; Sp Coll MacLehose 562].

47. Grant, Professor R. (1882), 'Observation of the Transit of Venus, 1882, Dec 6, made at Glasgow Observatory', *MNRAS*, 43, 62–3.

48. Grant, Professor R., F.R.S. (1888), 'Observations of Stars made at the Glasgow Observatory in connection with the Total Eclipse of the Moon of 1888, January 28', *MNRAS*, 48, 272.

49. Grant, Prof. (1861), 'Observations of Minor Planets made with the Meridian Circle at the Glasgow Observatory', *AN*, 54, 365–6.

50. Grant, Prof. (1871), 'Observations of Minor Planets and of Neptune made with the Transit Circle at the Glasgow Observatory', *AN*, 77, 225–30.

51. Grant, Professor (1867), 'Observations of the Meteoric Shower of 1866, November 13–14, made at Glasgow Observatory', *MNRAS*, 27, 29–34.

52. Ibid., 29–34.

53. Ibid., 29–34.

54. Grant, Prof. (1868), 'Observations of the Meteoric Shower of November 13–14, 1868, made at Glasgow Observatory', *MNRAS*, 29, 60–2.

55. Grant, Prof. R. (1885), 'Observations of the Meteoric Shower of November 27, 1885, made at Glasgow Observatory', *MNRAS*, 46, 64–5.

56. Grant, *Catalogue of 6415 Stars*, p. xxvii.
57. Ibid., p. lxvi.
58. Grant, *Catalogue of 6415 Stars*.
59. Ibid., p. xxix.
60. Grant, Professor (1883), 'The Proper Motions of Telescope Stars', *MNRAS*, 43, 120–46.
61. Grant, Robert (1884), 'II–The Proper Motions of the Stars', *Journal of the Philosophical Society of Glasgow*, XVI, 22–36 [GUL Res Ann I9-a.21].
62. 'Report of The Professor of Astronomy, April 28, 1864'.
63. Hutchins, *British University Observatories 1772–1939*, p. 142.
64. Grant, Professor R. (1888), 'Note on the Glasgow Star Catalogue', *MNRAS*, 48, 381–2.
65. Grant, Robert (1892), *Second Glasgow Catalogue of 2156 Stars*, Glasgow: James MacLehose & Son, Glasgow.
66. Hutchins, *British University Observatories 1772–1939*, p. 142.
67. Jack, Professor William (1893), 'Memoir of the Late Dr Robert Grant, FRS, Regius Professor of Astronomy in the University, and Former President of the Philosophical Society', *Journal of the Philosophical Society of Glasgow*, XXIV, 139 [Reprint in GUL Sp Coll Kelvin 136].
68. Copeland, R. (1892), 'Robert Grant', *Nature*, 47: 1,202, 36–7.

Chapter 8

Time is of the Essence

8.1 INTRODUCTION

Measuring star positions with instruments, such as sextants, allows location to be determined in terms of latitude and longitude, the latter being deduced only with reference to accurate time carried on board a ship. At any large port, it might be expected that there would be merchants offering sextants and marine chronometers for sale. In addition, it was necessary that Greenwich Mean Time (GMT) was available at the quayside so that mariners could set their clocks before departure on their sea voyage.

In the middle part of the nineteenth century, the business of moving people and transporting commercial goods was growing at an unprecedented rate. Ports around the world were being developed along rivers associated with large urban populations. This involved engineering by dredging river-beds and building shoreline cranes for offloading materials and goods to the newly constructed warehouses. As can be seen from reports such as that by Deas,[1] activity on the Clyde involved engineering projects on a grand scale to allow shipping to get further into the city, rather than relying on docks out along the estuary.

With mid-nineteenth century technology, the requirements for keeping GMT in any locality were a good quality transit telescope, with all its systematic errors regularly calibrated, and a system of accurate clocks, with known drifts, that were checked whenever the weather permitted observations. Local time obtained by recording stellar transits is relatable to GMT, using the knowledge of the observer's longitude. With the advent of telegraphic systems, signals could then be transmitted to stations where a time-ball or heavy gun was located. These visible and audible devices were triggered by electrical pulses from the local master clock; the shorter the telegraph wire, the less likely it was that there would be transmission delays, with errors in relaying the time, and line failures caused by accidents or bad weather. Time-balls appeared at ports around the world, with the 'drop' traditionally occurring at 1 p.m.; several remain today as tourist attractions, such as the simultaneous dropping the ball and firing the one o'clock gun in Edinburgh. The latter system, however, carries an inherent problem because of the finite speed of sound and the delay this introduces in hearing the gun at a distance.

The one o'clock signal was sufficient for mariners in port, but it had obvious limitations within a city and for the general public. It emerged that the ticking signals of regular seconds could be transmitted by wire, and these could be used to drive specially constructed clock dials with a large sweep finger pinned at the centre of the face and extending to the rim, so that GMT could be read accurately to the exact second at a reasonable distance; this design gave rise to the term 'seconds' clock'. Such clocks could be positioned at any strategic location if the appropriate wires were laid.

To answer the needs of mariners with regard to safe sea voyages, harbour masters around the world began to provide time at the waterfronts, initially by a 'one o'clock signal'. Despite the University Observatory being commissioned as early as 1845 to provide a time service, the Clyde River Trustees were relatively slow to establish a system for Glasgow in comparison with other places such as Edinburgh and Liverpool. Perhaps as a result of this, as well as through the experiences of other ventures elsewhere, not to mention having more time to consider a scheme, Glasgow benefited by providing a comprehensive system which, by its scale, was globally unique. This remarkable story is presented below.

8.2 THE BROOMIELAW TIME-BALL

Records[2] kept in the Mitchell Library in Glasgow show that the Clyde River Trustees had four clocks in operation in the 1850s for the purpose of providing time for the city and for shipping. They were located at the corner of Jamaica Street, at Finnieston Quay, at Court Miele and at the Police Office. These had been operating for a number of years, as a letter from Alexander Mitchell to George Knight Esq., of the River Trustees headquarters at 18 Robertson Street, suggests. In seeking a raise in remuneration for his work, he declared: 'These clocks I have kept for upwards of thirty years; for winding, keeping in repair, accountable for their correct timekeeping . . . 1/9 per week is too little'. It is unlikely that his plea would have been acknowledged, as the implementation of a time-ball was under consideration; seconds' clocks, controlled from the University Observatory, became established later.

The story of the painful path to establish the Glasgow time-ball, and of the short time it operated, has been well told by Kinns.[3] In the first place, the recommendation for operating such a device originated in Edinburgh, through Sir Thomas MakDougall Brisbane. After nearly being involved in a shipwreck in 1795, he developed his passion for astronomy and time-keeping. In 1808, he built an observatory at his home near Largs in Ayrshire, and is famous for establishing an observatory at Parramatta, near Sydney, Australia, while he was Governor of New South Wales. During an illustrious military career, he commanded a brigade which was heavily engaged in the battles of the Peninsular War from Vitoria to Toulouse. During this time, he continued to practise his astronomy so that, in the Duke of Wellington's words, 'he kept the time of the army', becoming affectionately known as 'Pendulum Tom'. He was involved in recommending Nichol to the Chair in Glasgow, and had political influence on matters of science in Scotland. In 1852, Charles Piazzi Smyth, Astronomer Royal for Scotland, who had worked previously as Assistant Astronomer at the Cape of Good Hope Observatory, established a time-ball in Edinburgh with the help of a local clockmaker, Frederick James Ritchie. Sir Thomas took a great interest in its operation and immediately recommended that similar ones should be erected in Glasgow and Greenock.

This suggestion was acted on, and *The Scotsman* of 6 June 1857 reported that 'A time-ball has been hoisted on the pole surmounting the Sailors' Home at the Broomielaw'. The Glasgow Sailors' Home at 150 Broomielaw was built in 1855–6 and opened on 26 January 1857. The architect was John Thomas Rochead, who also designed the building in Queen Margaret Drive, Glasgow, that was later used by the BBC. The 'Ball' was removed some time before the demise of the Sailors' Home, but the circular tower on which it was located remained on the skyline until its demolition in 1971 (see Fig. 8.3).

Although the exact date of the commencement of operations of the time-ball is uncertain, archived correspondence[4] between McGregor & Co., at 38 Clyde Place, and the Clyde River Trustees, shows that a contract had been formalised from 1 July 1858. This letter, with a letterhead embossed with a red emblem and bearing a Penny Red stamp, is dated 'Glasgow, 14th December 1858' and is addressed to 'Geo Knight, Esq., 18 Robertson Street'. It reads:

Geo Knight Esq.,
Dear Sir
Your favour of yesterday to hand, and as required beg to state that our duties with reference to the Time Ball are as follows —
1st To furnish the observations required in keeping the Time Ball regulator at Greenwich Mean Time
2nd To attend daily, and set the regulator at absolute Greenwich M.T.
3rd To furnish & maintain the chemicals required in forming the Magnetic connexion between the regulator & Time Ball
4th To attend daily for the purpose of putting the Battery in Action, Hoisting the Ball, seeing that it drops accurately, and generally to see that the whole Apparatus is kept in proper working order.
The foregoing we undertake to perform from 1st July 1858 till 1st July 1859 for the sum of Sixty pounds stg. payable Half-yearly.

We are Dear Sir
Your most Ob. Servants
D. McGregor & Co.

Another document from one year later reads:

D. MacGregor – Time Ball 1st July 59 to 1st July 60
38 Clyde Place
Glasgow

3rd December 1859

George Knight, Esq.,
Dear Sir
Your favour of the 2nd inst to hand and as requested beg to state that our duties with reference to the Glasgow Harbour Time Ball are as follows.
1st, etc as above.
The foregoing we undertake to perform from 1st July 1859 til 1st July 1860 for the sum of Forty pounds stg. payable Half-Yearly
We are Dear Sir
Your most Ob. Servants   D. McGregor & Co.

Over the page are the following notes:

Glasgow 1st July 1862 – We hereby agree that the foregoing Contract be continued in all respects for the year from this date till the end of June Eighteen hundred and sixty three

[Illegible signature] for the Clyde Trustees
D. McGregor

A further note added reads:

Glasgow 1st July 1863   until the end of June,
Eighteen Hundred and Sixty Four   D. McGregor.

Where the finance came from to install the time-ball, other than from the Clyde Trustees, is difficult to ascertain, as is the reason why a private company was awarded the contract to determine GMT and deliver the signal for dropping the ball. As early as 1855, an experiment was exhibited at the British Association Meeting in the Glasgow College demonstrating the triggering of a ball-drop mechanism by means of a telegraphic wire from Edinburgh Observatory (see the article, reproduced later in this chapter, in the *Glasgow Herald* of 12 December 1863). McGregor & Co. did not obtain GMT directly from the Royal Observatory, Greenwich, as might have been the case with the arrival of the age of telegraphy. Neither did they use 'time' from either the Glasgow or Edinburgh Observatories. The first item of their contract suggests that they themselves would furnish the observations required to provide GMT. They apparently maintained their own transit telescope for the purpose, as noted by Nichol[5] and Brown,[6] within their premises in Clyde Place; according to Nichol, the instrument was set up on the roof of McGregor's house.

All this contributed to a sense of dissatisfaction over the operation of the local time-ball. There were concerns over the accuracy of the 'ball drop', these underpinned by the lack of records available for scrutiny in respect of how GMT was obtained and in the way it was transferred to the ball-release mechanism. In addition there were problems in communication between various local parties, as well as ignorance as to what had been achieved elsewhere in the country and how the system should operate in Glasgow. In his letter[7] of 26 April 1859 to Sir George Airy, Nichol noted: 'I have been doing utmost to get them to put their Time Ball here in a right state. It is kept by a watch-maker, who has a rickety [observatory] in the top of his house'.

Included in the same letter to Airy was a piece from the *Glasgow Herald* by a Mr Allan, a member of the Clyde Trustees, containing several misinformed statements (see paragraph four of Nichol's letter of 30 April 1859 below, quoting some of Mr Allan's original statements). Some resolution to the arguments was achieved by a submission to the *Glasgow Herald* on 4 May 1859 by a Mr D. Dreghorn, a prominent local citizen and well-wishing intermediary; this included a letter written by Professor Nichol, with accompanying letters written in response to Nichol by Sir George Airy and a Mr Hartnup of Liverpool. The complete article reads as follows:

THE TIME-BALL ON THE SAILORS' HOME

To the Editor of the Glasgow Herald

Glasgow, 3d May, 1859.

SIR, – I send you a second communication I have received from Professor Nichol, called from him apparently by correspondence in your columns. As the matter is very important, I trust you will find room for it. I have recently learned that the subject has been attracting the attention of the directors of the Exchange. I am sure if we set ourselves to the work, there can be little difficulty in placing Glasgow—as far as her great navigation interests are concerned—under the system to which Mr. Hartnup alludes, as already existing and about to be greatly extended and perfected in London and Liverpool.—I am, Sir, your most obedient servant,

D. DREGHORN.

Nichol's covering letter to Mr Dreghorn reads:

Observatory, 30th April 1859.

My dear Mr. Dreghorn, – I had hoped that the last note I addressed to you, and which you sent to the *Glasgow Herald*, would have sufficed to remove misconception as to the matter of the time-ball. Permit me to intrude on you again, and to offer a few final remarks on the letter addressed by Mr. Allan, one of the Clyde Trustees, to the Editor of the *Herald*. In so far as Mr. Allan is concerned, my former letter was evidently useless.—There are three points to which I request your attention.

I. I repeat emphatically, in reference to the use of "electric currents," that the thing requisite, and which alone is of much moment, is this:—The working current ought to pass to the machinery of the ball directly from the transit clock, and be set in action, automatically, by that clock. No intervention of "a man with a watch" is now permissible as any part or stage of the process. The objection is to intervention of any kind; it is of no consequence at what part of the process such intervention may take place.

II. Statements are made and vouched for that an "adequate observatory" exists, and is in operation, within two hundred yards of the turret of the Sailors' Home. I have only to express a hope that the Clyde Trustees will verify this assertion. If the institution referred to is declared adequate by competent authority, the Trustees have an easy and comparatively inexpensive task before them. They have merely to lay a conducting wire through these two hundred yards; and to secure that records of the working of the institution be regularly kept, and remain open to examination.

III. You may remember that I referred, in corroboration of one of my main statements, to several well-known scientific gentlemen—among others, to Professor Airy and Mr. Hartnup of Liverpool. Mr. Allan takes up that reference, and writes as follows:—

"At some time, it should be known that Professor Hartnup of Liverpool does not recommend time-balls being dropped by electric currents transmitted from Greenwich—that is, not until proper arrangements shall have been made throughout the whole kingdom for this purpose. A time-ball was erected in Liverpool some time ago by the Electric Telegraph Company, and wrought from London, which had to be discontinued, as the signals were found not to be so correct as those of the ball dropped by hand in Mr. Hartnup's observatory. And, remarkably enough, the time-balls in Greenwich Observatory and at the Liverpool docks, are all at this moment dropped by hand, and not by electric currents; at least the former was so till very lately, if not so now, and the latter I know are still wrought in this way — so that, both in London and Liverpool it is the man with the watch who drops the ball after all. . . . [*the microfiche is undecipherable here*] . . . buildings in which these balls are to be placed, which clocks will be adjusted daily, precisely as is now done with the clock in the Sailors' Home, when the whole Liverpool arrangements will be as nearly as possible the same as at present in Glasgow."

Although perfectly aware that Mr. Allan had fallen into error, I enclosed copies of the foresaid extract to Professor Airy and Mr. Hartnup only receiving them to *trans[scribe?]* me to bring account of facts as they now stand. I have just received the following replies:–

From Professor AIRY to Professor NICHOL
Royal Observatory, Greenwich, April 27.

My dear Sir, —In the extract from a letter in a Glasgow newspaper which you have sent to me there are some inaccuracies which are important in reference to the question now before you.

1. The time-ball erected by the Electric Telegraph Company is in daily use at Liverpool, and is dropped with perfect accuracy (as regards the instance of time) by the current which originates at Greenwich. Occasionally, from defect of insulation, in damp weather, &c, it fails, but the failures are rare. 2. The time-ball at Greenwich is always dropped by the galvanic clock. It has not been dropped by hand for several years, except when the mechanism has been under repair or cleaning.

I do not know precisely what is contemplated at Liverpool in reference to the proposed new time-balls, but remarking that there is in the Liverpool Observatory a first-class transit instrument, used by an accomplished and long experienced observer, and remarking that the longitude of the Liverpool Observatory from Greenwich has been determined with great care, I should think it prudent (in order to avoid the difficulties the defital to a long line of telegraph) to drop the balls by galvanic current from Liverpool Observatory.

I should recommend a similar course at Glasgow—dropping the balls by current from the Glasgow Observatory. The distance of two or three miles offers no difficulty.

I am, my dear Sir, yours truly,

(Signed) G.B. AIRY.

(From Mr. HARTNUP to Professor NICHOL)

Observatory, Liverpool, April 27th, 1859.

Dear Sir, —It is quite true that I have recommended the new time-ball, about to be erected on the top of the Victoria Tower, to be dropped by the large turret clock in that tower by a mechanical arrangement. In the event of this being done the turret clock will, of course, be controlled by our normal clock at the Observatory, and thereby made to keep time throughout the day with the same degree of accuracy that we could drop the time-ball.

This method cannot, I think, be in operation at Glasgow, the Liverpool turret clock, on the top of the Town Hall, being, so far as I am aware, the only large clock at present controlled from an observatory. The practicability of controlling clocks by weak galvanic currents, is now attracting great attention in London, and the Astronomer-Royal, a few weeks ago, sent down the assistant who has charge of the galvanic department at the Royal Observatory, for the purpose of inspecting the arrangements by which the movements of the pendulum of the old clock in our Town Hall are controlled. It is quite certain that we now have it in our power to control all the public clocks in a large town, thereby making them all strike simultaneously, and show the same time throughout the day as the normal clock at the observatory by which they are controlled.

By a single wire you might control all the important clocks in Glasgow, and the Astronomer-Royal will doubtless soon do this in London.

The scientific public and intelligent ship-makers will never have confidence in either time-balls or clocks, which are not placed directly under the control of the offices of an astronomical observatory.

It was stated in a memorial by the president and officers of the British Association for the Advancement of Science in 1837, when recommending the Liverpool Town Council to establish an astronomical observatory in this port, that the Greenwich time obtained from various celebrated chronometer makers who had transit instruments of their own, had been found to differ from the correct Greenwich time sufficient to cause a wreck; and it was recommended by scientific gentlemen, who were consulted by the Liverpool Corporation, and particularly by the Astronomer-Royal, that an astronomical character should be given to the Liverpool Observatory, inasmuch as the accuracy which is indispensable in all astronomical observations, would thus be infallibly extended to all other departments of the establishment. — I remain, dear Sir, yours very truly,

(Signed) JOHN HARTNUP.

It is plain enough that public controversy on such matters can lead to no good result. It is really a technical and strictly scientific subject, and the state of information of the public mind on questions of this description does not offer adequate security against the acceptance of reiterated and plausible statements, how inaccurate soever they may be. I repeat my earnest hope that he will recommend the Trustees to request a visit and report from our Scottish Astronomer-Royal, Professor Piazzi Smyth of Edinburgh. There is no need of a *committee*. The appointment of a committee of scientific men would simply lose time. Professor Smyth's authority is acknowledged by every scientific man in Britain; and the visit of half a day will enable him to report definitely concerning the "observatory" and every other arrangement connected with the time-ball.

Other questions, I perceive, are being mixed up with this one. I am not indifferent to what has been written concerning the "Exchange clock" and "Chronometers" and shall take effective means to dispose of the averments referred to. It is fitting, however, that the time-ball question be kept by itself. – Ever faithfully yours,

David Dreghorn, Esq.   J.P. NICHOL.

The invoices of McGregor & Co. presented at the beginning of the chapter show that there was a formal contract from 1858 and that it lasted over the period when the concerns regarding competence were raised through the columns of the *Glasgow Herald*. There is no later reference to Piazzi Smyth taking an active interest in checking the efficacy of the overall system in Glasgow. His involvement with the establishment of time within the city four years later was more dramatic and contentious, as will be related later. The Glasgow time-ball operated for about five years. It was still in use during 1863, and the invoices of McGregor & Co (see above) suggest that it was maintained at least until June 1864. It is not recorded whether the signal was presented regularly and smoothly over the five-year period. On 17 February 1862, a beam for carrying Admiral Fitzroy's storm-warning signals was added to the tower of the Sailors' Home.

Figure 8.1 From left to right, the Marquis of Bute, Lancelot and Sultana are berthed on the Broomielaw. The one o'clock time-ball on top of the tower on the Sailors' Home can be seen in the distance. The right-hand image provides a magnified view of the tower carrying the Fitzroy arm, the ball, the compass arms and an emblem of a ship perched on top of the mast. The picture dates from around 1870.

The bustling Broomielaw of the late nineteenth century, as seen from the first bridge up the Clyde, is depicted in Fig. 8.1. A picture of the scene by David Small is shown in Plate 12. In the background, the tower of the Sailors' Home carrying the time-ball can be clearly seen. Other images of the Clyde with the time-ball tower within the scene can be found on the Virtual Mitchell Library website,[8] and images of time-balls around the world are available from the website[9] of the One o'Clock Gun & Time Ball Association. A view from the tower itself, looking eastwards up the Clyde, is shown in the top half of Fig. 8.2, allowing the location of the premises of McGregor & Co. to be highlighted. The lower half of this figure gives an image of the same view, but showing the time-ball itself and the Fitzroy arm.

In striking contrast, Plate 13 shows a modern view from the top of the new BT building, close to where the Sailors' Home once stood. Although the bridges match up, albeit with modified support pillars, and St Andrew's Church still stands towards the distant left, gone is the ground activity of people and horse-drawn carts. Gone, too, are the host of factory chimneys belching out their acrid smoke. Images of the Sailors' Home tower c. 1960 are shown in Fig. 8.3, the time-ball having been removed by this date.

The mechanisms associated with the time-ball have been described[10] by Brown in *Sketches on the Clyde*, in which it is noted that:

> On the Tower of the Sailors' Home, is the Harbour Time-Ball, which was erected in 1857, by Mr Alexander M'Kenzie mechanist, and has been worked, from the commencement, by the firm of M'Gregor & Co., chronometer makers.

Figure 8.2 Two views from the top of the Sailors' Home on the Broomielaw, with the distant spire of St Andrew's Church seen to the left in both images. The upper photograph of 1868 shows the location of the premises of McGregor & Co. in Clyde Place; the lower print includes the time-ball on the left of the mast with a ship at its top and the Fitzroy storm signal arm to the side.

Figure 8.3 The tower of the Sailors' Home c. 1960. The time-ball itself has already been removed. The left-hand image is from the cobbled James Watt Street with the Broomielaw running across the picture towards the top; the right-hand image shows the busy Broomielaw, with the dome of the Clyde Trust Building in the distance.

The Time-Ball is dropped daily, exactly at one o'clock, Greenwich mean-time, by an electric current from an astronomical clock, which is attached to the basement of the building; and a brief account of the mode of working it, may prove interesting, as many persons have been led to suppose that the ball is dropped by hand. The dial of the clock is cut through, above the figure 60, on the seconds-dial, and through the opening projects a thin plate of pure gold, which is inclined to the seconds-hand, also of gold, at an angle of about eight degrees. Concentric and revolving with the minute-wheel, is a wheel, notched out in three places, above which rests a lever connected with the gold plate or trigger. At a few seconds before five minutes to one o'clock, the lever drops into the first notch, allowing the gold trigger to fall into position for contact with the seconds-hand, which, as it completes the 60th second, touches the gold plate, and a minute bright spark is seen.

The signal is conveyed to the attendant, at the top of the Tower, and the ball is wound up half-mast high. The seconds-hand, after making the contact, pushes back the gold plate, which is very flexible, and continues its course; but before it completes another circuit, the trigger is lifted above the point of contact by the mechanism of the clock. At a few seconds before two minutes to one, the trigger again drops, the second contact is made, signalling as before, and the ball is wound up to the top of the staff; and when the seconds-hand completes the last second of the hour, it again touches the trigger and the ball instantaneously descends; and no one who ascends the Tower to witness the working of it, can fail to remark the unerring precision with which the ball is discharged by the clock below. The hands of this clock are never altered. It has a small losing rate, and a little before one o'clock, every day, the pendulum is accelerated, for a few beats, which brings it to the exact time.

It is gratifying to add, that Glasgow held a prominent place last year in the trial of chronometers at the Royal Observatory, Greenwich, as a chronometer, sent by Messrs. D. M'Gregor & Co., for trial, was so highly approved of, that the Lords of the Admiralty, offered the sum of L.50 for it, which was accepted.

The entire weight to the lifted is fifteen cwt., the ball itself being four cwt., and is five feet in diameter built of mahogany, and covered with zinc, nearly ⅟₁₆th in thickness. It rises fourteen feet, near to the model of the ship at the extreme point of the rod. The Tower, with the Time-Ball rod, measures 217 feet from the ground, and at the highest story, the view compensates the labour of the narrow ascent – the river in its windings, in its freights, in its bustle, and in its expanse, is seen and can be studied with advantage. In Edinburgh, where there is a time-ball on the top of Nelson's monument, Calton Hill, the apparatus, designed and erected by Messrs J. Ritchie & Son, is connected by a wire to a gun in the Castle; and at the same moment the sense of seeing is gratified, the hearing is also. At one o'clock P.M., the report of the cannon his heard in every quarter; and if Glasgow Time-Ball had such an apparatus, and if a similar sound were heard, the service to those interested would be tenfold – and why withhold it?

Further details of the building of the Sailors' Home under the auspices of the Trustees of the Clyde Harbour, and of the M'Kenzie devices for adjusting the pendulum prior to the dropping of the ball, with a figure depicting the gear wheels and contact mechanisms, are available in *The Practical Mechanic's Journal*,[11] in which the means of absorbing the fall of the ball by a pneumatic cushion is also described. Strong encouragement was also voiced for the establishment of a time-ball at Greenock on the Clyde estuary.

Some of the details as to the weight of the ball and the height of its drop have been challenged in a recent paper[12] by Clarke and Kinns. The account above certainly suggests that the ball was dropped by an electrical signal, but that the time reference must have been physically 'carried' from McGregor's premises in Clyde Place (see Fig. 8.2) to the astronomical clock in the basement of the Sailors' Home. According to Brown,[13] the company had an observatory on the south side of the river. He noted that it housed a transit instrument mounted on one block of polished marble, cut down centrally towards its base so as to allow the instrument to traverse the plane of the meridian.

As mentioned above, Nichol had described the observatory as 'rickety', and, in his letter of 4 May 1859, reproduced earlier, inverted commas were placed around 'adequate observatory', either indicating irony or suggesting that it was a quotation taken from someone in support of the competency of McGregor's company. Brown also notes there was intervention by hand to 'speed up' the clock just prior to the ball drop, a procedure that was anathema to Nichol. The time reference was transferred to the Sailors' Home in a cumbersome way by carrying a chronometer from one side of the river to the other, not in keeping with the latest development of tel-egraphic technology. Nichol's interventions in the form of letters submitted to the local newspaper must have established an antagonism between the University and McGregor's company that persisted after Nichol's death five months later and into the time of Professor Grant becoming Director of the University Observatory. The animosity burst out again in public during the saga of the one o'clock gun of 1863, as described below. It may be noted that an employee of McGregor's, a Mr Church, had contributed to the matter through correspondence in the *Glasgow Herald* (see p. 241), whereby he appeared to be au fait with the workings of the transit telescope. Mr Church commented on the arrogance of an assistant from the University Observatory and on an error the latter had made in carrying time from Horselethill to the city for comparison with 'McGregor time'. He also noted: 'At the time when the time-ball was first established, the Observatory, whatever its present position may be, was not in proper state of efficiency to maintain a correct standard of time'. Professor Nichol must have been sensi-tively aware of this. Probably to his embarrassment, he seemed to have made no offer of help, even though providing time to the city had been made a direct remit of the University Observatory on its establishment some fifteen years previously. By tarring the University in this way, it was likely that Mr Church's employer was sensitive to the threat of loss of income in respect of contracts for supplying time to the Clyde River Trustees.

At the end of the above excerpt from Brown's book (see previous page), reference is made to the firing of the one o'clock gun in Edinburgh, which commenced operation in 1861. An exhortation was given with the words 'and why withhold it?' that Glasgow should follow the same path. The call was answered not by a Glasgow initiative, but by one from Edinburgh. The ensuing noise from the opening rounds of fire reverberated into animosity between the astronomical institutions of the two cities for years to come.

8.3 THE DISTRIBUTION OF TIME

8.3.1 *Professor Grant's Remit*

Shortly after he took command in Glasgow, in his 1861 Report to the University Senate,[14] Grant summarised the role of any observatory. As related in Chapter 7, included in the itemised responsibilities was the requirement of 'Disseminating accurate time to one or more stations more central to any adjoining conurbation', the observatory master clock being out in the country at some distance from the population centre.

One of Grant's chief concerns, therefore, was the distribution of accurate time to the city. In the same Report, he noted that the Observatory had taken on board the transmission of 'correct time' to the Exchange and to the establishment of Mr D. McGregor, who had been charged by the Clyde Trustees with the daily dropping of the time-ball for the use of shipping. The 'transmission' was effected by dispatching once a week a person with a chronometer from the Observatory to the city. Grant recorded: 'I need scarcely express the opinion that such a mode of transmission is not in accordance with the advanced state of science in the present day'. According to Mr Church, who was employed by McGregor & Co., even this method was not infallible (see his letter below, taken from the *Glasgow Herald* of 5 October 1863).

At this stage, Professor Grant had been in contact with a Mr Bright of the Magnetic Telegraph Company, with a view to suggesting to the Glasgow public authorities the expediency of applying Jones's method to the regulation of some of the public clocks in the city. An estimate for laying down a wire from the Observatory to St George's Church 2¾ miles away was £120. From this point, a wire might be carried at moderate expense to other public clocks, all under the control of electric current from the Observatory. Grant planned to submit his proposals to the Town Council and other city authorities within a few weeks of his Report to the Senate of 17 April 1861.

8.3.2 *Glasgow's Big Bang of 1863*

In the late summer of 1863, Grant continued to have frequent interviews with the Chairman of the Committee of the Town Council who had been appointed to consider the proposals he had submitted regarding the University Observatory providing time to the city by means of telegraphic wires. According to his 1864 Report to the Senate,[15] very much to his surprise there then appeared in the local newspapers, at the close of September 1863, a statement to the effect that the Director of the Observatory had made arrangements, in concert with a commercial establishment actively employed in extending its business relations in Scotland, for firing a time-gun in Glasgow.

The trial for this was announced first in Edinburgh in *The Scotsman* of 24 September 1863, but with the Royal Observatory there providing the signal. The report reads:

> We are informed that experiments are likely to be made this week with a view to the establishment of a Time-Gun in Glasgow. The gun will be fired from the Edinburgh Observatory, in the same way as that now fired in Newcastle. Instead of an electric clock being used to pull the friction tube as in the Castle, a small

fuse filled with powder placed in the vent of the gun will be exploded by means of a spark evolved from Wheatstone's 'magnetic exploder' – the electric current being transmitted precisely at one o'clock from the Observatory through one of the Electric Telegraph Company's wires to Glasgow.

The announcement of the trials of the one o'clock gun in Glasgow was also published locally two days later in the *Glasgow Daily Herald* of Saturday, 26 September 1863. It reads as follows:

Glasgow Herald
1863 Sep 26

THE NEW TIME GUN

The arrangements for the new time gun experiment – the report of which came upon the community a day since with startling suddenness – are now progressing steadily, but there still remains so much to be done that the trial cannot be made for several days yet. The approaching experiment has originated with the Universal Private Telegraph Company, who have very spiritedly set to work to carry out their plans. Mr. Nathaniel Holme, the engineer of the company, has undertaken the superintendence of the arrangements, and the valuable co-operation of Professor Piazzi Smythe, Astronomer Royal for Scotland, has been obtained in furtherance of the scheme. In casting about for a suitable site for the gun, the attention of the Company was directed to a green which forms an eminence overlooking Sauchiehall Street and is entered from Renfrew Street, at the west side of the Corporation Galleries. This ground belongs to the City Bank, and the directors, on being applied to, generously granted it for the use of the experimentalists, while Mr. Long, at the back of whose gymnasium it is situated, frankly sanctioned the placing of this probably rather noisy neighbour in the immediate vicinity of his establishment, and on ground which he held as tenant.

The gun, a thirty-two pounder, has been given by Messrs. Napier & Son. It was intended, by the way, that it should have been sent to Mr. Napier's residence at Shandon, but on hearing of the projected experiment it was at once considerably put at the disposal of the Company. The proper charge of powder for the piece is 6 lbs, but owing to the present position of the gun in the midst of dwelling houses, not more than from 1½ lbs. to 2 lbs. will be used. As it is to be placed under the care of an experienced gunner, every assurance may be felt that no damage will be caused to the property in the vicinity, nor any unnecessary alarm occasioned to neighbouring residenters. The electric current will be passed from the Observatory at Edinburgh through a circuit granted for the purpose by the Magnetic Telegraph Company, who, with the same disinterested generosity, we believe similarly provided the wires for the Sunderland gun. Professor Piazzi Smythe has arranged his clock in the Observatory so that the current will be passed to the Glasgow gun simultaneously with a firing of the Newcastle, North Shields, Sunderland and Edinburgh pieces. It is expected that the test experimental gun will be fired on Thursday next the 1st proximo at one o'clock Greenwich mean time, and a plan has been prepared whereby those who are so far removed from the gun as not to hear it at the moment of firing may still calculate its report to a fraction of time. Taking the mean temperature of Glasgow at one P.M. at 54 Fahrenheit, the sound will travel at the rate of 1115 feet per second, and a map has been drawn out in which the city is divided into the circles, at determined distances from the gun, the space between each circle being traversed by the sound in one second. The citizens, therefore, must consult the map, and in calculating the moment at which the gun is fired make allowance for the distance at which they may happen to be placed from it. Thus, the sound will take 2 seconds to reach the Exchange, which is situated between the second and third circles, three seconds to reach the Broomielaw, which is on the third circle, 5 seconds to reach the College, 4 seconds before being heard at the Cross and so on according to the relative distances. These maps may be seen in Mr. Morrison Kyle's window, near the Exchange.

The site on which the gun is being placed is intended to be merely a temporary one. It was thought that the West End Park would have been too far west to admit of the gun being properly heard at the Broomielaw, where it is important that it should be distinctly heard, and the present position was selected as being elevated and central. But this very serious objection to the site chosen remains, that from the restricted charge of powder which will require to be used in this locality, the gun will not be heard at anything like the distance which it otherwise might be. At Newcastle, we understand, the time gun is heard within a radius of ten miles. If Gilmorehill could be obtained, and if a charge of 6 lbs. could there be used, with the gun directed towards the city, the report would be heard at a very much greater distance than can be the case under the present

arrangement. Its position in Sauchiehall Street, however, is, as we have said, purely tentative, and any suggestions which may be made as to an improved site will doubtless be considered. The great point, of course, to be gained is that the gun shall be so placed as to make itself perfectly heard at the Broomielaw and in the business parts of the city.

It is in contemplation to extend the scheme to Greenock, for the benefit of those residing in that busy port, and the public of Port Glasgow, Helensburgh, &c., if it should prove successful in Glasgow, and come to be regarded by commercial men and the community generally as a real public benefit. If it succeeds so far, the next step will probably be an endeavour to have the gun rendered permanent both here and at Greenock, as at Newcastle and North Shields, where the Tyne Commissioners have voted £ 200 a year for the maintenance of the guns.

It is interesting to note that maps had already been prepared and put on display, with marked distance-circles indicating the corrections required to allow for the sound of the bang to arrive at particular locations. It may be noted, too, that a possible site for the gun was Gilmorehill which became the current location of the University some seven years later.

From the letter immediately following the above article, it is obvious that Professor Grant had been given the courtesy of responding to the announcement, and the newspaper included his letter drafted on 25 September to a Mr Euing at the Exchange. It reads as follows:

PROPOSED TRANSMISSION OF GREENWICH TIME FROM GLASGOW OBSERVATORY.

We have been favoured with the following letter on this subject, addressed to our townsman, Mr. William Euing, by Mr. Grant, Professor of Astronomy in the Glasgow University.

Observatory, Glasgow, 25th Sept.,1863.

My dear Sir, On a recent occasion I had some conversation with you on the subject of transmitting current Greenwich time from the Observatory to the city of Glasgow. I pointed out how imperfect this object is effected at present, purely in consequence of the non-existence of an electric communication between the Observatory and the city,

and I mentioned that during the last two or three years I have been urging upon the Town Council the expediency of laying down a wire with a view, in the first instance, of controlling some of the city clocks by means of the normal clock of the Observatory, and ultimately of extending the method to other objects. The pendulum of the distant clock is maintained in a state of perfect control by a simple arrangement depending on the attractions of a magnet placed beneath it, the invention of which is due to Mr. Jones of Chester. When once the connecting wire is laid down the expense attending the application of the method is quite trifling. It may be employed in controlling a turret clock, in which case the first blow of the hammer will be the signal of the time for any hour of the day, or it may be applied to a clock with seconds hand, which will thereby indicate correct Greenwich time every instant; or lastly, it may be used in dropping a distant time ball at any specified hour. For several years past the town and port of Liverpool have enjoyed the advantages of these three distinct modes of obtaining correct Greenwich time by means of an electric wire connecting the Observatory of Liverpool with several public institutions, and in every instance the application of the method has been attended with the most complete success in support of this statement, I cannot do better than lay before you the following passages relative to the Liverpool Observatory extracted from the annual reports of the Astronomical Society of London for last few years.

[. . . . *Professor Grant then presented four extracts from the Liverpool Reports noting the successes and reliability in telegraphically driving the various City Clocks, including the large clock on the Victoria Tower with its six dials, each eight feet in diameter. He also noted that, on the 4th February, 1861, between 6 A.M. and 5 P.M., 1860 persons compared watches or chronometers with a Seconds' Clock placed in the office window of the Magnetic Telegraph Co. . . . .*]

It will be seen that nothing could be more satisfactory than the foregoing statements as regards the practical working of Jones' method. I trust that steps will soon be taken to connect this Observatory with several public establishments in Glasgow, so as to give correct indications of Greenwich time, in the way so admirably practised in Liverpool. The laying down of a wire two or three miles in length

cannot be regarded as a formidable object for a great city like this and I have already mentioned that the ulterior expense is trifling. I ought not to omit stating, that, in addition to the excellent time keeping which the Observatory had previously possessed, a first-class sidereal clock, by Mr. Charles Frodsham, of London, has been recently acquired for the express purpose of being employed in the transmission of correct time. I perceive, by the *Herald* of this morning, that steps are being taken to introduce a signal gun as at Edinburgh. This method no doubt has much of a sensational character, which cannot fail to recommend it to popular feeling, but, on grounds of real utility, the methods practised at Liverpool appear to be vastly preferable, more especially when the question relates to a great and rapidly extending commercial city, where the ear is assailed by the continuous din of the traffic of the streets, and the noise connected with a multitude of public works. However, if the public should desire a time-gun for Glasgow, it may be very easily introduced, in addition to one or all of the other methods to which I have already alluded.

In conclusion, I cannot refrain from again expressing a hope that steps will speedily the taken for connecting the Observatory with the city by an electric wire and that the public ere long will receive correct Greenwich time by the most approved methods from an institution to the instrumental equipment of which they have so manifestly contributed – .

Believe me, yours very truly.   R. GRANT.

William Euing, Esq., Royal Exchange.

By portraying the operations in the city of Liverpool as an example that might be followed, it can be seen that Glasgow very much lagged behind in the invaluable practice of distributing accurate time to the population and for shipping purposes. It may be noted that Professor Grant did not see the advisability of establishing a time-gun.

On 28 September, Grant also drafted a letter to the Lord Provost. It is plain to see that he was very sensitive to the intrusion of the initiative of Edinburgh Observatory in providing the one o'clock signal for Glasgow. From comments he made at the inauguration of the Ochtertyre Telescope and in later editorials in the *Glasgow Herald*, he appeared to have concerns over a developing political power associated with the science undertaken by Piazzi Smyth, who was possibly claiming Edinburgh as a 'National

Observatory' on the style of the Royal Greenwich Observatory. Grant strongly reiterated the nature of the commission he held in respect of the service that the Glasgow Observatory should fulfil in providing time to shipping on the Clyde. This letter was printed in the *Glasgow Herald* of 29 September and is transcribed below.

THE OBSERVATORY
AND
THE TIME GUN

We have been requested to insert the following letter which has been addressed by the Professor of Astronomy to the Lord Provost: –

Observatory, Glasgow, Sept 28, 1863.

My Lord. – You will no doubt have perceived, from a statement which appeared in the *Herald* of Saturday last, that arrangements are being made by the United Private Telegraph Company for firing a time-gun in Glasgow in connection with the Edinburgh Observatory. It would seem, also, that the originators of the scheme contemplate establishing the gun permanently, and placing similar guns on different points of the Clyde.

Permit me to inform you in reference to this matter, that by an express engagement entered into and with her Majesty's government, the University of Glasgow is charged, through the instrumentality of the Observatory established in connection with it, to afford all necessary facilities for supplying the shipping of the Clyde with correct time.

I need scarcely assure your Lordship that under no circumstances whatever will the University consent to forgo this engagement, or permit the usurpation by any other observatory, of the duties which it imposes.

The importance of placing the arrangements for the transmission of correct Greenwich time from this Observatory on a better footing than heretofore, has not failed to occupy the attention of the Professor of Astronomy, who, some time since, submitted his views on the subject to the consideration of the Town Council. I beg further, as a proof of the desire of the University to fulfil the obligation which it has contracted with the Crown in reference to this object, to call your attention to the enclosed copy of a memorial on the Observatory, which has been recently addressed by the Senatus Academicus of the University to the Lords Commissioners of Her Majesty's Treasury.

I would earnestly invite the Clyde Trustees to a consideration of the urgent necessity which exists for rendering the resources of this Observatory more effectually available to the shipping of the Clyde. Our instrumental means for the determination of correct time are unsurpassed anywhere, but they are rendered to the great extent powerless by the isolated condition of the Observatory, in regard to electric communication with the City of Glasgow and the Clyde. The Observatory will cordially receive from the Trustees any proposal in reference to this important object.

I have the honour to be, my Lord, your obedient servant.

(Signed) R.GRANT.

The Honourable Lord Provost of Glasgow.
Chairman of the Clyde Navigation.

The 'memorial' which is referred to in the letter above was written on 15 April 1862, the title[16] describing its nature. It comprised an introductory submission together with the document itself.

In presenting the history related to the Chairs of Astronomy in Britain, without specifically saying so Grant was referring only to the establishment of observatories under the chairholder's management. In this way he was able to claim that Glasgow's Chair was the first to carry the responsibility for undertaking observations. He states that there were four universities in Britain with Chairs of Astronomy, the first being the Regius Chair of Glasgow established in 1760. As related in Chapter 3, the Macfarlane Observatory was built in the grounds of the Glasgow College in 1757, and three years later Dr Wilson was appointed to the new Chair. In chronological order Grant then mentions the observatory established in 1774 by the Radcliffe Trustees in Oxford; this attracted a separate position for an 'observer', but he fails to mention the Oxford Savilian Chair of 1619. Similarly he records that an observatory at Cambridge was established in 1820, without mentioning the 1707 Plumian Chair of Astronomy and Experimental Philosophy. Finally, he notes that Edinburgh became established around 1834, probably referring here to the building of the observatory on Calton Hill. His noting that its management came about by attachment to 'another office through the Crown' suggests that he was referring to the establishment of a Chair. The date of 1834 in fact corresponds to the appointment of the first Astronomer Royal for Scotland, Regius Professor Thomas Henderson, the Chair in Edinburgh having been established in 1786. Grant also comments that the Glasgow Observatory had been purchased by the University in 1845 from an astronomical institution and occupied an honourable place with other similar establishments in both Europe and America. One of his pleas related to the fact that the government had supplied Edinburgh Observatory with two associates, and that the ancient University of Glasgow might reasonably expect that an observatory for which it had made such large sacrifices was also entitled to receive from the same quarter the small additional assistance requisite for placing the establishment on an efficient footing. In the covering letter, Professor Grant asked that 'their Lords & Gentlemen' might 'take into consideration the expediency of providing an Assistant at the Observatory of the University of Glasgow and appointing a small annual sum in aid in defraying the current expenses of the same establishment'. This submission appeared to fall on deaf ears.

On 28 September, a further announcement of the preparations for the Glasgow time-gun appeared in *The Scotsman*. The article is reproduced in Fig. 8.4.

Articles and letters then continued to appear in the *Glasgow Herald* concerning the experiments relating to the firing of a one o'clock gun and distributing 'time' across the city. On 30 September, the day before the first 'big bang', the following appeared:

GREENWICH TIME FROM THE OBSERVATORY: – We understand that steps are to be taken at once to connect, by an electric wire, the Observatory with one of the turret clocks in the city, for the purpose of controlling the latter by the normal clock of the Observatory, as practised so successfully at Liverpool. As each hour strikes, the first blow of the hammer will indicate the instant of Greenwich time, which may thus be obtained twenty four times in the course of a day. The controlling of a seconds clock by the same method (Jones') is also under consideration. It is obvious that, as soon as the connecting wire is laid down, the

GLASGOW — THE NEW TIME-GUN. — Several workmen are now busily engaged in making the arrangements necessary for the new time-gun experiment in Glasgow being tried at an early date. The approaching experiment has originated with the Universal Private Telegraph Company, Mr Nathaniel Holmes having undertaken the superintendence of the arrangements, while the valuable co-operation of Professor Piazzi Smythe, astronomer-royal for Scotland, has been obtained in furtherance of the scheme. The gun is being put into position in a green which forms an eminence overlooking Sauchiehall Street, at the west side of the Corporation Galleries, and it is intended to be fired on Thursday at one o'clock Greenwich mean time.

Figure 8.4 An article in *The Scotsman* of 28 September 1863 describes the continuing preparation of the Glasgow time-gun.

application of the method will be capable of indefinite extension. A sketch of this simple method of control will be placed for inspection in the Exchange, today, about noon. A similar sketch will be exhibited at the same time in one of the windows of Messrs. John Smith & Son, booksellers, St. Vincent Place.

Also on 30 September, a letter by a Mr F. G. Taylor appeared, which showed a measure of appreciation of the superiority of a seconds' clock display over the operation of a time-ball or gun signal. Although Taylor did not mention D. McGregor & Co. by name, he was somewhat scathing about the reliability of observations made by their transit telescope, readily identifiable by being in the new premises between Bridge Street and Dale Street.

TIME-GUN – TIME-BALL GLASGOW OBSERVATORY
To the Editor of the Glasgow Herald.

Glasgow, 28th September 1863.

SIR, – I see by your impression, dated the 26th inst., that a new time gun is about to be erected in Glasgow, a matter which must give general satisfaction to those who are practically acquainted with the present imperfect system of obtaining Greenwich time.

I have never personally inspected either the transit instrument or the methods of instrumental adjustment, two very important features for the determination of correct time; but one thing I certainly must say, if the place of observation (as I understand it to be) is situated in the new houses between Bridge Street and Dale Street (South Side) I would not value the results of such observations at a very high price.

In the first place, the most essential point to be observed in the erection of a transit instrument is stability; but the word stability, in its ordinary sense, is not the stability in astronomical sense as they are essentially of a different character. A transit instrument is not considered in a state of stability unless it has no other foundation but its own and that foundation too is solidly built, and the pillars upon which the horizontal axis rests so carefully and substantially constructed, that all the gales of wind that Glasgow is heir to cannot remove it from its meridional position.

Now, if the Town Council, or the Clyde Trustees, or whoever the gentlemen may be who sanction the dropping of the time-ball, could only visit the place of observation annually, or say half yearly to see that the instrument is in a state of efficiency, the instrumental adjustments and the general routine necessary for obtaining Greenwich time properly conducted, they would act very judiciously. This inspection shall take place not as a matter form, but as a matter of real utility and consequently should be superintended by the astronomer to the University – a gentleman who is really practically acquainted with these affairs, and who would conscientiously report when he considered the present place of observation in any way suited for the mounting of a transit instrument, and whether there is sufficient stability in the building itself to depend upon the instrumental error deducible from the observations (if any).

Unfortunately, however, another difficulty presents itself. In taking transit observations we observe the sidereal time and not the solar time of passage, and by certain formula we convert sidereal to solar time for our own meridian; and then by applying the difference of longitude between Greenwich and the place of observation, we obtain Greenwich mean time. The question then arises what opportunities have been afforded to obtain the true longitude? Answer suggests probably by chronometer observations with the Edinburgh Observatory. If so, first prove that the two places of observation are equally important or reliable as far as astronomical character goes, and then one can believe the ball drops punctually at one o'clock.

Only imagine that Glasgow, boasting, as it does, of its nearly half a million of inhabitants, is rendering itself conspicuous in astronomical history by allowing the time-ball to be dropped by an agency altogether independent of the Professor of its University. If I were a member of the Town Council, I would blush to think that a city like Glasgow, superior both in population and wealth to Edinburgh, should bow so humbly as to accept of the proposed scheme for giving us Greenwich mean time. What would be the natural conclusion arrived at by a person unacquainted with histories of the two cities? Why, that Edinburgh possesses facilities for determining Greenwich mean time which Glasgow was deficient of. But such is not the case. Glasgow has both a scientific institution generously equipped with instruments by its own citizens, and a Regius Professor possessing both zeal and abilities, and all the necessary qualifications for superintending time-ball regulations. Professor Grant states that the method of having the time by a signal-gun "has much of a sensational character, which cannot fail to recommend it to popular feeling, but on grounds of real utility and methods practised at Liverpool appear to me vastly preferable." Now this opinion must evidently be unanimous in the minds of those who give the least attention to this matter.

The Professor has given such striking proofs of the great control that the pendulum is held by Jones's method, in his extracts from the Royal Astronomical Society's Report, that Glasgow must not rest until it transmits its own time from its own Observatory, and by its own Professor. The Town Council have the choice of the time-ball, the signal-gun or the turret-clock. The turret-clock has a decided advantage over the other two, for this reason, that both the time-ball or gun gives the Greenwich time at one single hour of the day, and if the mariner happens to be absent from his ship at that moment of time he must lose that day's comparison; but if the turret-clock is placed in a public position, the mariner, at his own convenience, can take his chronometer and make comparison with the first stroke of the hammer at any hour of the day; or, if a seconds hand is applied to the turret-clock, every instant of this will be recorded, and the mariner can take immediate comparison and obtain his rate with precisely the same accuracy as the chronometers of her Majesty's Navy at the Royal Observatory, Greenwich. In conclusion, I only hope that the town Council will look before they leap and really investigate the merits of this matter. I feel sure that they will not allow sensationalism to predominate over good sense; and with full assurance that they will give their serious consideration in this case, and their hearty cooperation in that which really is most important. I respectfully subscribe myself, your obedient servant,

F.G. TAYLOR

On the following day, 1 October, it appears that the city authorities had responded to Professor Grant, as the *Glasgow Herald* featured the following article:

GREENWICH TIME FROM THE OBSERVATORY: – A plan of the proposed method of controlling the city clocks, to be employed at the Glasgow Observatory was exhibited yesterday in the Exchange. The system is very simple, and has already been shown elsewhere to be an effective one. The pendulum of the Observatory clock, at each beat, presses against a spring, so as to connect a galvanic battery with a wire which extends between the Observatory clock and the clock to be regulated. The current thus passing from the Observatory clock polarises once in each second the wire coil of the pendulum bob of the clock to be regulated. If the two pendulums beat in unison, so that the coil is exactly over the steel magnet (placed beneath the bob) on either side at the instant when the current passes through it, no effect is produced upon the pendulum of the distant clock, the influence of a magnet being exerted wholly on the direction of the pendulum rod. If the pendulum to be regulated lags behind it is at once pulled forward by the direction of a magnet. If on the other hand, it has a tendency to go too fast, so as to swing beyond the normal position above the magnet before the current passes through it, the direction of the magnet will pull it back. In this way the pendulum is bridled between the two magnets, and the vibrations will all be performed simultaneously with those of the Observatory clock.

In response to Mr F. G. Taylor's letter, one dated 2 October from Mr W. Church, an employee of McGregor & Co., appeared in the *Glasgow Herald* on 5 October. This letter contained a high level of anger, some of it directed at the University Observatory. It defended the company's ability to obtain accurate time from the transit telescope on the roof of their premises. Although it was not mentioned, in addition to operating the time-ball, McGregor & Co. had other revenue from regulating several clocks in the city, and understandably would not wish to see the loss of these contracts, particularly after investing in the establishment of a transit telescope.

THE GLASGOW TIME-BALL

To the Editor of the Glasgow Herald.

October 2nd, 1863.

SIR, – The firm of D. M'Gregor & Co. will not notice attacks upon their establishment, except where principals are concerned; but I, as being employed in the working of the time-ball, would request your permission to reply to some portions of Mr. F.G. Taylor's letter. I am not acquainted with the writer, but I infer from his letter that he possesses a very comfortable assurance of the value of his judgment and authority in matters relating to time-measurement, and that he shares a delusion, fostered by professional prejudice, that accurate time cannot be got or maintained outside the precincts of a public observatory. The firm of M'Gregor & Co., however, are not likely to attach much importance to his opinions respecting the transit observations, and they are certainly quite as well aware as he is of great importance of attending to the adjustment of a transit instrument, as, without such attention, it would be impossible to obtain true time.

What Mr. Taylor states concerning the stability of the instrument sounds very well in theory; and if we were compelled to take observations during those violent gales of wind which according to Mr. T., Glasgow is unhappily heir to, it might be sound expedient to plant the foundations of the instrument deep beneath the bowels of the earth, but no such necessity is imposed upon us, and the vibration due to the ordinary traffic of the streets during the day time cannot possibly affect the accuracy of sidereal observations taken at night, unless the meridianal position of the instrument had been disturbed thereby, but I can prove that it is not in the slightest degree perfected for its position, accurately determined by observations, can be readily verified by a well defined terrestrial meridian mark.

Mr. Taylor's assertions can never upset plain facts, a few of which I will submit. I have seen observations taking during a night of numerous stars, both high and low, in which the difference between greatest and least did not exceed a few tenths of the second, and Mr. Taylor, if he has any practical knowledge of observing, must know very well that such a result could not possibly be obtained unless the adjustments of the instrument were almost perfect. I can prove also that comparisons were made, scores of times, with the time of the Glasgow Observatory, showing nearly perfect coincidence each time, and thus affording proof positive that the firm possessed the means of obtaining true time; and the fact of their having been able, at any time, to check an error in the time sent from the Observatory was satisfactorily proved sometime ago in the detection of an error of several seconds in comparing the Exchange clock, which was owing to a gentleman having been present who had the true time from the firm, and although the assistant appeared to ridicule the idea of any one correcting the Observatory he was obliged to return the day following to rectify the error.

Mr. Taylor's question concerning the longitude might readily be answered by any intelligent schoolboy, that " it would be impossible to obtain true Greenwich time without possessing the true longitude of the place of observation ", but it is not consid-

ered essential to satisfy Mr. Taylor's curiosity respecting the mode of obtaining it.

Mr. Taylor's proposition to get the Professor of Astronomy to inspect the arrangement of the firm for obtaining time is simply absurd, for he must be perfectly well aware that the Professor has already given public expression to an antagonistic opinion.

Mr. Taylor expresses astonishment at the apparent anomaly of a time-ball being worked independent of the Observatory; but if he is really ignorant how the matter stands, the explanation is easily rendered. At the time when the time-ball was first established, the Observatory, whatever its present position may be, was not in a proper state of efficiency to maintain a correct standard of time; and the Clyde-Trustees, to whom the time-ball belonged, appointed the firm of M'Gregor & Co. to manage it, having, I suppose, sufficiently valid reasons for the confidence which they placed in them. I intend no illusion here to the astronomical instruments of the Observatory. Its transit circle might have been unsurpassed anywhere, but that could only have been used for the purpose of getting, but not maintaining, true time. The maintenance of a correct standard of time during intervals of bad weather, so frequent in our climate, must depend solely on the clocks of the Observatory, which ought to have been of the very first class, and sufficiently numerous for the purpose.

If the Observatory is now in a high state of efficiency – and we have Mr. Grant's assurance to that effect – by all means let it provide the time for the city of Glasgow; but I certainly consider that it is a very paltry mode of trying to attain this object, on the part of the advocates of the Observatory, by attempting to lower the credit and depreciate the services of other parties, and the Observatory might well exclaim "Oh! save me from my friends." Being itself not quite invulnerable, it has hitherto acquired no laurels in such a contest, the initiative in which has never been taken by the firm M'Gregor & Co., nor is it likely to do so in the present instance through the advocacy of Mr. F.G. Taylor; for, notwithstanding what he, or other parties, may assert, who possesses not the means of forming a judgment as to facts, I have no hesitation in saying that the time-ball has been, and is now, a standard of time sufficiently accurate for the purpose of rating chronometers the most important of all uses to which it can be applied – I am, Sir, your obedient servant.

W.CHURCH, Chronometer Maker.

As mentioned earlier, it is noteworthy that Mr Church believed that the University Observatory, at the time when Nichol was in charge, was not in a proper state for providing a reliable time service.

The time-gun project, of which Grant had been originally unaware, culminated with the first firing at 1 p.m. on 1 October 1863. He had already noted that the scheme had no element of permanence independent of the local support it might receive, and that it would not affect the thrust of the Observatory's aim to provide a control system for clocks at various locations in the city. The difficulties associated with undertaking the project on a permanent basis within the confines of an urban area, with the effects of the explosion being trapped within the tall stone buildings, were immediately apparent, and the voices of the citizens were heard. On 7 October, the *Glasgow Herald* featured the following article:

THE TIME-GUN

The present position of a gun in Garnethill being too confined to admit to a proper charge of powder, arrangements are being made to remove the gun to a more elevated position, from the immediate vicinity of the houses, so that the volume of sound from the gun can be increased to be audible over the entire City. It is expected that, about Wednesday next, the gun will be fired from its new position.

On 9 October, another letter was published in the *Glasgow Daily Herald* from an anonymous member of the public, no less than a T. Fugit, who had a troublesome concern.

THE TIME GUN
To the Editor of the Glasgow Herald.

SIR, – Without attempting to question the scientific merit of this experiment, I venture to call the new time-gun a nuisance if it is to remain longer where it now stands. For the first two or three days we were a little startled in this neighbourhood when we heard the one o'clock explosion, but for the sake of the Broomielaw and science, we did not care to complain. Today, however, the charge of powder has been increased, if we are to judge by the increased din. Now, I am a tenant in this locality, and I find my ceilings cracking, and in some places giving way altogether. That this is the result of the explosion there can be no manner of doubt, as, at one o'clock today one of the youngsters narrowly escaped a thump on the head from a yard or so of falling plaster. Nor is it all. The neighbourhood is surrounded by educational institutions, and I am told that some of the children attending them get quite sick when the gun is fired, and that, today, many of them got a greater fright than usual. On the whole, I think there is exhibited a woeful lack of common sense in placing the gun where it now is, more especially since, as I am informed, there be few at a distance that can hear it. Hoping to hear of its speedy removal, I am, &c.

T.FUGIT
Glasgow, 7th Oct., 1863.

This letter clearly describes the sheer disturbance caused by operating a time-gun within the city. On 12 October, another 'Letter to the Editor' appeared concerning this sudden public interest in the keeping of time. It reads as follows:

THE TIME-GUN AND GLASGOW OBSERVATORY
To the Editor of the Glasgow Herald.

SIR, –Mr F.G. Taylor and Mr W. Church having addressed you on this important subject, expose themselves and their widely different views to public criticism. Besides the thing is one of general interest. It will not be rude, therefore, to offer, and I hope you will have the goodness to tolerate, a few remarks on their respective letters.

There can be no doubt that Mr.Taylor, who took the initiative in this discussion, is familiar with the subject of which he writes. He shows up the difficulties which attend the system at present adopted for obtaining Greenwich time. His views, moreover, exactly coincide with those of the Professor of Astronomy, and their joint opinion on this matter should have much weight. In short, I can discover nothing objectionable in all he says. He is candid and consistent throughout, and, addressing himself not to persons but things, he writes with freedom and without fear. Mr. Church, on the other hand, appears to be in a passion. He defends a firm which Mr. Taylor never attacked, and replies to a question which Mr. Taylor never asked. Passing over, however, his somewhat warm language, Mr. C. referring to the transit instrument observes – 'What Mr. Taylor states concerning the stability of the instrument sounds very well in theory', implying evidently that *practically* the matter of stability is unimportant. But here is something more curious – 'The vibration due to the ordinary traffic in the streets during the day cannot possibly affect the accuracy of sidereal observations taken at night'; and then he speaks of his 'well defined terrestrial meridian mark'. Now, he acknowledges here that in the day-time at least a vibration exists; and when but in the '*day-time* can he take the observation of his meridian mark'. But, perhaps, the mark is illuminated. Mr. Church adds, that, in taking high and low stars with his transit instrument, the difference between greatest and least did not exceed a few tenths of a second. The phrase 'a few tenths', is very ambiguous, and in their interpretation of it I believe Mr. T. and Mr. C. widely disagree. I call 'Vega' a high star, and 'Spica' a low star; but if they did not produce the same clock error, within *one-tenth* or *two-tenths* to the extreme, I should be disposed to refer to my instrumental adjustment. The clock, of course, must be going mean time.

With regard to a periodic inspection by the Professor, I think it a capital idea, and regret much he should have expressed an 'antagonistic opinion'. He may, however, be prevailed upon to inspect under the instructions of the Board of

Trade. And is there anything unreasonable in the City Observatory dropping the ball? There may be no 'anomaly' a time ball being worked independent of the Observatory, still there is an incongruity akin to a hairdresser interfering with mechanics, or a dairy maid the politics.

Mr. Church is unacquainted with Mr. Taylor, and I with both. Nor have I the honour of being known to Professor Grant, so that my opinion, if it be despised, is at least disinterested. – Sir, most obediently yours.

TEMPUS VERUM.
Glasgow, 8 October 1863.

The battle between Edinburgh and Glasgow escalated in more vitriolic form, as revealed in a *Glasgow Herald* editorial on 30 October that read:

THE NEW TIME-GUN

More than a month has now elapsed since the intelligence that a Time-Gun was to be introduced into Glasgow burst so suddenly upon its inhabitants. An Edinburgh journal, in its impression of September 24, first gave currency to the report of the good things in store for Glasgow; for, up till then, the project had been matured with unaccountable quietness. It was stated, further, that the gun was to be fired in connection with the Edinburgh Observatory. In our impression of September 26 there appeared fuller details, by which the public were informed that the scheme had originated with the Universal Private Telegraph Company, which had secured the co-operation of Professor Charles Piazzi Smyth, Director of the Edinburgh Observatory—that the arrangements were all but complete—that a site had been procured for the firing of the gun, and that maps, to enable the public to take advantage more effectually of the new time signal had been *already prepared*, and might be seen in a bookseller's shop window. On the same day there appeared a letter from the Professor of Astronomy in our own University by which it appeared that he had, some time previously, submitted to the Town Council a scheme for the transmission of time, of a totally different character, but which had been practised with admirable success in Liverpool for several years past. It was plainly apparent from this letter that neither the University of Glasgow nor Professor Grant had received any information respecting the Edinburgh project until it was announced in the public journals; and this

was confirmed by a subsequent letter which the Professor addressed to the Lord Provost, in his capacity of Chairman of the Clyde Navigation. Referring to a special engagement with her Majesty's Government, by which the University of Glasgow, as represented by its Observatory, is pledged to furnish the shipping of the Clyde with correct time, he assures his Lordship "that, under no circumstances whatever, will the University consent to forgo this engagement, or permit the usurpation, by any other Observatory, of the duties which it imposes."

Clearly, then, the University of Glasgow and the Director of the Edinburgh Observatory are placed by the actual consummation of this project, in a position of antagonism. It becomes, therefore, an important public duty, to inquire into the circumstance whether Professor C.P. Smyth has been justified in the step which he has taken, and whether it is conducive to the public interest, or the progress of science that he should thus intrude into the sphere of duty occupied by our time-honoured University. We are enabled to notice this question more fully by availing ourselves of a report which Professor Smyth has recently addressed to the Board of Visitors of the Edinburgh Observatory, and which, by the way seems to be pervaded by a queer imitation of the form of language used in the annual reports of the Astronomer-Royal on the great national Observatory of Greenwich. If so, this must appear simply ridiculous to those who know the exact relation in which the Edinburgh Observatory stands to her Majesty's Government, and the exigencies of the nation. The Observatory of Greenwich was founded nearly two centuries ago, with a view to upholding of the interest of the country as a great naval Power. It is essentially identified with our maritime prosperity as well as the advancement of astronomy, and it has always constituted one of the scientific ornaments of our country. The Observatories of Glasgow and Edinburgh were purely academic institutions which have been established by the Crown in connection with their respective Universities. The office of Observer in the Glasgow University, conjointly with the Professorship of Astronomy, dates from the year 1760. The Edinburgh Observatory originated in a private society, called the Astronomical Institution, which, in the year 1834, transferred its Observatory to the Crown, upon the condition that it should continue to be maintained by her Majesty's Government. This was agreed to, and it was placed in connection with the Edinburgh University by the appointment of the Professor of Astronomy as its director. So far, it is clear that the two Observatories were placed precisely on the same footing, their respective functions being identical–namely, the diffusion of a knowledge of astronomy, and the advancement of the science by a course of observation. It is true that, while the Glasgow Observatory is only partially endowed, the Edinburgh Observatory is wholly endowed by the Crown. But it is important to bear in mind that the support conceded by the Government to the Edinburgh institution is exactly in accordance with its origin as an academic and not a national Observatory. In fact, the emoluments attached to the Professorships of Astronomy in the Universities of Glasgow and Edinburgh are as nearly as possible equal, the University of Glasgow supplementing the small salary which its own Professor derives from the Crown. We fully admit that both offices are underpaid, but we maintain that the claims in favour of the one are as good as those in favour of the other. It is true that the Professor of Astronomy in the Edinburgh University is styled her Majesty's Astronomer for Scotland, that this as we are informed, is a mere title like that of her Majesty's Painter for Scotland, or her Majesty's Jeweller for Scotland. There are

scores of such titles, which mean nothing, to be found in the west end of London, as well as at Windsor, the Isle of Wight, Edinburgh, Glasgow, and Aberdeen. Upon the ground, however, furnished by such a designation, the present occupant of the Chair of Astronomy in the Edinburgh University takes the title of a *bona fide* official—the title of Astronomer-Royal for Scotland—forgetful, we fear, of the respect due to the illustrious director of the Greenwich Observatory, who is the veritable Astronomer-Royal for Scotland, as well as for every other part of the British Empire. We repeat that the Edinburgh Observatory sprung from a private institution, and did not originate with the Government as a national necessity, like the Observatories of Greenwich and the Cape of Good Hope. As a natural consequence, it has never been recognised by the Government as a national Observatory, in which case it would have been placed under the Admiralty, with which it has notoriously no connection.

The public, however, will give themselves very little concern about what titles the Edinburgh Astronomer may be pleased to take, so long as they are not asked to pay for them. But it may be a very different matter should the absurd designation be made a pretext for obtaining additional support out of the national purse. It has been stated that as application for a grant of money on behalf of the Edinburgh Observatory is at present before the Government. The demand is said to include the proposal that a salary should be attached to the office of Astronomer-Royal for Scotland! an official who has not hitherto been recognised by Government in any other capacity than as Professor of Astronomy in the University of Edinburgh and Director of the Edinburgh Observatory. It would doubtless fortify any such claim to the countenance of the Treasury if it could be shown that the Edinburgh Observatory was the means of transmitting correct time to other parts of the country. But it does not by any means follow that it would be explained to the Government that time was thus sent to Glasgow without any application from the inhabitants, or their municipal authorities—that, even when it is sent, the seamen at the Broomielaw and the great bulk of the people never hear the gun;

and, finally, that our own University is not only competent, but is taking measures to supply accurate time to the public by means of the city clocks.

The inauguration of the Ochtertyre Telescope at the Glasgow Observatory, on the 30th of April last, will, no doubt, be fresh in the recollection of many persons. On that occasion Professor Grant delivered an address on Observing Astronomy, in which he showed that the notion of the Edinburgh Observatory being a national establishment, in the same sense as the Greenwich Observatory, was totally unfounded; that, in point of fact, the Glasgow Observatory, as an academic institution had the same claim upon the Government for support as the Edinburgh Observatory had; and he adverted to the anomalous circumstance that, while the Edinburgh Observatory was supplied with two assistants, paid out of the public funds, there was no provision at all for assistance at the Glasgow Observatory. Further, the University Commissioners, in their report printed in last June, had recommended the Glasgow Observatory to the consideration of Government. All this pointed clearly to the steps which the University of Glasgow has recently taken in memorialising her Majesty's Government on the condition of the Observatory. That, such a proceeding is not unreasonable, or uncalled for, may be inferred from the fact that, while the Regius Professor of Astronomy and the Observatory of Glasgow receive fifty pounds from the public funds, the Regius Professor of Astronomy in the University of Edinburgh and the Edinburgh Observatory receive just seven hundred pounds from the same source. But the very idea of the Glasgow Observatory presuming to take a position of equality with the Edinburgh Observatory is, to certain folks in the East, offensive. Still more alarming were the events clearly foreshadowed by existing circumstances. Professor Grant, after eighteen months of incessant occupation, had finally seen installed at the Glasgow Observatory the magnificent Ochtertyre Telescope. He had publicly stated, in his address on the occasion, that he had been for some time in communication with the Town Council on the subject of controlling the City Clocks from the Observatory, and that the munificence of the

subscribers to the Ochtertyre Telescope and enabled him to obtain from London a first-class sidereal clock, which was to be employed in carrying into effect that important object. It was clear, then, that matters were favourable to the claims of the Glasgow Observatory. But if the Edinburgh Observatory could quietly carry into effect its project of transmitting time to Glasgow, two important results would in all probability be produced—first, the Edinburgh Institution would then have some colour of a pretext for urging its claims upon the Government; and secondly, the Glasgow University might be effectually defeated in its application to the same quarter on behalf of its own Observatory. Nor were existing circumstances unfavourable to the success of the project. The Glasgow Observatory, it is well known, was not yet in electric communication with a contemplated field of its operations for the transmission of time, nor could an associated public company take in hand the city clocks, although it might co-operate in the firing of a gun.

Professor Smyth speaks of Glasgow having *applied* for a single gun; of its being the only Scottish city which had hitherto accomplished the object of its wish of its citizens having locally provided a cannon; of their strong common sense in perceiving the superiority of the new system, and of having vigorously adopted a new and more suitable locality for the firing of their gun. In reply to these statements, we have simply to say that in this city, of nearly half a million of inhabitants, there were probably not more than half a dozen of persons who were cognisant of the project of the Private Telegraph Company and Professor Smyth, previous to the announcement of it in the columns of the Edinburgh *Scotsman*.

It is not within our province to pronounce an opinion on the respective merits of the different methods of transmitting time which are now before the public. The University, we believe, are quite prepared to co-operate in adopting, according to the circumstances, whatever may be considered most suitable; and that, too, utterly irrespective of anything that may be done from any other quarter. It is understood that the preliminary permissions for laying down a wire from the Observatory to the College clock have been already obtained, and that actual operations will be commenced in the beginning of next week. We may expect, therefore, as in about three weeks hence the College clock will be maintained in perpetual control by the Normal Clock of the Observatory, and it will then be an easy matter to extend the system indefinitely.

Clearly the pretentious and autocratic role of the Astronomer Royal for Scotland was under attack as a result of the way he was riding roughshod over the commission that the Glasgow Observatory was under obligation to follow. This might be all the more surprising in that Piazzi Smyth was the son of Admiral Smyth, and it was Professor Grant who had been the mainstay of their collaboration in literary projects such as the translation of Arago's works. Surely the Astronomer Royal for Scotland must have known about Grant's obligations to the Crown. The attack on the title of Astronomer Royal for Scotland was indeed scurrilous, and a short article in the *Glasgow Herald* of 23 November 1863 put the matter right, but without giving an apology. That the title of Astronomer Royal for Scotland does exist was made clear. The article stated: 'We have been led into belief that no such title exists as that of "Astronomer Royal for Scotland"'. Details were then given of how it came into existence by 'Royal Commission' and with the 'Endorsement by the Treasury' in respect of the second title holder, Professor Piazzi Smyth.

Following the tirade of 30 October, the *Glasgow Herald* the next day published a 'Letter to the Editor' from Mr Nathanial Holmes, the engineer overseeing the exercise. In addition to presenting a validation of the experimental methods, it announced that an additional gun would be fired at Greenock. Holmes also accepted responsibility for establishing the whole project, declaring that no criticism should be directed to the Astronomer Royal. The last three paragraphs read as follows:

> It is in contemplation shortly to place a large experimental time gun at Greenock for the use of shipping at the Tail of the Bank, Port Glasgow, Helensburgh, Dunoon &c at which place its booming sound will distinctly be heard over the surface of the water.
>
> I feel it is unjust to attempt to throw blame upon the Astronomer-Royal for Scotland, and accuse him of forcing Greenwich mean time into Glasgow from Edinburgh Observatory. If there is any blame to be attached to the introduction of great public benefit to a great and wealthy community, it must be borne by myself as the most unfortunate individual.
>
> The comparative value of controlled church dials with heavy external pointers and audible gun signals as a popular means of accurately indicating true time can be discussed when the value of a permanent time signal for Glasgow and neighbourhood has been established – I am, your obdt. servt.
>
> NATH. J. HOLMES.

Throughout the experiments, open debate on the merits of time-balls, time-guns and the conversion of existing public clocks continued, particularly through the columns of the *Glasgow Herald*. On 6 November 1863, a spurious letter appeared concerning the problems of the delay in transmission of sound from the gun. It reads as follows:

THE TIME-GUN
To the Editor of the Glasgow Herald

Glasgow Nov 5

SIR, My attention having been directed to take in remarks on the time gun, allow me to say it must be borne in mind that an allowance requires to be made for the time sound takes to travel, and which is not 1100 feet per second as is generally supposed; but depends altogether on the strength of the sound – thus for instance, Sir John Ross, in the Artic regions heard the word of command given by the officer in charge of a gun after he heard the report of the gun, showing that weaker sounds travel much slower than stronger. It may also be observed that in fireworks, the great explosion is heard first, though the smaller coloured lights are known and seen to have exploded first. I am, Sir, yours &c.

PS

On 14 November, another letter, dated 10 November, from Mr Holmes was published, 'placing before the public' the relative merits of the various systems of presenting time. With reference to noting an exact moment and comparing it with a transportable chronometer, it was suggested that hearing the event was easier than using a visual reference. In any case, making modifications to church clocks was not thought to be an easy matter. A week later, on 20 November, a letter appeared from the patent holder of the method for electrical control of a pendulum, Mr R. L. Jones of Liverpool, who noted the great success of adapting the six dials of the Victoria Tower, each 8 feet in diameter. The 13-foot pendulum with a two-second beat had been replaced by a shorter one with a beat of one second.

Operations using the time-gun continued, with different sites being explored. The first experiment located the gun at Garnethill, close to Sauchiehall Street. In an article of 18 October 1865, in *The Mercury* of Hobart, Tasmania, it was noted that:

> The first Glasgow time-gun was supplemented by a second one in St. Vincent's Place on the 29th of October, and these two by a third at the Broomielaw, on the 10th of November, while a fourth gun was added to the system at Greenock on the 21st of November, all four being simultaneously fired through the agency of the electric current from the Observatory.

The article failed to mention that it was not the Glasgow Observatory that transmitted the signal.

A definitive moment occurred at a meeting of the University Senate on 27 November 1863, the minutes of which were published in the *Glasgow Herald* of 12 December. Some of the paragraphs of this long article have been condensed; these are presented in italics.

Glasgow Herald
1863 December 12

UNIVERSITY OF GLASGOW ON TIME SIGNALS

[. . . *The long letter, being a copy of a statement of the Senatus Academious of the University of Glasgow, is introduced by referring to the Report of 14th of October 1863, by Professor Piazzi Smyth to the Board of Visitors of the Edinburgh Observatory, containing statements having a tendency to injure the position, and detract from the usefulness of the Glasgow University Observatory . . .*]

[. . . *Also in the preamble, the letter sets the scene on the standing and responsibilities of the Observatory in supplying time to the Clyde; reference is also made to the recent establishment of the largest refractor in Scotland through support chiefly by the citizens of Glasgow . . .*]

[. . . *In running up to the announcement in an Edinburgh Journal of 24th September 1863, and more fully in a Glasgow paper of the 26th of the same month, for firing a time-gun in Glasgow with signal originating from Edinburgh. The committee commented on the problems in transmitting true time from the Glasgow Observatory three miles away to the City and the Clyde. An element of frustration with the slowness of the Town Council in responding to the meetings with, and submissions of, the Professor of Astronomy was apparent. . . .*]

It was not the object of the committee on this occasion to make any comment on the want of courtesy displayed by Professor Smyth in not communicating with the authorities of the University before presenting himself in a field in which it was well known that the Professor of Astronomy in this University was actively engaged, nor do they call attention to his proceedings from any apprehension in regard to the ultimate result of his operations for the transmission of time to the city of Glasgow and the Clyde. The committee desire merely to present the result of an inquiry which they have instituted into certain statements contained in Professor Smyth's report which affect the interests of the Observatory of this University, and which, if allowed to remain unnoticed, might exercise an injurious influence on the course of operations and the position of that establishment.

The following are the passages in the report to which attention is more especially called:–

"Several other cities have applied for the Edinburgh time-signal since then, and the local negotiations for the means of loading their respective guns are in different states of forwardness, but the only one amongst them that has succeeded in accomplishing its part is Glasgow, the Queen of the West. Glasgow had early perceived the advantage to astronomy for a division of labour amongst astronomical observatories, and as we had long since become a time-observing and time-signalling observatory, Glasgow preferred to get its time signal from the Royal Observatory, Edinburgh, rather than from its own very admirably planned Observatory, which was already fully taken up to the utmost of its strength with other branches of astronomy more immediately adapted to the nature of its establishment.

"With the full approval, therefore, of the British Association for the Advancement of Science, the Glasgow University, and the Glasgow Observatory too, as represented by its late popular director, the Edinburgh Observatory dropped a model time-ball daily, for a week, by electric means, in a public meeting room of the Glasgow College, in the year 1855. That work was successfully performed, besides its principle having been formally approved; but somehow there was not enough in the time-ball system, itself, to fully interest the practical inhabitants of the great western city. When, however, eight years had elapsed, and the inventions of Professor Wheatstone had enabled an electric current to explode a distant gun more easily as well as certainly than to drop a ball, the citizens of Glasgow had locally provided a cannon, and aided materially by the well known Magnetic Telegraph Company, did connect it by wire with the Edinburgh Observatory, and when the current from there did consequently discharge the Glasgow gun simultaneously with the several guns of Edinburgh, Newcastle, and Shields, the strong common sense of

Glasgow citizens immediately perceived the superior efficiency of the new system; for, after a week's experiment with from 2 lbs. to 2.5 lbs. of gunpowder a day, from a cannon temporarily placed, as will be seen in a schedule recently received from them, they are vigorously adopting a new and more suitable locality, with a noble charge of 8 lbs. of powder or nearly double what is used at any of the other stations."

These statements imply:–

1. That the dropping of a model time-ball at the meeting of the British Association held in Glasgow College in 1855, by means of a current of electricity from the Edinburgh Observatory, is to be regarded as a proof that the University of Glasgow assented to the permanent transmission of time signals from the Edinburgh Observatory to Glasgow.

2. That Glasgow preferred receiving its public time signals from the Edinburgh Observatory to receiving them from the Observatory of the Glasgow University, and that, in the course of last autumn, it had actually applied for the Edinburgh time-gun signal.

3. That, in reference to the project for establishing a time-gun in Glasgow in connection with the Edinburgh Observatory, the citizens of Glasgow provided a gun and connected it by wire with the Edinburgh Observatory.

4. That the citizens of Glasgow, in perceiving the superiority of the new system of time-gun signals, were vigorously engaged in carrying it into effect, and that they had actually forwarded to the Edinburgh Observatory a schedule, containing an account of their proceedings.

These different statements it may be proper to notice *seriatim*:–

1. At the meeting of the British Association in 1855, Professor Smyth was permitted to illustrate experimentally in the mechanical section, the dropping of a time ball by means of a current of electricity directed from a distant station. The association acted in this instance to Professor Smyth as they have invariably acted to any member of the association who might be desirous of explaining his views on any subject comprised within the province of their operations. The authorities of the University who had granted to the association the use of the apartments of the College, during the time of their meeting, took no cognisance whatever of Professor Smyth's experiment.

2. After a most careful inquiry, the committee have found it impossible to discover the slightest proof of Glasgow, as represented either by its municipal authorities, or by any collective number of its citizens, however small, having applied for the Edinburgh time-gun signal. The Lord Provost of the city has informed the committee that a gentleman representing himself as an official of the Universal Private Telegraph Company, waited on him, and requested, in the same of the company, permission to fire and experimental time-gun in a locality within the boundaries of the city that the applicant was told that the company could do so only at their own risk, that they would be obliged to make good any damage which the firing of the gun might occasion, and that if a complaint of the gun being the nuisance should be lodged with the police by any of the inhabitants of the neighbourhood, the authorities would require to give such complaint their consideration. The Lord Provost further informed the committee that such a complaint had actually been lodged against the company, but the gun, having been removed to another locality, the complaint fell to the ground. The owners of the property on which a site for firing the gun was first obtained, have informed the committee that the requisite permission was asked solely on behalf of the Universal Private Telegraph Company, and that no allusion whatever was made to the inhabitants of Glasgow as being in any way connected with the experiment. Information to the noise effect has been obtained from the Forth and Clyde Canal Company, who granted permission for the second site of the gun. Finally, the committee would refer to a letter from Mr. Holmes, the engineer of the Universal Private Telegraph Company, published in the *Glasgow Herald* of 31st October, 1863, in which an account is given of the proceedings connected with the time-gun experiment in Glasgow, and not the remotest allusion made to the inhabitants of Glasgow as having applied for a time-gun signal from

the Edinburgh Observatory. The committee have been unable to discover the slightest proof that Glasgow has preferred to receive its time signals from the Edinburgh Observatory rather than from the Observatory of the University of Glasgow.

3. The committee have been unable to ascertain that any of the inhabitants of Glasgow have co-operated with Professor Smyth and the Private Telegraph Company in the time-gun operations which are being conducted in this city, beyond the simple fact that Mr. Napier, the engineer, upon becoming acquainted with the experiment, kindly lent a gun for the occasion. As regards the connecting of the gun by wire with the Edinburgh Observatory, which the report attributes to the citizens of Glasgow, the operation neither originated with the inhabitants of the city, nor was it carried into effect by them. The connecting of the gun with the Edinburgh Observatory was entirely accomplished by the Universal Private Telegraph Company, who have been the active associates of Professor Smyth in all these proceedings. The committee would direct the attention of the Senatus to the significant fact that while the Electric Telegraph Company, and the Magnetic Telegraph Company are thanked in the report for their co-operation in transmitting the Edinburgh time-signals the Universal Private Telegraph Company, to whom the practical realisation of the scheme is mainly due, is passed over in silence; and in so far as Glasgow is concerned, the citizens of Glasgow, who have had nothing to do with the transaction, are substituted.

4. It has been stated that the citizens of Glasgow neither originated the time-gun scheme, nor co-operated in connecting the gun with the Edinburgh Observatory. The committee have been unable to discover the slightest trace of their having engaged in any of the subsequent operations, and are at a loss to understand what was the nature of the schedule said to be forwarded to the Edinburgh Observatory containing an account of their proceedings.

It may therefore be asserted as the result of the inquiry which the committee have instituted with a reference to the statements affecting the Observatory of this University, contained in Professor Smyth's report, that

the University of Glasgow did not in 1855 assent to the transmission of time signals from the Edinburgh Observatory to Glasgow, or take any cognisance whatever of the matter; that Glasgow did not prefer receiving the time signals from the Edinburgh Observatory to receiving them from the Observatory of this University; that it did not apply for the Edinburgh time-gun signal; that the citizens of Glasgow did not connect by wire the time gun recently introduced here with the Edinburgh Observatory, or co-operate in any of the subsequent operations connected with the firing of the gun; finally, that they have not in any way associated themselves with the project. The committee would further remark that the statements contained in the report are at variance with the letter of Mr. Holmes, the engineer of the Universal Private Telegraph Company. Mr. Holmes makes no allusion whatever to the citizens of Glasgow as having in any way participated in the operations connected with the firing of the time-gun. It is worthy to remark also that, while Mr. Holmes states that the firing of the gun had hitherto been merely experimental, the report, which is of an earlier date, asserts that Glasgow had consented permanently to receive its time signals from the Edinburgh University.

In bringing the foregoing statement to a close, the committee deem it unnecessary to enter into any discussion of the relative merits of the various modes of transmitting time signals, but, without disparaging in any way the method by firing a gun, they feel themselves warranted in stating that this method may prove imperfect and very objectionable in a crowded city like Glasgow, whereas the method of regulating clocks by electricity possesses several advantages, among which it is only necessary to mention its giving the most accurate possible indications of the time at every instant of the day, and is very low-cost, while, at the same time, should it be desired, the same apparatus may be applied to give the additional signal of dropping a ball, or firing a cannon.

The committee know that the Senatus will at once disclaim the intention of discouraging the transmission of accurate time signals from any extraneous quarter, and they feel sure that neither the people of Glasgow nor the

University can be fairly accused of any selfish feeling if they consider it to be most natural and proper that Glasgow should prefer to receive is correct time from the Observatory of its own University, even if the results had only equal advantages with those offered from a distance. They hope, therefore, that the University may look with confidence for the support of the citizens in the present effort to render their Observatory practically useful to the city and district with which it is associated.

[. . . *As well as being involved with the production of time signals, the committee also believed that it was the duty of the University Observatory to diffuse a sound knowledge of the principles of Astronomy drawn from local observations and experience.* . . .]

The committee cannot refrain from expressing their deep conviction of the beneficial consequences which have already ensued, and which they are persuaded will hereafter ensue, from the circumstance that the office of Regius Observer is this University is placed conjointly with the Professorship of Astronomy in organic connection with, and under the direct control of, the University. To this circumstance they attribute the strenuous efforts made by the University to supplement the endowment of the chair, and promote the instrumental efficiency of the Observatory. They are firmly convinced that to the same cause is due the munificent public and private aid which the University has received in furtherance of the same object. This concurrence of the Crown, the University, and the public has resulted in obtaining for the West of Scotland an astronomical observatory, with which, in regard to commodiousness and instrumental resources, no other establishment of the kind in this part of the British Isles can be compared, and which may fairly take rank in both these respects with the most complete observatories of Europe. The committee feel assured that the Senatus would not have considered themselves warranted in memorialising her Majesty's Government on the Observatory of this University if they had not been enabled to support their prayer for the endowment of an assistant, by an appeal to the actual successful efforts of the University and the public in contributing towards the same object.

WILLIAM THOMSON, LL. D., F. R. S., L. and E.,
Professor of Natural Philosophy.

ALLEN THOMSON, M. D., F. R. S., L. and E.,
Professor of Anatomy.

HUGH BLACKBURN, M. A., F. R. S. E.,
Professor of Mathematics.

ANDERSON KIRKWOOD,
Professor of Conveyancing.

[The above statement was approved by minute of the Senate of the University of Glasgow, dated 27th Nov., 1863.]

The above document clearly relates the displeasure felt regarding Professor Smyth's insensitive intrusion and the wanton interference by Edinburgh on matters that should essentially be ascribed to Glasgow. In his Report to the Edinburgh Observatory Board of Visitors Committee, Professor Smyth had been free with his grand presumptions in relation to Glasgow citizens and with the permissions he assumed. Suggesting that 'Glasgow' as some kind of authoritative body had asked Edinburgh to supply the system, and that it had been duly appreciated by its citizens, was a complete distortion of the facts. The *Glasgow Herald* columns showed strong concern regarding Edinburgh's intrusion, and there was dissent on the efficacy of the system and on the general disturbance it caused. What is of interest, though, is the record that the telegraph system had been previously tested out at the British Association Meeting held in Glasgow in 1855, with a time signal dispatched from Edinburgh, which Professor Smyth clearly thought provided evidence that permission had been given by the 'Glasgow authorities' for the system to be applied to the time-gun. It does seem a little strange that such a system had not been discussed at the time of the establishment of the Glasgow time-ball in 1857, but that this scheme of signal dispatch was resurrected only for the 1863 time-gun experiments.

Immediately, but while the experiments with guns continued, the University authorities took the initiative of laying down a wire connecting the Observatory with the Old College University Buildings. The problem required the installation of a wire that would necessarily have to pass over the roofs of the houses along the whole length of its course. On 24 December 1863, the turret clock of the College was placed under the electric control of the Standard Mean Time clock of the Observatory at Horselethill, some 3 miles away. There were some initial problems, but these were soon solved with the help of Professor William Thomson. In addition, a small seconds' clock was fitted at the entrance into the Inner Court of the College and buildings. With this feat accomplished, Professor Grant was able to submit a full proposal to the Lord Provost, who was Chairman of the Clyde Trustees, giving options as to how the time service might be presented, that is by operating the time-ball and/or the time-guns at 1 p.m., or by 24-hour dial clocks. A copy of the formal letter of 28 December 1863 forms Fig. 8.5. It was proposed that Jones's method of controlling clocks by electrical signals from a master mean time clock at the Observatory should be adopted. This method had been employed with complete success at Liverpool, where it had been introduced a few years earlier by John Hartnup, Director of the Liverpool Observatory. A committee of the Town Council was appointed to enquire into, and report upon, the proposal submitted to them.

Reference to this letter, and to the confirmation that the College clock was displaying time in direct response to signals from the Observatory, appeared in the *Glasgow Herald* of 5 January 1864. It reads as follows:

CONTROLLING OF THE PUBLIC CLOCKS BY AN ELECTRIC CURRENT FROM THE OBSERVATORY OF GLASGOW UNIVERSITY.

We understand that the University of Glasgow has submitted to the Town Councils of Glasgow, Paisley, Port Glasgow, and Greenock, a plan for the controlling of all the public clocks in the four towns, by means of a current of electricity directed from the standard mean time clock of the Observatory of the University, in accordance with the ingenious invention of Mr. Jones of Chester, which has been for several years in practical operation at Liverpool, where it has been attended with the most complete success. If this proposal be adopted, the result will be that the pendulums of all clocks under control will vibrate in perfect unison with the pendulum of the Observatory clock, and the first blow of the hammer for the successive hours will indicate Greenwich meantime, and will, in the case of every clock, occur at the incident when the seconds' hand of the Observatory clock points to sixty. It may be interesting to the public to know that this beautiful result has been already realised in the most satisfactory manner in the case of the University turret clock, which was connected by an electric current with the Observatory clock about ten days ago. This will be apparent from the following letter addressed to Professor

OBSERVATORY,

GLASGOW, *28th December*, 1863.

MY LORD,

THE Observatory of the Glasgow University being now connected with the College and with the City of Glasgow by an electric wire, the occasion seems proper for suggesting to the authorities of the Clyde Trust, the expediency of co-operating with the University in establishing upon a satisfactory and permanent footing, a system of arrangements for the transmission of correct time from this Observatory for the use of the Shipping in the Clyde.

If the Trustees should decide upon availing themselves of the resources of this Establishment, in connexion with the object mentioned, it will afford much pleasure to the authorities of the University, and to myself, individually, as Director of the Observatory, to assist in promoting the practical realization of their views. Without attempting to pronounce definitively upon the plan of operations which, in that case, it might be ultimately found most convenient to adopt, I would propose, in the meantime, for the consideration of the Trustees, the following suggestions:—

1. The erection of a Turret Clock, with large dials, on some commanding position of the Broomielaw, the said Clock to be furnished with a Jones' magnetic-electric pendulum, and to be controlled by an electric current directed from the Standard Clock of the Observatory.

2. The erection of a small Seconds' Clock, similarly controlled from the Observatory, but supplied with a Seconds' hand, by which the correct time may be ascertained at any instant of the day, and placed in a situation where it may be conveniently consulted by the Shipping public.

3. The dropping of the Time-Ball on the Sailors' Home by a mechanism acted upon electrically from the standard Mean-Time Clock of the Observatory.

4. The firing of a Gun from some central position on the Broomielaw.

5. The establishment of an Office for the rating of Chronometers, to be placed under the control of the Clyde Trust, and to be supplied with special facilities from the Observatory for ascertaining the correct time.

For carrying out satisfactorily these various suggestions, the Observatory of the Glasgow University is equipped in the most complete manner with instrumental appliances, while at the same time the Director of the Establishment is aided by an effective observing force.

I have to add in conclusion that the University of Glasgow is bound by an express engagement entered into with Her Majesty's Government, to afford all necessary facilities for the transmission of correct time from this Observatory, for the use of the Shipping of the Clyde. The University is prepared to carry into effect the terms of this engagement, and it relies on the friendly co-operation of your Lordship and the Clyde Trustees, in its efforts to discharge an important public duty which it cannot, with honor, consent to forego.

I have the honor to remain,

MY LORD,

Your obedient Servant,

R. GRANT,
Professor of Astronomy in the University of Glasgow.

The HONORABLE the LORD PROVOST of GLASGOW,
~~Chairman of the Clyde Navigation.~~

Figure 8.5 Following the establishment of the control of the University clock from the Observatory 3 miles away, Professor Grant set out his proposals on 28 December 1863 to the Lord Provost of Glasgow.

Grant by Mr. Macfarlane, the senior assistant to Professor William Thomson, who very obligingly undertook to compare daily the University clock with the indications of the Observatory clock. In order to appreciate the exact force of Mr Macfarlane's observations it may be well to mention that at every second of the minute, except the thirtieth second, a current or pulse of electric fluid is transmitted from the Observatory clock to the University clock. This effect is indicated at the University clock by a corresponding deflection of the needle of a galvanometer placed in the circuit, occurring at every second except the thirtieth, when, there being no current, the needle will for an instant stand still in a vertical position. Now, if the two clocks beat simultaneously, this position of the needle, indicative of the absence of a current, ought to occur invariably at the instant when the seconds'

hand of the University Clock (which, we may remark, is not seen from the outside of the College buildings) points to thirty. How far this result has been realised will be seen from Mr Macfarlane's letter.

College, Glasgow, Jan 4, 1864.

"Prof. Grant,

"Sir, – I have regularly observed the working of the College clock several times a day from Saturday the 26th of December last till today, with the exception of Friday and Saturday last when my place was supplied by Mr. Tatlock, and we have invariably found that the *no current* beat of the Observatory clock to coincide exactly with the 30th second of each minute of the College clock. – I remain, Sir, your obed. servt.

DONALD MACFARLANE.

A few days later, on 13 January 1864, an editorial in the *Glasgow Herald* reported the following:

Glasgow Herald
13th January, 1864

For the last two or three months the curiosity of the inhabitants of our good City has been kept alive by the untiring efforts of our friends of the Edinburgh University on the one hand, and of the Manager of the Universal Private Telegraph Company on the other, to supply the City and Port of Glasgow with correct time. It must be admitted that during all this time they have been complete masters of the situation. The means of supplying the City and Port of Glasgow with correct time had hitherto been confessedly imperfect, and were altogether inconsistent with the advanced state of science in the present day. No one had insisted more strongly or more perseveringly on this point than the Professor of Astronomy in the Glasgow University, who, it is well known, was actively negotiating with the Town Council in reference to a project of connecting electrically the Observatory with the City of Glasgow, with a view to the controlling of the public clocks by a current of electricity directed from the standard meantime clock of the Observatory, and for other collateral purposes. But all this was still in abeyance. The

scheme necessitated in the outset a pecuniary expenditure of some amount; the University had no funds for such an object, and corporate bodies, including—it must be feared—our own Town Council, are proverbially slow to move in the direction of giving an order upon their Treasurer. It may possibly, therefore, have been imagined by the gentlemen to whom we have first referred, that, by a timely exercise of energy, the Observatory of the Glasgow University would be fairly checkmated, and would be compelled to accept, as a *fait accompli*, the efforts of its rivals, if we may so term them, before it extricate itself from the situation of isolation in which the force of circumstances had hitherto confined it. If this belief was really entertained by the projectors of the new scheme, they must have been convinced by this time that they were labouring under a delusion. While the gentlemen of the Edinburgh scheme have been announcing themselves to us by the daily crashing of their guns, the Observatory of the Glasgow University has been quietly but resolutely making his preparations; and the day has at length arrived when both parties are placed upon the same footing. We confess this is exactly the position which we wished to see established; for any contest, be it what it may, loses its true gusto to the onlooker, if he finds that one of the parties in the strife is fighting with an undue advantage over his opponent.

We had occasion recently to announce that the Observatory of the University is now connected electrically with the City of Glasgow, and that the Turret Clock of the University has been definitively placed under the control of the standard mean time clock of the Observatory. This, we understand, is but the commencement of an extensive system of operations of a similar nature which is under contemplation, and which is destined to place the City of Glasgow, and the neighbouring towns, on a footing unsurpassed anywhere, in so far as the advantages of correct time are concerned.

We have, then, submitted to us two projects for supplying the City and Port of Glasgow with correct time—the one emanating from our ancient University, and having the source of operations included within the extreme boundaries of the City; the other offered from a distance, under the joint management of Professor Piazzi Smyth, of the Edinburgh Observatory, and of Mr. Nathaniel Holmes, of the Private Telegraph Company. Dismissing all extraneous considerations, it may not be an unprofitable task to cast a glance at the relative advantages of the two schemes. The Observatory of our native University proposes to introduce perfect time by the controlling of clocks, whether of great or small dimensions, the dropping of a ball, or the firing of a gun, or by a combination of two or all of these different methods. The circumstance of the Observatory forming, so to speak, part of the City, and its comparative proximity to the other towns upon which it is contemplated to operate, render the application of any of these methods perfectly practicable. In this case, then, the public have the means of selecting the method or combination of methods which may appear, according to circumstances, to be most desirable, while at the same time they may be said to have it completely under their own control, so closely is this City identified by a thousand friendly associations with the venerable University of the West of Scotland.

The projectors of the Edinburgh scheme propose to supply us with correct time by the daily firing of a gun or guns in connection with the Edinburgh Observatory. The distance of the source of operations from Glasgow precludes the idea of including the control of clocks in their programme. They offer us, therefore, merely one of the methods which our own University offers, with the disadvantage of the current being transmitted from a distant station, along the wire of a public Company, the remuneration of whose services must be chargeable upon Glasgow. It is further to be borne in mind that casualties arising from storms to a wire fifty miles in length, and placed wholly beyond our own control, are of a much more serious nature than those which may affect a wire belonging to ourselves of only three miles in length. We conclude then, that we can do much better in applying ourselves with correct time by means of our own resources at home, than by introducing the desideratum from an observing station fifty miles distant. The home project gives us the choice of several modes of indicating the time, instead of confining us to one. It is more economical, less liable in its practical operations to casualties from storms, and would be, to all intents and purposes, under our own control. Surely, under such circumstances, our friends in the East must excuse us if we decline to hesitate for a moment in making our selection with respect to the two projects before us. We have been hitherto in the habit of relying upon our own resources and our own institutions in supplying this City with every object of importance which may be conducive to the health, the comfort, and the general interests of the public. We see no sound reason why we should deviate in this instance from the course which we have heretofore uniformly pursued.

Instead in abiding by this decision we should be only adopting a line of conduct which a wise judgment and a proper feeling of self-respect have prescribed under similar circumstances in other places. When, two or three years ago, the citizens of Edinburgh, by a spirited movement, got up a subscription to provide themselves with a time-gun, they naturally connected their enterprise with the Observatory on the Calton Hill; and they would doubtless have felt very indignant if there had been submitted to them a project for effecting the same object in which their own Observatory was ignored, and they were offered the firing of the gun from some observing station fifty or a hundred miles off. We hold that upon such an occasion as this it is a duty incumbent upon the inhabitants of our City, and upon the ruling authorities more especially, to take matters into their own hands, and not to subject us to the inconvenience of being indebted to strangers for an improvement which we can work out much more effectively and more economically by ourselves. About two months ago there was a good deal said respecting the rapid progress of the time-gun system at Newcastle, Sunderland, and North Shields, under the joint management of Professor Smyth and Mr. Holmes. Recently, however, as we are informed the tie which connected each of these three "affiliated cities" with their elder sister had been severed. Newcastle, while welcoming any useful suggestions from without, has reserved to itself the rights of managing its own affairs in the way in that may seem most conducive to its own interests, and neither of the gentlemen just mentioned have now any connection with the gun which is fired daily in that town. We would desire to speak with the utmost respect both to Professor Smyth and Mr. Holmes—and indeed, we thank them for the

interest which their operations have excited in this quarter—but we cannot help thinking that if Newcastle, which has no Astronomical Observatory of his own, has been enabled to provide itself with correct time independently of either of these gentlemen, surely so can Glasgow, which is so much more favourably circumstanced in respect to astronomical advantages.

It would be noted that this important subject engaged the attention of the Town Council at its last meeting when a memorial, signed by the most eminent men in the City, was presented, respectively requesting that the Council would "take such steps as may appear most expedient for co-operating with the University in tendering the admirable instrumental appliances, and the resources in general, of the Observatory practically available to the public," for the purpose of conveying correct time to the City with which it is so closely connected. The memorial was very favourably received, and we cannot doubt that, under the patronage of the Corporation, and the other public bodies, the very moderate amount of funds necessary will be readily provided, and all the City clocks will soon be controlled, and a time-gun, or guns, fired by the instrumentality of our own Observatory.

At the meeting of the Clyde Trust on 2 February 1864, the decision was made in favour of Professor Grant's proposals, and the intervention of the Edinburgh consortium was turned down. As it turned out, the availability of 24-hour clocks was so successful that the thought of using signals for the time-ball and time-guns was abandoned. The announcement of the decision was made a day later and appeared in the *Glasgow Herald* of 3 February as follows:

CLYDE TRUST

The usual meeting of the Board was held yesterday – The Lord Provost presiding.

TIME SIGNALS

The Sub-Committee of the Harbour reported that in accordance with the remit from the Committee of Management, they had considered a communication from Professor Grant regarding certain proposed arrangements for the transmission by telegraph of correct time for the use of the shipping of the Clyde and a communication from Mr. Holmes in relation to time-gun signals. The committee were of the opinion that the Trustees should not entertain Mr. Holmes' proposal as to time-gun signals; and, with regards to Professor Grant's comment as to the transmission of correct time by telegraph, they had remitted to the Chairman and Messers Allan Gilmont and W. Allan to consider the proposal and report.

In the same issue of the *Glasgow Herald* the engineer, Mr Holmes, included the following announcement:

DISCONTINUANCE OF THE TIME-GUNS
To the Editor of the Glasgow Herald

SIR, – I desire through your columns to inform those interested in the establishment of correct time signals for Glasgow, Greenock, and the surrounding parts, that the four time-guns hitherto fired daily at 1 P.M., Greenwich mean time, will cease firing on Saturday the 6th instant. The experiment I had the honour of introducing to the city has proved successful; and if it is desired to have guns – having laid the matter before the several authorities – the guns can be resumed as soon as the necessary pecuniary arrangements can be made. Leaving, therefore, the question of guns or no guns in the hands of the public, I am, yours obediently

NATH. J. HOLMES.

Thus, the five-month battle and bitter dispute was over, largely the result of the determined perseverance of one man, Professor Robert Grant; his scheme won the day to provide accurate time to the city and along the Clyde, although it must be admitted that he had practicality on his side. In no way could the once-a-day dropping of a time-ball or the firing of a one o'clock gun compete with the striking of the hammer of a clock each hour, or the continuous display of the movement of a seconds' finger. The telegraphically controlled slave clocks gave time continuously over 24 hours, not just singularly at 1 p.m. The guns became silent, and although there is no reference to the demise of the Broomielaw time-ball, its operations would also seem to have been abandoned in the early part of 1864.

It is obvious from the articles and letters that appeared in the *Glasgow Herald* in the autumn of 1863 that many feathers were ruffled by the testing of the system and the firing of a gun. In the first place, Professor Grant's nose must have been put out of joint by the Astronomer Royal for Scotland, Charles Piazzi Smyth, who seemingly went behind his back by telegraphing a 'one o'clock signal' from Edinburgh to operate the firing of a gun in Glasgow, particularly as he had been in discussion with the local authorities for several months about establishing a superior system of seconds' clocks, so that GMT would be available continuously rather than just once every 24 hours. Professor Grant was also guarding the responsibilities of the Observatory to the Crown. As presented earlier, spurious letters were also submitted by the public, triggering a response from Mr Church, an employee of McGregor & Co. Although a working relationship existed between the Observatory and the McGregors, this letter hinted at a lack of confidence in the standards of time-keeping at the Observatory, as reference was made to a mistake made by one of the time-couriers from Horselethill; or was it that the company, which had invested in a transit telescope, could see the loss of a contract and of revenue from the Clyde Trustees?

Grant considered that all the extra work in relation to the transmission of time to the city was inconsiderable compared with the value of the attained object. Although the actual operations for the transmission of time proved to be anything but onerous, the negotiations and arrangements for effecting the systems had been very time-consuming. Grant commented that it did not form part of the duties of an astronomical observatory to prescribe to the public the particular mode of indicating correct time that it might be most desirable to adopt. It was for the observatory simply to supply the time. Accordingly, in conducting the negotiations for the transmission of correct time from his Observatory to the city, Grant felt that he was only carrying out the wishes of the University, leaving it entirely to the public to decide for themselves how to apply the generated signals to present 'time' in the most useful way.

In his 1865 Report to the University Senate,[17] Grant reiterates:

> It is of importance to bear in mind that the office of Astronomical Observer in the University is not a mere accidental appendage to the Chair of Astronomy. On the contrary, it was instituted by the Crown as an Office co-ordinate with, and distinct from, the Professorship, and on this ground is equally entitled to the watchful consideration of the authorities under whose control it is placed. I may be permitted, therefore, to express a hope that the University will not relax its efforts to secure for the Observatory such a measure of public support as is consistent with its excellent instrumental equipment, its general activity, and its well founded claims as one of the most ancient University Observatories in the British Isles.

His persistence in persuading the city authorities and in negotiating support from the telegraph companies is clearly apparent in all of his later Reports to the Senate, as well as in other documents and papers.

8.3.3 The Distribution of Regular Time

In February 1864, a seconds' clock belonging to the University was placed in a window of Mr Muirhead's establishment in Buchanan Street, in order that the public might have the opportunity to satisfy themselves on the accuracy with which true time might be transmitted from the Observatory. Applications were received from several quarters to have further seconds' clocks established. Grant noted that on completion of the operations in progress, there would be ten clocks in the city of Glasgow under the Observatory's control. Proposals related to telegraph links were also submitted to the Clyde Trustees and to the authorities of Greenock, and a desire for cooperation with the University from these parties was recorded.

Professor Grant's Report of 1864[18] noted the progress made with the authorities of the Exchange, who had decided to run three clocks at their buildings, including the turret clock. By the beginning of March 1864, momentum had grown for the formal acceptance of the electrical signals and a note in the *Glasgow Herald* of 5 March 1864 declares this as follows:

TRANSMISSION OF TIME FROM GLASGOW OBSERVATORY. ELECTRIC OPERATIONS.

We understand that the directors of the Exchange, at a meeting held on Tuesday last, decided upon taking immediate steps for introducing into the Exchange the system of controlling clocks by electricity, which has been recently established in this city, in connection with the Observatory of the Glasgow University. The wire extending from the Observatory to the College being already attached to the building of the Exchange, the clocks of the latter may be connected with the system at a trifling expense. It is proposed to place under control the great public clock and the clock in the interior, which is above the entrance into the building. A conspicuous clock, in place of the present small clock, showing the time to seconds, will also be fitted up in the great room, and will be similarly placed under control. This clock will furnish an interesting test of the exactitude of the method of control, inasmuch as the jump of the second hand at the sixtieth second, corresponding to each successive hour, must invariably coincided with the first blow of the hammer of the bell of the turret clock. This result has been verified in the most satisfactory manner at the College, were a controlled seconds clock in Professor William Thomson's class-room may be seen to beat in perfect unison with the great clock in the tower of the building. At the thirtieth second of each minute a no-current signal is transmitted from the Observatory clock, which is indicated by the needle of a galvanometer placed in the circuit standing still for an instant. This circumstance furnishes a ready means of testing the coincidence of the controlled clock with the standard clock, for if a galvanometer be placed in the circuit beside the controlled clock, the needle—which alternately oscillates from side to side—will be observed to come to a momentary stand still at the thirtieth second of each minute, provided the control be perfect. We understand that the preliminary operations in connection with the clock at St George's Church are nearly completed, and that it will be forthwith placed under control. In this case the system of control will be applied in a somewhat different manner, inasmuch as an arrangement is being made by which the electricity will act upon a two seconds pendulum, in accordance with the method devised by Professor Wm. Thomson. Hitherto, it has been considered desirable, when a turret clock has been placed under control, to replace the two seconds pendulum by a seconds pendulum, beating in unison with the pendulum of the controlling clock. Several obvious advantages will be gained if the control of a turret clock can be affected, while at the same time the two seconds pendulum is retained. No doubt is entertained of the perfect practicability of the method proposed in the present instance. We trust ere long to witness the rapid extension of this system of indicating correct time, in which the truths of astronomy, electricity, and horology are so beautifully blended together. We understand that applications have already been made to Professor Grant from several establishments in the city with a view to its adoption upon their own premises. The extra work required at the Observatory in connection with the transmission of time to the city, is of an extremely simple character. It does not, in fact, occupy more than an hour and a half daily of the time of an assistant; and it deserves to be borne in mind that the same expenditure of time will suffice for the controlling of an indefinite number of clocks, and, we may add, for the firing of any number of guns which the public may desire to place in connection with the Observatory. The University of Glasgow has freely placed the resources of the Observatory at the disposal of the city, with a view to the establishment of a complete system of correct time indications; and we have no doubt that not only the various corporate bodies, but also the inhabitants generally, will reciprocate this act of courtesy by cordially supporting the University in its present exertions. We want no officious meddling from without. We feel particularly competent to perform our own work, and we are confident that, in the present instance, it will be accomplished in a manner consistent with the dignity and importance of our ancient city.

The conclusion of the above article once again emphasises Glasgow's independence in supplying its own time system without the interference of outside parties. At about the same time, connections to the St George's Church Clock, made by Thwates of London in 1809 (see Plate 14), had been completed, and preliminary controlling experiments were underway. The 1865 Report to the Senate,[19] reveals that, including this church clock, there were then ten clocks in the city of Glasgow established with an electrical connection to the Observatory and maintained effectively at an average distance of 3 miles from the master source.

In a celebration of Professor Grant's success, the editor of the *Glasgow Herald* encouraged him to write three articles on how time was determined at the Glasgow Observatory; these appeared in the editions of 14 March, 19 March and 26 March 1864. The descriptions were substantive, with tabulated details of the measurements of some stars. Grant also explained how observations of stellar transits were made to an accuracy of one-tenth of a second, and how local sidereal time was converted to GMT, according to the longitude of the Glasgow Observatory. He mentioned the especially good weather of February 1864 when 329 stellar transit recordings were made.

In his 1864 Report to the Senate,[20] Professor Grant noted that the Bryson sidereal clock had been removed to the new building to be used with the recently acquired Ochtertyre Telescope, and that the new sidereal clock by Frodsham was employed for observations with the transit circle. The Muirhead mean time clock was used as the standard clock, maintained free of error, for transmitting the electrical signals from the Observatory to the city. This latter clock was used for this purpose for many years to come. It was a handsome piece by James Muirhead of 90 Buchanan Street, Glasgow, with a mercury compensated pendulum, supported in an oak case with castellated top. It is depicted in Plate 15 as it stands within the University, in the Melville Room, which was named after the sixteenth-century Principal who reformed the syllabus, emphasising the requirements of astronomy as a standard subject (see Chapter 1, p. 16).

Around this time, while attending a meeting of the Royal Astronomical Society, Professor Grant described[21] the success of distributing time across Glasgow. He considered the most suitable displays for a large city to be clocks with a seconds' pendulum and a dial around 3 feet in diameter, showing the time to hours, minutes and seconds. Clocks of this construction had been set up in the public thoroughfares of Glasgow and found to be exceedingly useful. The signalling was effected by transmitting a pulse every second, with a break once every minute. Grant reported that there were eleven clocks of various forms in Glasgow under control of the master clock of the Observatory. This number was due to be increased to seventeen or eighteen, with a system gradually extending all over Glasgow. He commented that the scheme operated equally well with turret clocks using a heavy pendulum that oscillated once every two seconds or a more standard form of the one-second pendulum.

The 1866 Report[22] described the continuing success of the transmission of time signals to the city and the port of Glasgow. The project for providing time to the river had received support from the Clyde Trustees and from several of the most eminent engineering and shipbuilding establishments along the route of the wire, and the operations for bringing it into effect were in a satisfactory state of progress. At this stage, the wire, nearly 7 miles in length, had been laid down, and several clocks had already been placed under electric control along the river.

Arrangements were made under the auspices of the Astronomer Royal, Sir George Airy, for a scheme whereby the transits of groups of stars were timed first at Greenwich, and then minutes later in Glasgow, with the pairs of meridian passage being recorded simultaneously at both stations by means of electrical impulses. The chief engineer of the Electrical and International Telegraph Company, C. F. Varley, arranged the use of the cable between the two cities. For a four-week period in 1865, set aside for the collaboration, there were four nights (1, 2, 22 and 25 May) when the skies were simultaneously favourable at the two sites. Mr Plummer was the observer in Glasgow. As a result of this study, it was concluded that the transit circle at the Glasgow Observatory was located at a longitude 17′ 10″.55 W of Greenwich, about 1″.25 less than previously thought, 'moving' the city by about 300 metres relative to its then mapped position. By using this ingenious method of making observations at the two locations and sending the signals of the star transits both ways, the effects of line time delays could be removed. A paper[23] describing the experimental details and success of the exercise was later published.

In his 1867 Report to the Senate,[24] Grant listed the various clocks within the Observatory which formed so important a part of the instrumental equipment. He noted that the transit clock continued to give much satisfaction. The same was said with regard to the Muirhead mean time clock, used as the 'normal clock' for transmitting the time to the city and port of Glasgow. The Muirhead had been fitted with output terminals (see Plate 15) for providing electrical pulses to the telegraph system. The sidereal clock by Frodsham, which at the time of the previous Report was in the hands of its maker, had been recently returned. The sidereal clock employed in connection with the Ochtertyre Equatorial, and which was electrically connected to the transit clock, continued under perfect control. A clock by Shelton with a gridiron pendulum was used for educational purposes in connection with a small transit instrument bought from Troughton and Simms in 1863. This latter piece may be the one that remains in the possession of the current University

Observatory (see Fig. 12.9), and used by students up until the 1980s. The aforementioned Shelton clock must then have been of considerable age, as the heyday of this manufacturer was in the mid-sixteenth century. According to Gavine,[25] a Shelton clock was ordered in 1754 by Robert Dick, who was at that time the Professor of Natural Philosophy. It is referenced as being stored in the upper dome room in 1863,[26] cleaned and altered by Ritchie in 1894 and located in the transit room in 1906.[27] The clock maker is famed for providing five astronomical regulators for the Royal Society in connection with the timing of the Transits of Venus in 1761 and 1769, the latter event forming a basis for Captain Cook's expedition to Tahiti.

It is likely that, at the time of the move from the High Street to the new site at Gilmorehill, Professor Grant took the seconds' clock from the College yard and re-sited it in the West Quadrangle. According to his 1871 Report to the Senate,[28] a branch wire was laid in March 1871 to the new University buildings, and the clock was reconnected to its master at Horselethill. An additional clock was also installed in the circuit of the Clyde wire. The responsibility for maintaining the batteries of the clock system was transferred from the British and Irish Magnetic Telegraph Company to the Authorities of Her Majesty's Post Office, and this was also noted in the 1871 Annual Report.[29] From the early initiation of the time service for the Clyde and the city, Professor Grant went to considerable efforts to chivvy the Town Council and Clyde Trustees into effecting their part in the cooperation. Some thirteen years after the first telegraphic signal was transmitted from the Observatory, Professor Grant still found it necessary to remind[30] the authorities of the importance of providing this important service to commerce. Even while his work on the production of the Glasgow Star Catalogue must have occupied most of his time, dedication to maintaining the city's time service[31] was very important to him. The seconds' clock from the Old College remains on view, although inoperative, in the south-east corner of the West Quadrangle, as shown in Fig. 8.6. It seems ironic that the Muirhead clock that controlled it from the Horselethill Observatory, first at a distance of 3 miles across the city, and then from 1 mile from across Byres Road, is now found just a few metres away, one floor above. Also the mechanism (see Plate 14) of the turret clock, the first to be controlled from the Observatory, is not much further away at the entrance doors of the Bute Hall.

Returning to the Report[32] of 1867, it was recorded that operations for transmission of true mean time from the Observatory to the city and port of Glasgow continued to expand. Professor Grant noted that on 5 February a public seconds' clock, which had been erected by the Clyde Trustees for their workshops at Govan, was definitively placed under electric control. This clock was inspected regularly by one of the assistants from the Observatory, and was invariably found to be perfectly coincident with the normal mean time clock at Horselethill. As it fronted the river, and the dial was distinctly visible from on board the multitude of vessels passing up and down, Grant

Figure 8.6 The Seconds' Clock, formerly at the Old College, remains in the south-east corner of the West Quadrangle of the University Building at Gilmorehill. The mechanism, requiring regular winding, with the hollow-bob pendulum controlled by an electromagnet and beating to the seconds dispatched from the Muirhead Clock, is in a cupboard behind the clock face, inside the Quadrangle building.

commented that it could not fail to be a great public utility as the standard exhibitor of GMT. It was also reported that four clocks attached to the ship-building establishments on the river had been recently placed under electric control by connecting them to the Clyde wire. Three clocks besides, which were intended to be connected electrically to the Observatory, were in the hands of the clock maker, undergoing the preliminary process of rating.

At the time of Grant's death, the public clocks were still under control, the upkeep being part of the daily duties of the Observatory. Professor Becker continued observations with the transit telescope, and the University Observatory provided the source of standard time to Glasgow for over fifty years.

8.4 DUNCAN MCGREGOR & CO.

As it turns out, the site of the McGregor business at 38 Clyde Place is the only building of the nineteenth century in current occupation along that part of the waterfront. All the others have been demolished or are in a poor state of repair. A photograph of the building is presented in Plate 16 and matches up with the detail in the upper part of Fig. 8.2. There are no signs of there ever having been an observatory there. In the same picture, a copy of

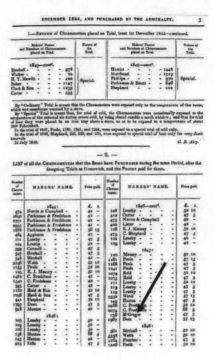

Fig 8.7 The Record of chronometers, submitted for trial at Greenwich from 1 December 1844, lists one by D. McGregor & Co., subsequently purchased by the Admiralty for £47 5s.

the McGregor company emblem is shown. The motif of an ensign holding a sextant was a common feature of the time and was used by many marine instrument makers.

It is gratifying to note that in addition to the local business operated by the McGregor company, Glasgow also featured prominently in the trial[33] of chronometers at the Royal Observatory, Greenwich in the early 1840s. A marine chronometer sent by Messrs D. McGregor & Co. for trial was so highly approved of that the Lords of the Admiralty offered the sum of £47 5s for it, which was accepted (see Fig. 8.7). This was noted in Brown's article[34] presented earlier in this chapter.

With this endorsement, many such chronometers were sold, including one, No. 3529, which now finds itself in Australia in the Museum Victoria, within the Melbourne Observatory Collection. It is described as a 'marine chronometer by D. M. McGregor & Co., Glasgow and Greenock, Scotland', and was once used at Melbourne Observatory for recordings involving the noting of time. It was still in use around the period 1945 to 1949, as it has notes inside its lid that indicate the gaining rate of the piece (see Fig. 8.8).

Within the correspondence in the *Glasgow Herald* related to the time-gun, McGregor & Co. received a recommendation from a very satisfied customer

Figure 8.8 A chronometer by D. McGregor & Co. of Glasgow, formerly used at Melbourne Observatory, is on display at the Melbourne Museum. The records within the lid show that it was gaining about three seconds each day.

as a result of a repair to his pocket chronometer which he had unfortunately dropped. Mr James Smith wrote of his approbation on 2 November 1863, and this appeared in the paper two days later:

THE TIME-GUN

To the Editor of the Glasgow Herald

SIR, In consequence of the announcement in the Herald of Friday last, that a gun would be fired in St Vincent Place, on the same day at one o'clock, I was waiting, having previously compared my pocket chronometer with the Exchange room, and immediately after with that in Mr Muirhead's shop. The following were the results. My watch (accidently of course) agreed with the time indicated by the gun fire to a second, but the Exchange Clock differed from it by three seconds, and Mr. Muirhead's by five. Such minute deviations from true time, though of little consequence for the common purposes of life, are of essential importance when the object of observation is to obtain the rates of marine chronometers, and, I think, must satisfy any person accustomed to consider the subject how essential it is that in a great commercial port like Glasgow the instruments of regulating time should be in immediate connection with an established observatory I have such perfect confidence in the extreme accuracy of my learned friends, Professors Smyth and Grant, that I shall be well contented to have the time regulated by either still better by both; but what ever is done, I hope there will be a time-gun in the harbour, for it appears to me that by on other

Figure 8.9 An invoice for winding the Harbour Clock at Greenock, over the period of a year from September 1915, shows that Glendinning & Co. had taken over the operations of McGregor & Co.

means can the same degree of precision be acquired in the comparison of time keepers. When I have said that the perfect agreement between my chronometer and the gun fire was accidental, I by no means wish to under value its performance. About three years ago it met with such injury from a fall as to have been rendered useless I put it in the hands of Messrs MacGregor, who I understand, regulated the time-ball at the harbour. Since it received the necessary repairs by the chronometer maker, corrected with their house, I can say with truth that I have never seen a pocket chronometer which kept time with greater accuracy. I am &c.

JAMES SMITH

Jordanhill 2d November

As already mentioned, there appear to be no records of the exact date when the Broomielaw time-ball ceased to operate. What is known, however, is that by 1916, McGregor's business was either on the decline or already defunct. According to a bill written on the headed notepaper of the Greenock branch, the operations were in the hands of Glendinning & Co. (see Fig. 8.9).

8.5 OTHER ARCHITECTURAL ARTEFACTS

An inspection of the west elevation of the City Chambers in George Square clearly shows the presence of an astronomer. The building dates from

Figure 8.10 The entrance to the City Chambers carries figures representing various merchants, and includes an astronomer with his telescope and a surveyor with a globe.

1883–8, and the figures in the frieze over the main entrance are by George Anderson Lawson, with the arrangement loosely based on the Arch of Constantine in Rome.[35] The subjects looking out to the west celebrate 'the peaceful victories of art, science and commerce'. The man-made artefacts relating to sculpture and architecture are to the left; the technical aspects of human activity such as astronomy and chemistry are on the right. A surveyor with his globe and an astronomer behind him with a telescope can be clearly seen in Fig. 8.10.

Of more recent interest is the appearance of an abstract illuminated wall sculpture dating from 1993. It was designed by Michelle de Bruin and Callum Sinclair, and is mounted high on the south-east gable of a modern tenement building at the junction of Byres Road and University Avenue. The piece was commissioned by Hillhead (now Glasgow West) Housing Association, and it remains in their ownership. The materials used were slate, black granite, onyx, glass and halogen bulbs.

According to McKenzie,[36] its dimensions are approximately 1.2m × 2.4m, and it comprises a central disc of black granite with arms radiating at irregular intervals, suggesting a solar eclipse, and the phases of the Moon and a variety of Celtic knot designs inlaid in slate in the background masonry. As can be seen in Fig. 8.11, there are six arms, which may represent the five ancient planets that can be seen by the unaided eye, with the

Figure 8.11   The apex of the modern tenement building at the junction of Byres Road and University Avenue commemorates a total solar eclipse.

addition of Earth. They are designed to be illuminated under the control of a light sensor. Rather than representing lunar phases, the cutaway circles may relate to the partially eclipsed Sun, seen on either side of the total eclipse.

References

1. Deas, James (1876), *The River Clyde: An Historical Description of the Rise and Progress of the Harbour of Glasgow, and of the Improvement of the River from Glasgow to Port Glasgow*, Glasgow: James Maclehose [GUL DX660 DEAS; GUL Sp Coll Mu25-d.17; Deas, James (1893), *Clyde Navigation: Description of 130-ton Crane Seat and Steam Crane on Finnieston Quay, Glasgow Harbour Officially Tested 3rd May 1893. – Pamphlets of Speeches given at the Institution of Civil Engineers* [GUL Sp Coll Kelvin 160].
2. Glasgow Mitchell Library: Letter from Alexander Mitchell to George Knight Esq. [T–CN 12/516].
3. Kinns, Roger (2010), 'Time Balls, Time Guns and Glasgow's Quest for a Reliable Local Time Service', *Journal of Astronomical History and Heritage*, 13, 194–206.
4. Glasgow Mitchell Library: Invoices from D. McGregor & Co. [T–CN 12/535].
5. Nichol, J. P. (1859), Letter to Sir G. B. Airy, dated 26 April 1859 [RGO 6/615, Leaf 75].

6. Brown, James (1862), *Sketches on the Clyde. No. I. Gourock. No. II. West Kilbride. No. III. Hamilton. And Guide to the Watering Places and Chief Resources of the Clyde*, Glasgow: n.p. [GUL Sp Coll BG57-e.2].

7. Nichol, Letter to Sir G. B. Airy.

8. The Virtual Mitchell Library. Available at: http://www.mitchelllibrary.org/ See images C2590 and C1635.

9. The One o'Clock Gun & Time Ball Association. Available at: http:// www.1oclockgun.org.uk/glasgow_tb.html

10. Brown, *Sketches on the Clyde*.

11. (1858), 'The Glasgow Harbour Time Ball', *The Practical Mechanic's Journal*, III (second series), London: Longman, Brown, Green, Longman and Roberts, 179–80. Available at: http://archive.org/details/practicalme2318581859glas

12. Clarke, D. and Kinns, R. (2012), 'Some New Insights into the History of the Glasgow Time Ball and Time Guns', *Journal of Astronomical History and Hertiage*, 15, 59–67.

13. Brown, *Sketches on the Clyde*.

14. 'Report of The Professor of Practical Astronomy in The University of Glasgow, Addressed to the Committee, and read at the Visitation of the Committee, on the 17th of April 1861' [GUL Res Ann I15-d.5].

15. 'Report of The Professor of Astronomy in The University of Glasgow, Addressed to the Committee of the Senate Appointed to Visit the Observatory, April 28, 1864' [GUL Res Ann I15-d.5].

16. 'MEMORIAL on Astronomical Observations in general, with especial reference to the Observatory of the University of Glasgow, addressed to the Commissioners appointed by Her Majesty to enquire into the condition of the Universities of Scotland, by Robert Grant, MA, FRAS, Regius Professor of Practical Astronomy and Observer in the University of Glasgow' (15 April 1862) [GUL Res Ann H25-y.11].

17. 'Report of The Professor of Astronomy in The University of Glasgow, Addressed to the Committee of Senate Appointed to visit the Observatory and Read at the Visitation of the Committee on the 24th of April, 1865' [GUL Res Ann I15-d.5].

18. 'Report of The Professor of Astronomy, April 28, 1864'.

19. 'Report of The Professor of Astronomy, 24th of April, 1865'.

20. 'Report of The Professor of Astronomy, April 28, 1864'.

21. Grant, Professor (1865), 'The Time-signalling Operations at Glasgow', *Monthly Notices of the Royal Astronomical Society* [hereafter MNRAS], 26, 66–7.

22. 'Report of The Professor of Astronomy in The University of Glasgow, Addressed to the Committee of Senate Appointed to visit the Observatory and Read at the Visitation of the Committee on the 29th of March, 1866' [GUL Res Ann I15-d.5].

23. Grant, Professor (1865), 'On the Determination of the Difference of Longitude between the Observatories of Greenwich and Glasgow by Galvanic Signals', *MNRAS*, 26, 37–44.

24. 'Report of The Professor of Astronomy in The University of Glasgow, Addressed to the Committee of Senate Appointed to visit the Observatory, and Read at

the Visitation of the Committee on the 29th of March, 1867' [GUL Res Ann I15-d.5].

25. Gavine, David M. (1981), 'Astronomy in Scotland 1745–1900', unpublished PhD thesis, Open University, p. 55.

26. Glasgow University Manuscript 117/27400. Manuscript Notes on b) The University Observatory Dowanhill [Glasgow University Archives].

27. Glasgow University Manuscript 117/27409. Catalogue of Apparatus in the Department of Astronomy – 1947 [Glasgow University Archives].

28. 'Report of The Professor of Astronomy in The University of Glasgow, Addressed to the Committee of Senate Appointed to visit the Observatory, and Read on the Day Visitation, 20th of April, 1871' [GUL Res Ann I15-d.5].

29. Glasgow Observatory (1872), 'Report for 1871', *MNRAS*, 32, 153.

30. Glasgow Observatory (1877), 'Report for 1876', *MNRAS*, 37, 169.

31. Glasgow Observatory (1875), 'Report for 1874', *MNRAS*, 35, 189; Glasgow Observatory (1876), 'Report for 1875', *MNRAS*, 36, 168; Glasgow Observatory (1880), 'Report for 1879', *MNRAS*, 40, 221.

32. 'Report of The Professor of Astronomy, 29th of March, 1867'.

33. Hay, J. H. (1849), 'The Record of chronometers submitted for trial from the 1st December 1844, and purchased by the Admiralty; with reports on Mr Loseby's mercurial compensation for chronometers'.

34. Brown, *Sketches on the Clyde*.

35. McKenzie, Raymond (2000), *Public Sculpture of Glasgow: An Illustrated Handbook*, Liverpool: Liverpool University Press [GUL Fine Arts G267 MACKE2].

36. Ibid.

Chapter 9

The Turn of the Century

9.1 THE APPOINTMENT OF PROFESSOR BECKER

Professor Ludwig Wilhelm Emil Ernst Becker was born in Wesel, Germany, in 1860. He was educated at the University of Bonn, eventually being awarded a PhD there for the calculation of the general perturbations of the minor planet (72) Feronia. After being employed by the *Berliner Jahrbuch* for two years, during which time he played a considerable part in the calculations of the motions of the minor planets, he was recommended to the Earl of Crawford and Balcarres, who owned a large private observatory at Dunecht, near Aberdeen, and he came to Scotland in 1885. At the time, Dr R. Copeland was in professional charge at Dunecht. In the autumn of 1888, Lord Crawford donated the whole of his equipment, as well as his astronomical library of 18,000 volumes, to the Lords Commissioners of the Treasury, for a new Royal Observatory, shortly to be built on Blackford Hill in Edinburgh. As part of Lord Crawford's arrangement, Copeland was appointed to the Regius Chair in Edinburgh in 1889, becoming the Astronomer Royal for Scotland; Becker was invited to join him as an assistant at the Royal Observatory, where he remained from 1889 to 1893.

In collaboration with Copeland, Becker published a paper[1] in 1892 relating to a nova that appeared in the constellation of Auriga. A year later, in 1893, he was appointed to the Regius Chair in Glasgow and held the post for forty-two years until his retirement in 1935. His letter of application[2] refers to his work at Dunecht, which produced six papers, two of which were published. The accompanying testimonials included short letters of support from Professors Copeland and Tait of Edinburgh, Sir William Huggins, the celebrated spectroscopist, Professor F. Tietjen from the University of Berlin and Professor F. Kuestner from the University of Bonn. A mid-life portrait of Professor Becker is presented in Fig. 9.1.

Following his appointment, Becker tidied up the work undertaken earlier with Professor Copeland and published a paper[3] on the stellar transit observations made at Dunecht between September 1886 and May 1889, with some supplementary observations providing coordinates for 217 nebulae. He published another paper,[4] this one related to the determination of the

Figure 9.1   A mid-life portrait of Professor Ludwig W. E. E. Becker.

absolute positions of stars around the pole, the motion of the pole and the constant of aberration, independent of the change of latitude. The work was frustrated by the lack of clear nights, but he endeavoured to continue this theme and intended to make observations at Glasgow.

Professor Becker's time at Dunecht showed that he had practical skills in several observational areas. He made visual records[5] in 1888 of Comet Sawerthal, measuring the boundary of its coma. From the data of observations transmitted from Harvard Observatory, he redetermined[6] the orbital elements of Comet Temple which he had calculated a number of years previously. In 2005, this same comet was the target of the Deep Impact space mission, and had a hole blasted into its rocky core. This work on comets formed part of his publication list in his application for the Glasgow Chair; at the time, other cited papers had not been formally published. Becker made use of a remarkable spectrometer mounted on the top of Barmerkin Hill, near Dunecht, to observe the Sun around dusk and dawn during three consecutive summers. He catalogued 3,637 absorption lines in the solar spectrum, together with 928 telluric lines produced by absorption in the Earth's atmosphere, with wavelength positions determined to an accuracy of $\pm$ 0.02 Å. His report[7] earned him the Makdougall-Brisbane Prize of the Royal Society of Edinburgh, and a readily available summary of his work and its worth was reported[8] by the Council of the Royal Astronomical Society.

9.2 EARLY YEARS AT GLASGOW

On his arrival in Glasgow, Professor Becker's first task was the renovation of the instruments. To carry out this work he was fortunate to receive a grant of £1,000 from the Bellahouston Trustees.[9] The 6-inch Ertel transit circle was reconditioned, and the 20-inch Breadalbane Reflector was overhauled and provided with a large spectrograph, a highly powerful instrument for that time. According to Gavine,[10] the 20-inch reflector, which Professor Grant had mothballed, was dismantled and found to have an inscription stating that it was Ramage metal, reworked by Grubb. He contemplated having the mirror refigured, but Hilger advised against it and Howard Grubb warned that his father had had trouble working with its thin metal. Grubb gave an estimate of £80 for re-grinding and £90 for a new mirror; Hilger offered a similar mirror for £200. In 1895, new reflector mirrors from an unknown maker were installed, together with dome shutters and sundry equipment costing £307. Other items included clocks at £70, a chronograph and electrical apparatus at £82 and an anemograph at £25; a Feuss chronograph costing £23 was also purchased to support the intended transit work.

A report[11] written in 1895 on the activities of the Observatory, soon after Professor Becker had taken up post, confirmed that an overhaul, with modifications by Troughton and Simms, had been made to the 6-inch Ertel transit telescope, then housed in a moveable hut. Becker hoped to mount the instrument on two old pillars to the east, and to recalibrate the meridian by stars he had measured at Dunecht. One interesting point was that small electric lamps of one candle power, one for each of the ten microscopes, had been installed to illuminate the circles, the light being conducted to the scales by means of glass rods polished at their ends. This was a further technical advance on an earlier improvement (see Fig. 7.16).

Experiments for controlling and working electric clock dials continued at the instigation of the Town Council. It is not certain if the previous systems had gone into disrepair, or whether this was a new enterprise. The intention was to control eight clocks placed at various points in the city, and these in turn were to work around three hundred street dials on twenty independent circuits.

Becker started observing as soon as the instruments were ready. With the transit circle, his programme was to investigate polar stars and, in particular, a pair of stars close to the pole, so that they would always be within the telescope's field of view, with the instrument remaining static. By observing the same stars at intervals of several hours, he was able to determine the position of the pole relative to the instrument, and then the positions of the stars. His intention was to determine accurately the constants of aberration and nutation, over the 18.6-year lunar cycle. Although hampered by the unsuitable climate and the proximity of a large industrial city, he carried on the observations single-handed for ten years, at which point the project was abandoned.

The results of the work were published in the *Memoirs of the RAS*, and a final value for the latitude of the Glasgow Observatory was given[12] as: 55° 52' 42".09 ± 0".05. The quoted uncertainty is equivalent to about ± 1.5m on the ground, a latitude determination better than that provided by current GPS devices, although the time taken to determine the value was considerably longer than is the case today. According to Grant,[13] based on measurements of circumpolar stars in 1860, 1861 and 1862, the latitude was determined as 55° 52' 42".8 N; the systematic difference of the order of 0".7, corresponding to ~20m on the ground, could be explained by the siting of Becker's telescope in a hut within the Observatory garden, away from the original transit room. Becker also provided a new determination of the constant of aberration of 20.454 ± 0.025 arcsec. For his spectroscopic observations, inclement weather was a most frustrating factor, but very long exposure spectrograms were made of a number of objects, such as novae, comets and nebulae. Photographic spectra of the major planets were also obtained.

His first success with the large spectrometer attached to the 20-inch telescope was the observation of Nova Persei, with the experimental arrangement more fully described in a paper published in 1904.[14] He concluded that the records were in the form of line spectra but with bright broad bands, especially across the Hβ, Hγ and Hδ lines; the degree of line broadening was proportional to the line wavelength and independent of the element.

Professor Becker made the acquaintance of John Franklin-Adams, a pioneering amateur astronomer who later produced a remarkable photographic atlas (*Franklin-Adams Charts*) covering the entire sky. It consists of 206 sections, each of 15° square, showing stars as faint as seventeenth magnitude on a scale of 15mm per degree. Becker assisted him in setting up his instruments at Machrihanish, in Argyll. He and Franklin-Adams observed the total eclipse of 30 August 1905 in Tunis, Becker making photometric observations of the corona.[15] Also in 1905, on 8 February, Professor Becker delivered a Centenary Lecture[16] to the Royal Philosophical Society of Glasgow, entitled 'Progress of Astronomy in the Nineteenth Century'. It was highly comprehensive, comprising an overview of planetary motions, the continuing discovery of asteroids, stellar problems and the nature of nebulae and galaxies.

Over the first thirty years of the new century, Becker had the support at the Observatory of James Connell, who had been employed in the latter days of Professor Grant's tenure, in particular supporting the production of the *Second Glasgow Catalogue*. Connell was able to apply his experience and knowledge to the ongoing observational work and could also respond to issues related to Grant's catalogues. An example of the latter can be seen in a letter of enquiry from Professor Ristenpart of Berlin, relating to some of the entries in the catalogue. The original data were re-examined by Connell, and comments prepared for a reply. A copy of the original letter and Connell's notes are presented in Fig. 9.2.

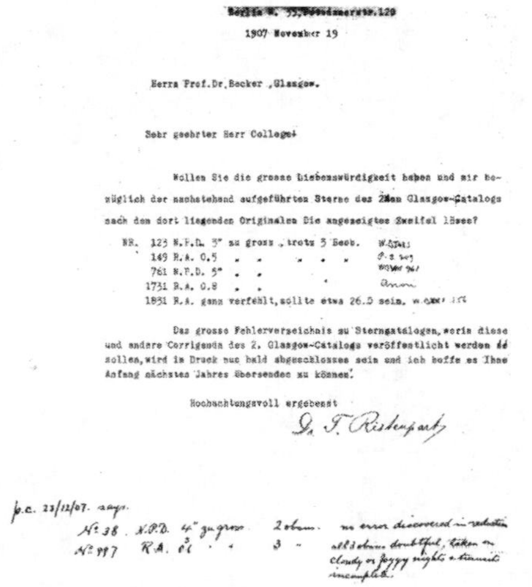

Figure 9.2 The top image is a letter dated 19 November 1907 from Professor Ristenpart of Berlin, in which he enquires about some of the entries in the Glasgow Star Catalogue. The data related to this and other enquiries were re-examined, and rough note comments prepared. A comparison of the handwriting on the postcard depicted in Figure 9.3 shows that the re-calculations were performed by James Connell.

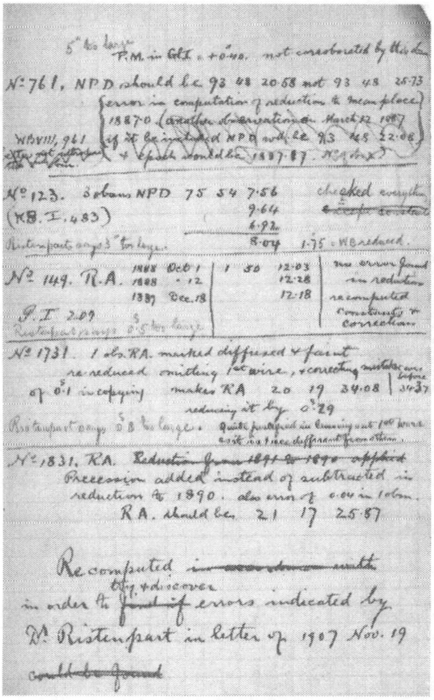

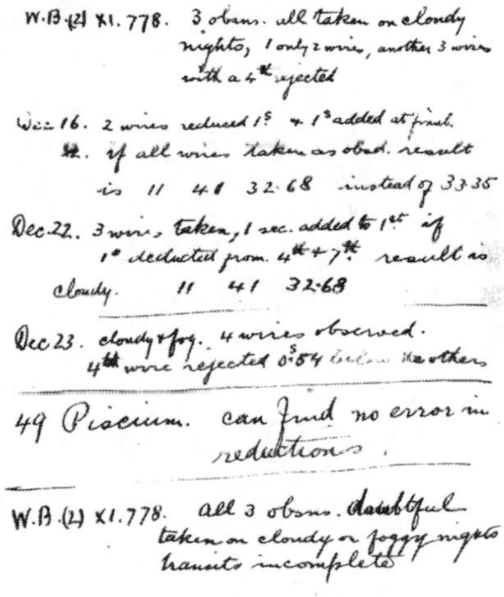

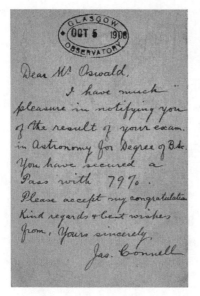

Figure 9.3   Student counselling was apparent in 1908, with James Connell sending the good news of examination success.

Figure 9.4   A 1904 postcard sent to R. Becker in Hildersham, Germany, gives details of a family trip to Loch Lomond. The photograph provides a rear view of the Observatory with its garden and drying washing. The line of writing at the bottom notes that the view is from the tennis court.

In addition, James Connell contributed to the teaching of the use of instruments and engaged in interaction with students. The tradition of the University having a cordial relationship with the student population is apparent in a postcard from the Observatory sent in 1908, communicating the good news of an examination success; a copy of it is depicted in Fig. 9.3.

At the turn of the century, family life in the Observatory house and gardens appeared to be happy and without troubles. Connections with parents in Germany were maintained, as a postcard of 1904, shown in Fig. 9.4, suggests. It was addressed to 'Land Surveyor R. Becker' in Hildesheim, probably Professor Becker's father, and was written either by Becker's wife or one of his daughters. A translation reads:

Dear Father, I have received your letter. We are very pleased that our dear Grandmother is getting better each day. My arm is quite better. I got two cards from Mother. Yesterday we were at beautiful Loch Lomond. Warmest greetings to dear Grandmother and yourself. From your loving Elsa.

9.3 RESEARCH FRUSTRATIONS

The refurbished instruments, particularly the re-mirrored Breadalbane Telescope, supported Becker's research for the first ten years or so, although he had to battle against the worsening conditions of the site at Horselethill. Comparison of the Ordnance Survey maps of 1859 and 1895, as presented in Fig. 9.5, vividly reveal the extent of the urban sprawl. The major effect of this would have been haze and smoke dramatically reducing the atmospheric transparency, rather than the more modern problem of light pollution. His frustrations are clearly registered in the Annual Reports over the period from 1900 to 1910,[17] in which he expressed his hope that the University might provide more substantial research facilities at an improved site. In 1905, Becker made a proposal (see Figs 9.6 and 9.7) to the University Court

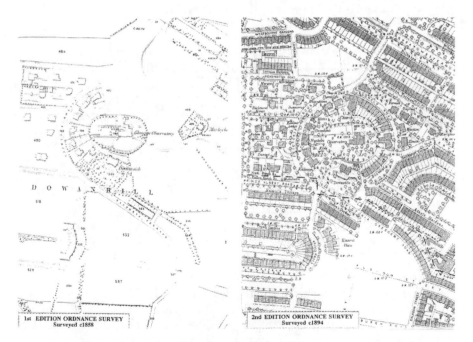

Figure 9.5   The left-hand image displays the 1859 Ordnance Survey Map with the Horselethill Observatory at Dowanhill and the disposition of local properties at the time of Professor Grant's appointment. Some thirty-six years later, the OS map of 1894 graphically displays the creep of urban development while Professor Grant had been in office.

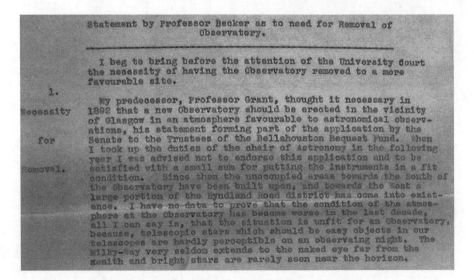

Figure 9.6 The introductory part of the Memorandum of 1905, written by Professor Becker to the University Court, describes the Observatory as being unfit for purpose.

3.

Suggestions.

,a refractor
of 7 inches
aperture,

The conflicting demands enumerated above can be satisfied by two Observatories - a research Observatory within travelling distance from Glasgow and a small Observatory in the University grounds with a South aspect where students can practise observing and the time service and the meteorological work can be attended to. The Students' Observatory would contain the instruments now used by students and for time observations, viz., a small transit instrument of 2⅞ inches aperture, a small reflector, the Sidereal and Mean Time Clocks and the meteorological Instruments except the anemometers, which require to be in an exposed position, say somewhere on the roof of the University buildings. My present assistant, who is now 20 years at the Observatory might be placed in charge of it under my superintendence: he would attend to the time service and give the students the required assistance, both being duties which he performs at present.

Figure 9.7 Part of Professor Becker's Memorandum of 1905 containing the section dealing with the proposal of running a teaching observatory at the College in addition to an outlying research facility.

that a new research observatory might be established well away from the city, but at the same time noting that the required teaching facility for students might be established in the College grounds at Gilmorehill; Horselethill could then be closed.

The first two paragraphs of Becker's Memorandum, as they were presented in the original document, are shown in Fig. 9.6. The fruitlessness of his research and the frustrations it caused can be gleaned from the paragraphs immediately following, which read:

One of the researches I have been carrying on has for its object the determination of the position of the pole from measurements with the Transit Circle of the star of the 9.3 magnitude close to the pole. I require a clear sky for at least four hours and I have often spent several hours uselessly at the telescope, because increased haziness prevented me from securing the corresponding observations in the fourth hour. In the seven months, September to March, which are especially valuable for astronomical research on account of a long duration of darkness, I had on an average nine nights on which I was able to observe the star for four hours, and eight nights for six hours or more. Our experience with the reflector is similar. For instance, in two winters I endeavoured to photograph the spectrum of the Nebula in Andromeda, the necessary exposure being about 100 hours, (my assistant guiding the telescope while I was engaged with the Transit Circle). In 1899-1900, 24 hours of exposure were secured in three months, and in 1900-1901, 39 hours. Both plates contained nothing and the work was consequently lost.

Owing to the haziness of the atmosphere we have, as a rule, to expose at least twice or three times as long as would be necessary in a purer atmosphere and the efficiency of the fine instrument is thus greatly reduced. On average there are about 150 nights a year on which bright stars can be seen for a shorter or longer period in the telescope. In my opinion it is less the cloudiness than the haziness of our atmosphere which shuts off the light of the stars.

Observations of the transparency of the atmosphere at day-time support this view. During the past four years I let my meteorological clerk note three times a day distant objects he can see from the Observatory roof. From November to February the Aberfoyle hills (distant about 30 miles) were visible on 5 per cent of the occasions, and on 68 per cent the range of visibility did not extend beyond the University tower or the high ground at Maryhill, (distant about 1 mile), from April to August there is not so much dense haze, the corresponding percentages being 21 and 32, and for July, the month in which the Fair holidays fall, 23 and 23. Towards the Clyde valley, Paisley, Govan, and especially towards the City, the atmosphere is generally thick, and it is only on the finest days that the chimney stalks in the East end of Glasgow can be discerned from the Observatory.

His proposal for the move considered the contracted formal requirement to provide time to the city, as well as the requirement to have a teaching facility near the College buildings. Also, in the interests of continuity he planned to keep the meteorological instruments in situ as before thanks to an annual grant from the Meteorological Council of the Royal Society, supplemented by funds from Glasgow Corporation and the Clyde Trustees. He suggested that a research observatory should be established within travelling distance of the city, to include the housing of the Cooke refractor and the Breadalbane Telescope; with the overhaul of these, plus a coelostat, the budget for the entire system was around £27,000. A new small observatory near the College should house a newly acquired 7-inch refractor and a transit telescope for teaching purposes, with a reliable assistant in charge. Gavine[18] has noted that

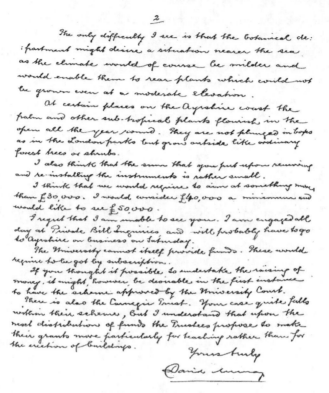

Figure 9.8   The second page of David Murray's letter of reply concerning Professor Becker's Memorandum of 1905.

the 7-inch refractor had been gifted by Cameron Corbett, MP, in about 1900. From the Annual Report for 1910,[19] it may be noted that this telescope had been established under a canvas-covered dome and used by students.

Becker dispatched a copy of his memorandum on 28 April 1905 to David Murray of Maclay, Murray and Spens, and received a reply dated 11 May (see Fig. 9.8). Murray advised that twenty acres was insufficient, and that one hundred acres should be sought in Ayrshire; £100 per acre was the likely full price, but there was a possibility of finding a cheaper plot. He suggested that the Department of Botany might be approached to consider a joint venture, although a milder climate nearer to the sea might be required, so that palms and other subtropical plants could be grown. He also thought that Becker should aim for more than £30,000; £40,000 would be the minimum, and £50,000 a good figure. It was noted that the University would be unable to provide funds itself, and that monies might be raised by public subscription, or possibly by approaching the Carnegie Trust, although that body was disposed to provide grants more for teaching than for the erection of buildings. In the first place the scheme should be approved by the University Court.

Professor Becker's proposal was turned down, but three years later he submitted a more modest plan, this dated 10 January 1908. Following his interest in solar research previously engaged in at Dunecht, he had in 1905 proposed the construction of a large siderostat, with a 36-inch mirror feeding a concave mirror 33 inches in aperture and 100 feet in focal length, to undertake systematic observations of solar phenomena. In the new proposal he noted that during the previous three years, the International Union for Cooperation in Solar Research had been founded, which embraced many of the large national observatories as well as the most prominent observatories of America; the Carnegie Institution had erected and endowed a special observatory for such research on Mount Wilson, California, in unrivalled atmospheric conditions. It was obviously inordinate to consider competing with an installation of this kind, which would be supported by proper infrastructure and manpower.

He had also realised that the Ertel transit circle and the Breadalbane Reflector, although recently refurbished, were old and insufficiently modern to be able to compete with research elsewhere, no matter where they were sited. Consequently, the idea of moving them was abandoned. It was suggested that a five-acre site might be purchased for a research station housing a new photographic transit instrument, and that the Observatory at Horselethill be kept with the old equipment; the thought of using them for any ambitious research was abandoned. The cost of the overall scheme was estimated at £12,300, with two-thirds of this figure being set as an endowment to provide monies for yearly upkeep and salary costs (see Fig. 9.9).

University of Glasgow.

STATEMENT

BY

PROFESSOR BECKER

REGARDING

THE OBSERVATORY.

10th Jan. 1908

5. My proposals now are as follow: To leave the Observatory at its present site and to erect in the country a small Research Station with a new photographic instrument, the larger scheme as embodied in my first report being postponed until for other reasons it may become desirable that the Observatory grounds be abandoned.

I recommend for the Research Station a photographic transit instrument capable of registering stars brighter than the tenth magnitude. Since the stars fainter than this magnitude have already been photographed there will be no risk of the new instrument becoming obsolete owing to the want of light-gathering power of the lens.

6. Estimate of probable cost of buildings, equipment and maintenance of the Research Station.

a. Buildings,		
5 Acres of land, - - - - - - - -	£500	
House and piers for photographic transit instrument, -	700	
House for assistant, with office, and 2 rooms for Professor, - - - - - -	1000	
		£2200
b. Instruments,		
New photographic transit instrument, - - - -	1500	
Clock installation, - - - - - - -	300	
		1800
c. Upkeep of station,		
Yearly salary of assistant, - - - - - -	150	
Yearly upkeep, - - - - - - -	100	
	250	
Capital sum necessary to give £250 interest, - - -		8300
		£12,300

According to the above estimate two-thirds of the sum of £12,300 are set down as endowment, without regard to the possibility of yearly contributions being obtainable from other sources.

L. BECKER.

Figure 9.9 Professor Becker's cover page for his Statement of 10 January 1908 to the University, together with his summary proposals at the end of his Page 2.

Again, Professor Becker's recommendations were not acted upon. Later, on 18 October 1909, a third statement was handed to the Principal with a proposition that the suggested research station might be even smaller than previously mooted, with the supporting house perhaps rented and the instrument mounted in a hut within the leased property. This was considered by the Observatory and Laboratories Committee, on 30 November 1909, and again it received a negative response.

From then on, Becker's ambitions to undertake a serious line of observational work were completely thwarted, although he continued to make useful, but casual, records of a few objects until the onset of World War I.

9.4 THE BEST OF A BAD JOB

In a brief Report[20] for 1911, Professor Becker commented that he had been using the 20-inch reflector to photograph the spectra of planets, and that every available night was used in an unsuccessful attempt to obtain a large dispersed spectrum of Brooks' Comet 1911 c, with a total exposure time of 15.3 hours; using a 3¼-inch 15° objective prism[21] bought for £32 and a telescope with a focal length of 30 inches, with the arrangement attached to the tube of the 20-inch reflector, he managed to obtain five spectra at low dispersion. In the same year, he used both the 20-inch reflector with spectrometer and the 3¼-inch objective prism to observe Nova Geminorum. The resulting paper[22] described the envelope shape of the spectra and the behavioural details of the lines, particularly those of hydrogen. The objective prism remains in the safekeeping of the current Observatory, and a photograph of it is shown in Fig. 9.10.

The same spectrometer arrangement of the 20-inch reflector was used between 1909 and 1911 to obtain spectra of Jupiter and Saturn. The records[23] showed peculiar bands covering 6175–6207 Å for Jupiter, and 6137–6216 Å

Figure 9.10   The small objective prism used by Professor Becker to record spectra of Nova Geminorum in 1912.

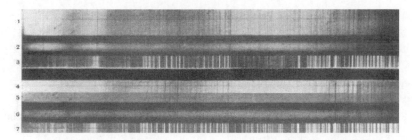

Figure 9.11 Spectral records obtained by Professor Becker; no. 1 is of the Moon, no. 2 of Jupiter and no. 3 of an iron-sodium arc. Numbers 4 to 7 are the same material but with the reversed lunar spectrum superimposed at no. 5 and no. 6.

for Saturn. At the time, water was supposed to be present in these planetary atmospheres, but the supposition was not borne out by these observations. Becker's photographic records are reproduced in Fig. 9.11.

From then on, Becker spent more time conducting theoretical work on a variety of topics, and the Breadalbane Telescope fell into disuse. In 1912, however, he was experimenting[24] with a new form of photographic transit telescope, the basic instrument purchased from the Cambridge Scientific Instrument Co.

9.5 WORLD WAR I

According to the *Ordinances of the Universities of Scotland* published in 1915, the Professor of Astronomy's salary was £600; this sum took into account the fact that he had an official house. As a result of ill-feeling at his being German by birth, Professor Becker was forced to vacate the Observatory on Horselethill during World War I. His treatment on this matter must have caused him great distress. Although he had become a naturalised British citizen in 1892, before his appointment to Glasgow, this fact failed to save him from unmerited suspicion. Initially, the Principal and the Secretary of Scotland defended him, but they eventually succumbed to the clamour and had him temporarily removed. Becker had insisted that such action was not legally permissible, because he held the sovereign's commission. Hutcheson and Conway[25] noted that he was interned, or at least restricted in movement, in a house near Crieff. According to Smart,[26] public opinion forced his temporary retirement to Aviemore in the Highlands, where he lived in seclusion until the end of hostilities and when anti-German sentiment had subsided. The Annual Reports[27] during this period euphemistically note that the Director was 'on leave of absence', with the Observatory's ever faithful assistant, James Connell, in charge. Becker did, however, manage to submit a few papers which were published.

Figure 9.12 The modification of a sextant by the addition of an 'artificial horizon'.

Although it is not certain when or where he performed the study, his deliberations on a geometric analysis of an image of the famous spiral nebula M 51, obtained at Yerkes Observatory, were published[28] in 1917; the discussion related to an interpretation of the motion of the spiral arms. One work that was dispatched from Aviemore involved a description of the design of a modified artificial horizon for a standard sextant. The paper[29] was submitted in June 1916, but was not published until the end of the war. An addendum to the paper noted that 'Publication was deferred by the RAS Council till after the termination of the war'. Whether this was because of Professor Becker's status at the time, or whether it was thought that the described modification might be of interest to the military, is an open question. An image of the modified sextant is presented in Fig. 9.12.

Becker's return to office appears to have been quietly effected soon after the end of the war. An anecdote related to his return has been recorded[30] by Robert Hutcheson as follows:

> In 1920, I was sent to the Observatory to return the petty-cash book which was in the care of a lecturer called Connell, the first assistant. On arrival, I found an ill-kempt man labouring in the garden. My progress was barred and business challenged. I said 'I have come to see Mr Connell about the petty cash'. His reaction stunned me. 'I shall take the book to Mr Connell, I am Professor Becker'. This was the first indication to the University that Becker had been released from interment.

It may be noted that James Connell (1867–1930) served Professor Becker and the Observatory with great dedication. He had been appointed towards the end of Professor Grant's reign and deserved special mention in the production of the *Second Glasgow Catalogue*. While Becker was away from Glasgow, he took care of the day-to-day running of the site and, as can be gleaned from the various mentions of him in this chapter, played a part in teaching

Figure 9.13 James Connell at the eyepiece of the 9-inch Ochtertyre Refractor.

and maintaining the equipment until his death. A photograph of him at the eyepiece of the 9-inch Ochtertyre Refractor forms Fig. 9.13.

9.6 THE LATER YEARS

A change of tack, in terms of Becker's research, was immediately apparent on his return to the Observatory; age was beginning to catch up with him and he had lost the incentive to struggle with night-time observations, using inadequate equipment in a poor climate. At this point he had developed an interest in stellar dynamical problems and, in 1920, he wrote two papers[31] on capture theory. His main proposition was that a binary star may be disturbed by an approaching third star, such that one of the components may be replaced by it, this being dependent on a set of special conditions associated with the initial binary system and the approaching star. According to Smart,[32] this work was taken up by others some thirty years later, in theories associated with the formation of planetary systems. In 1927, he published two papers related to the behaviour of the major axis of planetary or satellite orbits. The first[33] treated the integration of the differential equations for the major axis of a satellite's orbit, as disturbed by the Sun, and of the orbit of a planet, as disturbed by other planets. The second[34] calculated the motions of the planets with the consideration that gravitational forces travel at a finite speed. Predictions of orbital behaviour were explored, particularly in respect of perehelion precession. One of the conclusions was that the precession of the perihelion of Mercury could not be explained by this thesis without the velocity of gravitation being a large multiple of the velocity of light.

Just as in Wilson's times, the Observatory was a target for burglary. An inventory[35] shows that on 24 July 1924, two stopwatches were stolen in addition to a direct vision diffraction spectroscope by Thorpe. Although there is little reference to the continuation of transit observations for the purpose of local timekeeping, as purchase[36] had been made of 'apparatus for

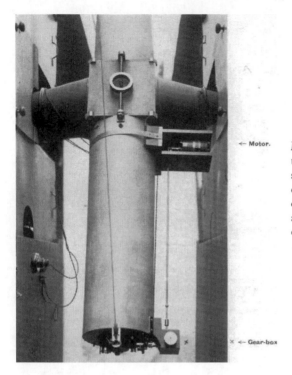

← Motor.

× ← Gear-box

Figure 9.14 The modification to the Ertel transit telescope shows the drive rod running outside the tube from a motor close to the telescope rotation axis down to the micrometer eyepiece.

reception of Wireless Time Signals' in October 1925, an innovation to the eyepiece micrometer of the Ertel instrument was made in 1928. The paper[37] describing the motorising of the movement of the cross-wires provides the only available photograph of the Ertel transit telescope, as reproduced in Fig. 9.14. The new system replaced the traditional hand-operated method, which was a delicate and tedious process; instead, the screw was adjusted by an electric motor at a rate in excess of the star's image movement and the rate was adjustable according to the declination of the star. A one-night trial suggested that the equipment represented a substantial improvement over previous experiences of basic hand-controlled eyepiece wires.

Becker's last submitted paper[38] involved the presentation of graphical maps, allowing values to be determined for solutions to formulae related to spherical triangles. In his latter years, he had intended to produce a text-book on spherical astronomy, but the high cost of reproducing the many diagrams, so important to the presentation, seemed to prevent publication. The typescript with manuscript corrections has been archived,[39] and a page from the draft text is presented in Fig. 9.15.

Throughout his office, the Observatory continued as a meteorological station, and regular records of temperature, pressure, wind speed and direction, and so on were taken. Again, however, Becker[40] had to endure a lack of support for his activities as the annual grant from the Meteorological

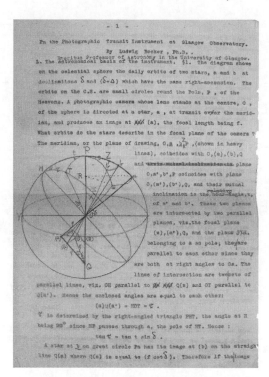

Figure 9.15 A leading page from Professor Becker's draft typescript entitled 'Spherical Astronomy'.

Committee of the Royal Society was withdrawn in 1913, causing him to dispense with the second assistant at the Observatory and to abandon instruments that had been in operation for forty-five years. An analysis of the data through the nineteenth century, and up to 1925, was published[41] by the Meteorological Office. The title page and photographs from the document are presented in Fig. 9.16.

9.7 ASTRONOMY TEACHING

Becker commenced teaching immediately on his arrival in Glasgow, as can be seen from the University calendar of 1893–4. The first-level class used Barlow and Bryan's *Elementary Mathematical Astronomy*, a much-used text in English universities. This was accompanied by J. F. W. Herschel's *Treatise on Astronomy* or Newcomb's *Popular Astronomy*, and *The Nautical Almanac*. The senior class extended its reading to Todhunter's *Spherical Trigonometry* and Brünnow's *Spherical Astronomy*. The course was broad, covering spherical astronomy, precession, nutation, aberration and parallax effects on coordinates, explanations of *The Nautical Almanac*, theory of the theodolite and transit circle, determination of latitude and time, geodetic measures, triangulation and theoretical astronomy. Becker also offered to instruct advanced students in orbit determination for no charge. During Grant's

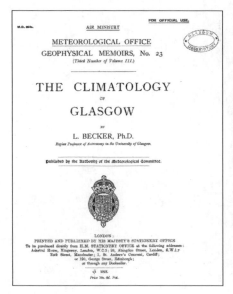

Figure **9.16** Becker's Meteorological Report of 1925 included photographs of three anemometers on the roof of the Observatory and the thermometer screen attached to the north side of a detached building.

time, degrees in natural philosophy were awarded as an MA, at Ordinary or Honours standard; optional papers in astronomy were set for the Ordinary level (see Chapter 7). As the study of the sciences expanded, BSc degree subjects simply comprised engineering, geology and biology. Later, natural philosophy changed to BSc degrees. According to the University calendar of 1895–6, astronomy was available for the final BSc, as one of three subjects at about the level of Honours MA. This course covered celestial mechanics, the two-body problem, including Hamilton Jacobi's treatment, the n-body problem, with an introduction to perturbation, precession, nutation, interpolation and mechanical quadrature. A modified course was available for the majority of students preparing for the BSc in Engineering Science, using the first eight chapters of Barlow and Bryan. A first course for the MA, also covering the introductory work for the BSc, included topics related to the elementary celestial sphere, description of instruments, determination of latitude and time, precession, nutation, aberration, occultations and calculation of comet orbits. The course used Barlow & Bryan's text and C. A. Young's *Manual of Astronomy*, and could be adapted to the needs and attainments of students. The class met three times weekly at the Observatory,

Figure 9.17 Some student astronomy classes of years past. The upper left is from 1910–11 and shows Professor Becker as a young man at the centre, with James Connell standing to the right; the upper right is from c. 1922, again with James Connell on the right; the lower left picture shows Professor Becker in profile with James Connell standing on the left; the lower right photograph presents the class of c. 1934.

which had a small students' department containing a 36cm circle theodolite in a revolving hut and several sextants. In 1899–1900, Becker was teaching the same courses, with additional lectures on the method of least squares and on advanced geodesy for engineers.

After his return to post in 1920, teaching flourished, with Becker covering physical and dynamical astronomy to classes by far the largest in the United Kingdom. Photographs of the participants of some of the student classes are presented in Fig. 9.17.

Becker's lectures did not concern popular or descriptive astronomy, but were geared to the mathematics of the celestial sphere and to dynamical aspects. He presented a sound theoretical development of the subject to a sufficient depth for practical usefulness. Students computed the actual ephemeris of a planet from the orbital elements. For practical work, they had to learn to make observations with various instruments and to determine clock error, latitude and azimuth.

Many of Becker's students later made their mark in the scientific world.

Prof. Ludwig Becker
(Astronomy).

Figure 9.18   Professor Becker is caricatured joyfully riding his telescope.

Have you heard of Ludwig B.,
Professor of Astronomy?
He lives up in the college tower,
And rings the hour-bell every hour.
And if you look up to the spire,
You'll see him hanging by a wire.
And when the rain is coming down,
He catches it upon his crown.
But when the lightening flashes out,
He tears his hair and jumps about!
And when the wind is blowing high,
You'll find him soaring to the sky.
He puts his fingers to his nose,
And spreads his coat and up he goes.
When other folks are safe in bed,
He walks about upon his head!
Yet here's a mark to know him by,
He holds a spyglass to his eye!

According to Hilditch,[42] Becker's enthusiastic teaching methods had a profound influence on John Jackson, who graduated in Glasgow in 1907, directing him into a career in astronomy. In 1933, Jackson was appointed His Majesty's Astronomer at Cape Observatory in Cape Town, South Africa. Becker also guided Alexander Thom in Glasgow in the calculation of orbits and other astronomical skills. Thom, who became Professor of Engineering at Oxford, was later to champion the surveying of megalithic sites (see Chapter 1, p. 3) and to advocate that many of them carried information on various astronomical alignments. Professor William Smart, who eventually returned to Glasgow to take up the Regius Chair on Becker's retirement, also graduated under his tutelage. Professor Becker appears to be fondly remembered by his students, as the cartoon in Fig. 9.18 suggests. It appeared, together with the attached poem, in the *Glasgow University Magazine*[43] of 15 February 1899.

The reference made in the poem to the tower and to Becker ringing the hour-bell may be related to the control and running of the seconds' clock in the West Quadrangle (see Fig. 8.6). Whether or not it also once controlled the ringing of the bell in the tower at Gilmorehill is unknown.

Professor Becker was a most approachable man and easily made friends with his students. According to John Jackson,[44] mentioned above, soon after the commencement of term, he would give his students a key to the Observatory and tell them that it was open from 'noon till noon'. This interval of time defined the astronomer's day, according to the system of

Greenwich Mean Astronomical Time (see Chapter 7, p. 208) which was in use prior to 1925.

Professor William M. Smart, his successor, relates in his obituary[45] that Becker was an inspiring teacher. John Jackson, one of his earliest students, and mentioned above, noted[46] that Becker's wife and family spent most of their time in Germany and that he remained in Glasgow for financial reasons. After retiring in 1935, he lived in Lagundo in northern Italy, but his latter years were saddened by World War II and the anxieties concerning his children that it brought. He died in Merano on 11 November 1947, in the presence of his wife; all his children were separated from him by closed frontiers. He was survived by his widow, four of their five children and eight grandchildren. Becker was one of the gentlest of men, modest and retiring in disposition and, whatever the outside world thought, respected and esteemed by his students both for his learning and for his kindly qualities.

The appointment of his successor, Professor Smart, saw a change in research direction; it became limited to theoretical studies, with no intentions of pursuing local observational matters. As a consequence, it was thought better to bring the practical teaching requirement closer to the University site, and a new small observatory was built at University Gardens. The grounds of the decaying Horselethill Observatory, which operated for nearly 100 years, were sold for about £9,000 to the Roman Catholic Church, and the site is currently the location of Notre Dame High School at 160, Observatory Road.

A view of the entrance gates to the Observatory in around 1905 is shown in Fig. 9.19. An aerial view of Horselethill towards the end of its life as a

Figure 9.19 The entrance gates of the Horselethill Observatory in 1905 show the encroachment of residential properties.

Figure 9.20   A 1930 aerial view of Horselethill shows the magnificence of the main Observatory buildings with the large dome at the right-hand side of the rectangular block. Lower to the left, on the slope of the garden, is another small dome and a hut. The Botanic Gardens are to the upper right of the picture.

Figure 9.21   The large dome below the main building of Horselethill Observatory, c. 1927.

functioning Observatory is depicted in Fig. 9.20; it can be seen that there were several domes within the grounds. A photograph of the larger one immediately behind the main building is presented in Fig. 9.21.

Perhaps the last photograph of the mature building is that shown in Fig. 9.22. It was probably taken in the early 1930s, and the figure may be James

***Figure* 9.22** The Horselethill Observatory c. 1930, just prior to its demolition, showing some of the meteorological instruments with the dome behind. A feature of the entrance door is the emblem of the 'snake, triangle and star', the carved stone being incorporated in the later observatories.

Connell, who was Professor Becker's assistant over the greater part of his incumbency in the Chair. A feature, clearly seen in the photograph, is the emblem over the lintel of a snake, a triangle and a star. Its origin remains an enigma. Although it is similar to the styles used by the Theosophical Society and the Masons, its appearance is unique to the Glasgow University Observatory. The snake eating its own tail, or the Ouroboros, is a very old symbol and is taken to represent the cyclic nature of the Universe, which recreates itself out of destruction. It may have been chosen and designed by Professor Nichol when Horselethill was built, as it perhaps relates to the 'nebular hypothesis' that he described and advanced in his writings. The whole design, however, may be just a simple representation of astronomy being the measurement of stars in a recycling Universe. It was likely that the same emblem was incorporated at the rear of the building as two of them are extant, one in the Observatory at Acre Road and another at the entrance of the Kelvin Building, the current home of the School of Physics and Astronomy within the University. The only remaining pointer to indicate the former glories associated with Nichol, Grant and Becker is the road leading from Byres Road up to Horselethill. A photograph of Observatory Road's nameplate is shown in Fig. 9.23.

Figure 9.23 The last tangible reminder of the location of the Glasgow Observatory of 1841–1939: the junction of Observatory Road and Byres Road.

References

1. Copeland, R. and Becker, L. (1892), 'On the new star in the constellation Auriga', *Astronomy & Astro-Physics*, 11, 593–602.
2. (1893?) Testimonials in Favour of Ludwig Becker, Ph.D., Candidate for the Chair of Practical Astronomy in the University of Glasgow, and the Directorship of the University Observatory [GUL Sp Coll Kelvin 152].
3. Becker, L. (1902), 'Observations of 217 Nebulae made with the Transit Circle at Dun Echt Observatory', *Annals of the Royal Observatory Edinburgh*, 1, 1–46.
4. Becker, L. (1902), 'Positions of Stars within one degree from the North Pole, and of Three Fundamental Pole-Stars', *Annals of the Royal Observatory Edinburgh*, 1, 47–70.
5. Becker, L. (1888), 'Note on Comet *a* (Sawerthal)', *Monthly Notices of the Royal Astronomical Society* [hereafter *MNRAS*], 48, 380–1.
6. Becker, L. (1891), 'On the Orbit of the Periodic Comet 1867 I', *MNRAS*, 51, 475–94.
7. Becker, L. (1890), 'Low Sun Spectrum', *Transactions of the Royal Society of Edinburgh*, 36, part 1, 99–200.
8. See the section entitled 'The Solar Spectrum at Medium and Low Altitudes' in 'Report to the Council of the RAS Seventy-Second AGM' (1892), *MNRAS*, 52, 294–6.
9. See the section entitled 'Glasgow Observatory' in 'Report of the Council to the Seventy-fifth Annual General Meeting' (1895), *MNRAS*, 55, 217.
10. Gavine, David M. (1981), 'Astronomy in Scotland 1745–1900', unpublished PhD thesis, Open University, 88.
11. See the section entitled 'Glasgow Observatory' in 'Report of 1895' (1896), *MNRAS*, 56, 232.

12. Becker, L. (1917), 'Note on the Latitude of Glasgow Observatory', *MNRAS*, 77, 662–3.

13. Grant, Robert (1883), *Catalogue of 6415 Stars for the Epoch 1870*, Glasgow: James MacLehose & Sons; London: Macmillan & Co, p. xi.

14. Becker, L. (1904), 'On the Spectrum of Nova Persei and the Structure of its Bands, as Photographed at Glasgow', *Popular Astronomy*, 12, 603–8.

15. Becker, L. (1908), 'The Distribution of Blue-Violet Light in the Solar Corona on August 30, 1905, as Derived from Photographs Taken at Kalaa-es-Senam, Tunisia', *Philosophical Transactions of the Royal Society of London*, 207, 307–39.

16. Becker, L. (1904–5), 'Progress of Astronomy in the Nineteenth Century'. Centenary Lecture, *Proceedings of the Royal Philosophical Society of Glasgow*, 36, 136–83 [GUL Sp Coll MS Gen 1756/8/37].

17. See the sections entitled 'Glasgow Observatory' in 'Report of 1900' (1901), *MNRAS*, 61, 225; 'Report of 1905' (1906), *MNRAS*, 66, 199; 'Report of 1907' (1908), *MNRAS*, 68, 272; 'Report of 1908' (1909), *MNRAS*, 69, 274.

18. Gavine, 'Astronomy in Scotland 1745–1900', 88.

19. 'Glasgow Observatory', in 'Report of 1910' (1911), *MNRAS*, 71, 299–300.

20. 'Glasgow Observatory', in 'Report of 1911' (1912), *MNRAS*, 72, 278.

21. Glasgow University Manuscript 117/27409. Catalogue of Apparatus in the Department of Astronomy – 1947. See the section for 1939 [Glasgow University Archives].

22. Becker, L. (1912), 'Observations of the Spectrum of Nova Geminorum', *MNRAS*, 72, 709–14.

23. Becker, L. (1917), 'The Spectrum of Planets Jupiter and Saturn', *MNRAS*, 78, 77–8.

24. 'Glasgow Observatory', in 'Report of 1912' (1913), *MNRAS*, 73, 242–3.

25. Hutcheson, Robert T. and Conway, Hugo (1997), *The University of Glasgow 1920–1974*, Glasgow: Glasgow University Library [GUL James Herriot Lib – Education ST71 1997-H].

26. Smart, W. M. (1948), 'Obituary – Prof. Ludwig Becker', *Nature*, 161: 4,083, 161.

27. See the sections entitled 'Glasgow Observatory' in 'Report of 1916' (1917), *MNRAS*, 77, 330; 'Report of 1918' (1919), *MNRAS*, 79, 247; 'Report of 1919' (1920), *MNRAS*, 80, 378–9.

28. Becker, L. (1917) 'On the Spiral Nebula, M 51, Canum Venaticorum', *MNRAS*, 77, 655–62.

29. Becker, L. (1919), 'On a new Artificial Horizon for Sextants', *MNRAS*, 79, 593–604.

30. Hutcheson and Conway, *The University of Glasgow 1920–1974*.

31. Becker, L. (1920), 'On Capture Orbits', *MNRAS*, 80, 590–7; Becker, L. (1920), 'The Capture Hypothesis of Binary Stars', *MNRAS*, 80, 598–603.

32. Smart, 'Obituary – Prof. Ludwig Becker'.

33. Becker, L. (1927), 'On the Perturbations of the Major-Axis of the Orbit of a Satellite or Planet', *MNRAS*, 88, 88–93.

34. Becker, L. (1927), 'The Effect of Finite Velocity of Transmission of Gravity on the Motion of the Planets', *MNRAS*, 88, 93–97.

35. Glasgow University Manuscript 117/27409. Catalogue of Apparatus in the Department of Astronomy – 1947.
36. Ibid.
37. Becker, L. (1928), 'The Motor-Micrometer of the Glasgow Transit-Circle', *MNRAS*, 88, 604–6.
38. Becker, L. (1930), 'Graphical Solution of a Spherical Triangle', *MNRAS*, 91, 226–30.
39. Becker, L. (1947), 'A Textbook of Spherical Astronomy' (Becker's draft typescript with manuscript amendments, 240 sheets) [GUL Sp Coll MS Gen 1723/2].
40. 'Glasgow Observatory', in 'Report of 1913' (1914), *MNRAS*, 74, 299.
41. Becker, L. (1925), *The Climatology of Glasgow. Meteorological Office – Geophysical Memoirs, No. 23*, London: His Majesty's Stationery Office.
42. Hilditch, R. W. (1995), 'A Century of Scottish Astronomy 1894–1994 and Future Perspectives', *Quarterly Journal of the Royal Astronomical Society*, 36, 11–27.
43. Hutcheson, R. T. and Oakley, C. A. (eds) (1951), *The Fleeting Years: A Collection of Drawings of Incidents and Personalities at Glasgow University*, Glasgow: Glasgow University Graduates Association [GUL Education S277 1951-H; GUL Sp Coll Mu Add. 148].
44. Jackson, J. (1948), 'Obituary Prof. Ludwig Becker', *MNRAS*, 108, 41–3.
45. Smart, 'Obituary – Prof. Ludwig Becker'.
46. Jackson, 'Obituary Prof. Ludwig Becker'.

Chapter 10

The Astronomical Society of Glasgow

10.1 EARLY DAYS

Around the time of Professor Becker's appointment to the University, in 1890, the British Astronomical Association (BAA) was formed. The organisation was established to fulfil the interests and needs of amateur astronomy around the country in a coordinated way. Soon after its formation, in 1894, a group got together in Glasgow to form a West of Scotland branch, and it is from there that the organisation eventually developed into its current form of the Astronomical Society of Glasgow, or ASG.

Two episodes have already been described in which the groundswell of interest in astronomy within Glasgow society resulted in the establishment of an observatory. The first, in 1808, through the Glasgow Society for Promoting Astronomical Science, saw the erection of an ornate Egyptian-style building at Garnethill, which had its formal opening in 1810. The activity there answered the need of the Society for a short while, and then financial problems soon arose. Approaches were made to the University to buy out the enterprise, but nothing came of these. The Society was disbanded in 1822; some of the instruments are known to have been sold off, while others simply disappeared. Around 1830–2, the building was demolished.

A second wave of enthusiasm for astronomy appeared in the mid-1830s, as described in Chapter 6. In 1836, a number of gentlemen in Glasgow and the surrounding district formed themselves into an association styled the Astronomical Institution of Glasgow. Their objectives were better defined than those of the previous venture; their aim was to erect an observatory, with part of its direction the pursuit of astronomical observations for the purpose of providing accurate time for shipping on the Clyde. An agreement was established whereby the Institution would provide an observatory and the University would supply the requisite instruments and an observer. A new observatory was built on Horselethill and opened in 1841. The Institution obtained a substantial grant from the government for the project, but as a result of Professor Nichol's extravagance and other financial problems, the scheme fell into distress. Following a couple of years of difficult negotiations, the University agreed to buy out the project in 1845 and to take over the Observatory, accepting responsibility to the Admiralty of establishing time for local shipping. The shareholders of the Institution received a reasonable fraction of their original investment, and other creditors were paid off. For another ten years or more, however, the aims of the enterprise were not fulfilled. They only emerged through the efforts of Professor Grant following his appointment, as related in Chapter 8.

10.2 OTHER GLASGOW ACTIVITIES

Outside the precincts of the University, in the latter half of the nineteenth century, science lectures were regularly delivered in the City Hall. As part of this, astronomy continued to be of interest to the general population. For example, on 30 January 1879, in the City Hall, Professor Balfour Stewart[1] delivered a lecture entitled 'Suspected Relations Between the Sun and Planets'. In his talk he presented an overview of the nature of sunspots and their cyclic behaviour. The main theme suggested possible connections between solar activity and the positions of the planets in their orbits, although the idea now has no import whatsoever. The magnetic nature of sunspots was not known at the time. In the main thesis of the talk, Professor Stewart explored the possible correlations between sunspot numbers and the configurations of the planets, particularly the juxtaposition of the inner planets, Mercury and Venus. In reference to Fig. 10.1, he says in his transcript:

> Now regarding the earth simply as the point from which phenomena are viewed, we see by this diagram that whenever Venus or Mercury is nearly between high earth and the centre of the sun sun-spots behave as in the 1st and 5th figures – that is to say, as they are carried round by rotation near to the influential planet they become less, and as they are carried away from it may become greater. Secondly, when Venus or Mercury is at the extreme right of the sun the spots diminish in size all away across, as in figures 2 and 6. Thirdly when Venus or Mercury is on the other side of the sun, exactly opposite the earth, spots have their maximum at the centre as in figures 3 and 7. Lastly, if Venus or Mercury is near the extreme left the spots augment in size all the way across, as in figures 4 and 8.

Needless to say, such proposals were very much in the form of conjecture, based on the interpretation of selected and insufficient data. One of his 'magic lantern' slides was of local interest, as it gave a dramatic picture of the nature of the Wilson effect. It is reproduced in Fig. 10.2 and shows the cavitous nature of sunspots, albeit greatly exaggerated.

A phrase used by a Dr Chalmers of Glasgow, and quoted by Professor Stewart in his lecture, continues to have resonance in respect of the present-day expansion of our knowledge and understanding of the Universe: 'The greater the diameter of light – the greater is the circumference of darkness'.

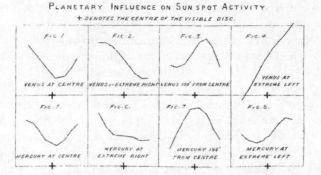

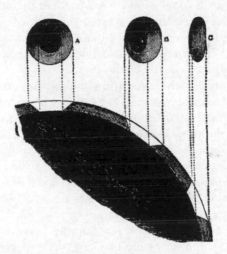

Figure 10.1 Data presented by Professor Balfour Stewart in his lecture of 30 January 1879, used to provide evidence that sunspot activity was related to the positions of Mercury and Venus in their orbits.

Figure 10.2 A slide from Professor Balfour Stewart's lecture of 30 January 1879 dramatically explains the Wilson effect in terms of depressions on the solar surface.

10.3 THE SCOTTISH BRANCH

The West of Scotland branch of the BAA was formed in Glasgow in the autumn of 1894, four years after the foundation of the parent body. The inaugural meeting took place on 23 November of that year, at which members were addressed by E. W. Maunder, founder of the BAA and editor of their journal. His name is associated with an inactive period of the Sun, from 1645 to 1715, which he discovered by examining historical records of sunspot observations. His lecture was entitled 'In Pursuit of a Shadow', and was an account of a recent eclipse expedition.

The area covered by the West of Scotland branch was defined as comprising the counties of Lanarkshire, Renfrewshire, Ayrshire, Argyllshire, Dunbartonshire and Stirlingshire, and membership was initially confined to members of the parent association. Some of the first meetings were held in the Church of Scotland Training College, Dundas Vale, on New City Road in Glasgow. For some years afterwards, the branch met in the Athenaeum in Buchanan Street. Early membership was comparatively small. In spite of the enthusiasm within the group, there appears to have been some anxiety about its affairs. After considerable negotiation with the parent association, arrangements were made in 1903 whereby the branch could enrol associates not directly attached to the parent body, who would be entitled to all privileges of the branch, except for restrictions on eligibility for election to the council. The principal office-bearers and two-thirds of the entire local council were required to be full members of the BAA. Power was also acquired by the branch to enrol members resident in any part of Scotland. The new arrangement resulted in a rapid increase of membership, from about thirty-five in 1904 to about one hundred and ninety in 1907, of whom some one hundred and forty were associates. The increase in membership radically changed the character of the branch. From being a small group whose main function was 'the association of observers', the main purpose of the parent association, the branch became a society with the aim of keeping its members informed on matters pertaining to astronomy, with a high standard of presentations covering a broad range of subjects. In addition to the majority of members who considered astronomy a passive pursuit of simply attending lectures, there were some who regarded their telescopes as their choicest possession, and yet others who were prepared to construct their own. As a result of the increasing membership, the venue of the meetings was transferred to the new buildings of the Royal Technical College in October 1905. This association has happily been maintained, and, to this day, the Astronomical Society of Glasgow regularly meets within the University of Strathclyde.

One of the key people in establishing the branch was Bailie John Dansken. Born in 1836, he was a surveyor by profession and played a part in many of the professional, political and educational bodies within Glasgow. He was a life member of the British Association for the Advancement of Science, and of the Society of Arts, and was a Fellow of the Royal Astronomical Society. In addition, he was a Vice-President of the Glasgow Geological Society, and was the first President of the West of Scotland Branch of the BAA. A portrait of him is shown in Fig. 10.3, together with the minutes of the branch meeting of 23 November 1905 announcing his death. At his home, at 2 Hillside Gardens, Partickhill (the address has since been redesignated), he had a private observatory in the attic of the coachhouse. Two of the telescopes donated to the branch no doubt were once used in it. The property still stands in a renamed thoroughfare, but, from the outside, there is no

The Technical College,
Glasgow; 23rd Nov. 1905.

The second meeting of the twelfth session was
held here this evening, Mr Robert Robertson, B.Sc., in the
chair. The President referred to the great loss which
the Branch had sustained through the death of their
senior vice president, and moved 'That this Society
begs to place on record an expression of the feelings of
deep sorrow with which they learned of the sad and
sudden death of Mr John Dansken, F.R.A.S. Mr Dansken
was connected with this Branch of the British Astronomical
Association practically from its foundation, and he all
along took the warmest interest in the affairs of the
Society having, on several occasions, filled its highest
office with the greatest satisfaction to its Members,
and credit to himself. Mr Dansken was associated with

Figure 10.3 A portrait of John Dansken, a founder of the West of Scotland branch of the British Astronomical Association, together with the minute of 23 November 1905 announcing his death.

Figure 10.4 Observatory Cottage, once the private observatory of John Dansken, retains the dome support rotation ring.

evidence, other than its name, of an observatory structure ever having been there. Inside, however, the ring support for the dome makes an interesting feature (see Fig. 10.4).

Memorable meetings held under the auspices of the branch during the first two decades, were *conversazioni* held on 20 April 1900 and 31 March 1909, largely thanks to the generosity of two Presidents, Bailie John Dansken and Major John Cassells, respectively. These gentlemen are also remembered in connection with their gifts of instruments for the encouragement of astronomical study. Dansken donated a 13-inch reflector and 4-inch refractor to what was then the Royal Technical College, and Cassells presented a 4-inch refractor to Albert Road Academy. It was also Bailie Dansken who inaugurated the collection of lantern slides owned by the branch. At a later date, a library for the benefit of members was established.

It is also worth mentioning the observational studies of the planets made by Henry McEwan from various residences in Glasgow. He was President of the West of Scotland Branch of the BAA for a while, but his main contribution to amateur astronomy was his position as Director of the Mercury and Venus Section of the BAA, which he held for sixty years (1895–1955). He was very much concerned with determining the rotation of Venus and engaged in correspondence with many of the professional astronomers around the world regarding the mapping of the planets. He made occasional use of the 7-inch refractor at the University's Observatory, but generally preferred to conduct his observations with his own instruments at home, particularly a 5-inch refractor. His biography has been thoroughly presented in two papers[2] by McKim; these include many of McEwan's drawings which appeared at various times in the BAA's journal.

The first of the branch visits was to the then new Royal Observatory at Blackford Hill, in Edinburgh, where the members were conducted personally through the buildings by the Astronomer Royal for Scotland, and received from him the details of the telescopes and working instruments. The branch celebrated its silver jubilee in September 1919; although World War I was over, railway restrictions still prevailed and this prevented a visit to Edinburgh. The Society continues to have annual outings to places of astronomical interest and has returned on a number of occasions to the Royal Observatory.

In 1937, authority was obtained from the parent association to change the name from 'West of Scotland Branch' to 'Scottish Branch', although there appears to have been no corresponding widening of the area of branch activities; meetings continued to be held in Glasgow only. At the same time, the restriction of only allowing one-third of the membership of the council to be associates was removed.

Records of the branch have been kept in three tomes since 1897, four years after its inception, as Minutes of Meetings, Minutes of the Council and a Cash Book. The first ever records of the minutes are reproduced in Fig. 10.5, and it may be noted that those of the council meeting of 13 October 1897 bear John Dansken's signature.

During World War II, the activity of the branch was greatly restricted, particularly by the blackout. In compensation, the sky became visible for the first time to thousands of city dwellers and, as a result, interest in astronomy was stimulated. The early period of the war affected the membership and attendance at meetings, but throughout this difficult period there was no interruption to the branch's programme. Accounts and minutes of the meetings held by the current society make interesting reading. For example, Professor A. D. Rowes of the University of Western Australia, while on a visit to the UK, gave a paper on star groups. By contrast, a novelty was the reading of a paper by Mr J. R. Simpson on references to astronomy in the poems of Robert Burns. Again, Mr David Sinden and his brother, Mr

Figure 10.5 The Minutes of Meetings, Cash Book and Minutes of the Council have been kept from 1897 onwards. The first two entries date from 13 October 1897, with that from the Council Meeting signed by the President, John Dansken.

Frederick Sinden, presented the 1996 O'Neill Lecture, entitled 'The Stars o' Robert Burns'.

The fiftieth anniversary of the branch occurred at an opportune time. With the close of the session 1943–4, the branch completed fifty years of useful function and this was celebrated, amongst other things, by the re-election of Professor Smart to the Jubilee Chair. By 1944, the danger of invasion had passed, and the war was near its turning point. It was decided to celebrate the date on the largest possible scale, an extremely courageous decision on the part of the office-bearers, considering the very limited resources and what was noted as a comparatively small membership of about 150. Details of the proceedings appeared in the local newspapers, and Fig. 10.6 reproduces a cutting from the *Glasgow Herald*. The guest speaker, Sir Harold Spencer Jones, the Astronomer Royal, noted how the wartime blackouts had allowed city dwellers to become conscious of the stars being more readily visible in the dark skies. A photograph of the dignitaries attending the accompanying reception for the branch and its guests, hosted by the Lord Provost of Glasgow, is presented in Fig. 10.7.

ASTRONOMY IN WAR-TIME

Scots Society's Jubilee

"So far as astronomy is concerned, the war is not going to be unproductive of good," said Sir Harold S. Jones, Astronomer-Royal, Greenwich Royal Observatory, yesterday at a civic reception in the City Chambers, Glasgow, on the occasion of the jubilee celebrations of the Scottish branch of the British Astronomical Association.

"One thing we all want," he went on, "is the end of the war and the end of the black-out, so that we can again enjoy street lighting and shop-window lighting. But the black-out has had one compensation. So many of us have been on duties in the Home Guard, in fire watching, and as wardens in the black-out that we have been able to see the stars in all their glory as we never saw them in peace-time, and it has been possible for us to appreciate the splendour and the beauty of the heavens and to see something of the changing face of the skies throughout the year."

The membership of astronomical societies, said Sir Harold, was increasing, and he hoped that the jubilee celebrations of the Scottish branch would lead to an increase in Glasgow and Scotland and that there would be a renewed interest in what had been aptly called the queen of sciences, astronomy.

Figure 10.6 A report from the *Glasgow Herald*, describing the fiftieth anniversary celebrations of the Society, notes the advantage of having a dark sky in wartime.

Figure 10.7 In 1944, the Scotland Branch of the British Astronomical Association celebrated its golden jubilee with a public lecture and civic reception in the City Chambers, hosted by the Lord Provost. Professor W. Smart is to the left and Sir Harold Spencer Jones, the Astronomer Royal, is to the right.

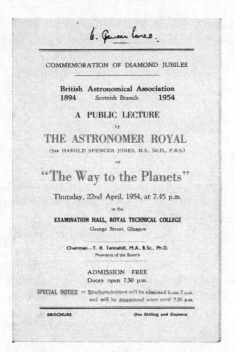

COMMEMORATION OF DIAMOND JUBILEE

British Astronomical Association
1894   Scottish Branch   1954

A PUBLIC LECTURE
by
THE ASTRONOMER ROYAL
(Sir HAROLD SPENCER JONES, M.A., Sc.D., F.R.S.)
on

"The Way to the Planets"

Thursday, 22nd April, 1954, at 7.45 p.m.

in the
EXAMINATION HALL, ROYAL TECHNICAL COLLEGE
George Street, Glasgow

Chairman—T. R. Tannahill, M.A., B.Sc., Ph.D.
President of the Branch

ADMISSION   FREE
Doors open 7.30 p.m.

SPECIAL NOTICE — Brochureholders will be admitted from 7 p.m.
and will be guaranteed seats until 7.30 p.m.

BROCHURE   One Shilling and Sixpence

Figure 10.8 The announcement of a lecture by the Astronomer Royal, Sir Harold Spencer Jones, duly autographed.

The society's diamond jubilee was also celebrated in style. The main event was a public lecture (see Fig. 10.8), again delivered by Sir Harold Spencer Jones, this time entitled 'Life on the Planets'. The presentation was made possible by the generosity of the Royal Technical College governors, always good friends to the branch, who chose the occasion to restart the David Elder Lectures, which had been discontinued during the war. The lecture proved to be exceedingly popular, the audience filling the examination hall of the Royal Technical College, with many unfortunately turned away through lack of space in the auditorium. As a consequence of the talk and its associated publicity, there was an immediate increase in membership.

10.4 THE ASG

The period around the diamond jubilee marked an important stage in the history of the society. For many years, the branch had functioned as an almost independent organisation. Its constitution clearly carried restrictions to its operation as a branch of the parent association, particularly with regard to limitations on the choice of office-bearers, making the conduct of its affairs more difficult than they needed to be. With the full assent and cooperation of the parent association, it was decided that at the end of the 1953–4 session, on 30 April 1954, the branch should be wound up, with the establishment adopting the status of a completely independent society and changing its title to one with more local significance. The celebration of the sixtieth anniversary commemorated not only the climax to six decades of fruitful activity as part of the premier organisation for amateur astronomers, but also the beginning of a new and independent society: the Astronomical Society of Glasgow, or ASG.

For the first session in its new form, there were two honorary Presidents, Professor W. M. H. Greaves, Astronomer Royal for Scotland, and W. M. Smart, Regius Professor of Astronomy at the University of Glasgow; the President was Dr T. R. Tannahill, a Lecturer at Glasgow University, and A. E. Roy acted as Secretary. The wording of the society's title was deliberately chosen not to be 'The Glasgow Astronomical Society', partly because of problems that its acronym might have caused. The changeover of the accounts under the new umbrella of the ASG is noted in Fig. 10.9.

The centenary of the local Astronomical Society was again celebrated with a civic reception hosted by Glasgow Council in the City Chambers on 23 November 1994. On this occasion, the wit and humour of Malcolm Kennedy, the then Secretary, was very apparent in the first recitation of his poem marking the announcement that the IAU had named Minor Planet No. 5805 'Glasgow', in honour of the Society's contribution to the furtherance of astronomy. It reads as follows:

A Rosey, Rosey Glow
I was just a wee asteroid in space all alone
and did not have a name –
A nonentity known just as five eight zero five
Until GLASGOW I became.
Now I'll orbit in the wintertime
and the summertime also;
And my presence will let members of the ASG
Be surrounded by a rosey, rosey GLOW!

While taking part in a mercy mission carrying aid to eastern Europe, Malcolm Kennedy died in a road accident on 18 November 1997. He was born in New Zealand but lived in Scotland, working as a civil engineer. As secretary of the ASG, with his tall presence he organised the meetings with panache, showing a mastery of wordplay and puns. To honour his memory, asteroid No. 7166, originally designated 1985TR, was named 'Kennedy' on 8 August 1998. As with asteroid 'Glasgow', No. 7166 was discovered by Dr Bowell at the Anderson Mesa Station associated with the Lowell Observatory in Flagstaff, Arizona.

The BAA held its Out-of-London meeting in Glasgow on 17 September 1994, as the announcement presented in Fig. 10.10 describes. Thirteen years later, in 2007, the BAA Out-of-London weekend was again held in Glasgow, from 31 August to 2 September, with the local events organised by the ASG. A civic reception was provided and hosted by the Lord Provost; around

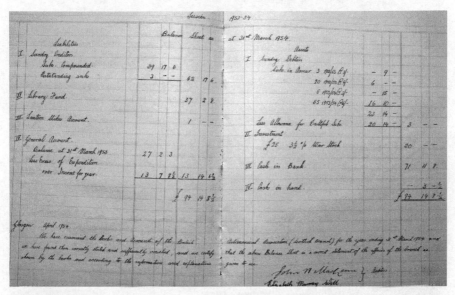

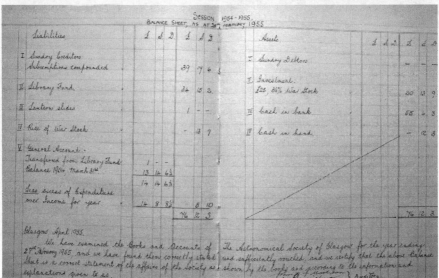

Figure 10.9 The last account of the Scottish Branch of the BAA in 1954 and the first account of 1955 of the Astronomical Society of Glasgow show the level of financial activity of the Society at the time of its reform.

one hundred and twenty people attended the day lectures, and sixty people visited the University Observatory.

The Smithsonian Astrophysical Observatory (SAO) launched an amateur science programme in 1956 called 'Operation Moonwatch', which became

Figure 10.10 An announcement of the BAA giving notice of its Out-of-London meeting in Glasgow on 17 September 1994.

part of the International Geophysical Year of 1957. It involved spotting artificial satellites before professionally manned tracking stations were available. Its initial goal was to enlist the aid of amateur astronomers and other citizens, who would help professional scientists spot the first artificial satellites. Even after the introduction of more professional equipment, the project ran until around 1975. Dr A. E. Roy procured a set of rugged elbow telescopes from the SAO and enlisted the help of the ASG to make observations at the Observatory in University Gardens. Some photographs of this activity form Fig. 10.11.

Lecturers of the highest order are invited each year to keep the membership informed on all manner of astronomical topics. A new highlight of the annual programme is the Tannahill Lecture, given in January each year, in memory of Dr Thomas Russell Tannahill, the first President of the newstyled society, who bequeathed a substantial legacy on his death in 2006. The first Tannahill Medal was awarded at the inaugural lecture in 2009 to Mr Tom Boles, an amateur observer and the discoverer of the greatest number of supernovae. As a teenager fifty years previously, he had attended talks given by Dr Tannahill in Glasgow.

The Society's annual programme also involves making use of facilities at the current University Observatory at Acre Road. In addition, regular observing nights are organised at Mugdock Park, north of the city, where the skies are darker. Keen to undertake outreach activities, the Society also runs general educational programmes for the public at large. This is achieved

Figure 10.11 The Astronomical Society took part in the SAO Moonwatch Programme organised by Dr Roy at the Observatory, in University Gardens. Small telescopes were positioned on concrete pillars, with recordings made of the passage of artificial satellites. In the picture on the right, Margaret I. Watson (now Mrs Morris) is seen collating the various observations.

by using their personal equipment to show daytime live images of the Sun and, on cold winter nights, by setting up their telescopes to provide views of the Moon, planets and other celestial objects. Several thousand people have now taken advantage of events such as 'The Sun over the Botanics' and 'The Moon over the Botanics', held in the Botanical Gardens. Details of these events, and other activities, together with the annual programme of meeting nights and lectures, can be found on the ASG website.[3] With a society of this nature and pedigree, the spreading of up-to-date knowledge of astronomy is well served within the city.

References

1. Stewart, Professor Balfour (1879), *Suspected Relations Between the Sun and Planets*. Glasgow Science Lectures: Series 1878–79–80, Manchester: John Heyward [GUL Gen Sci C12 1880-G].
2. McKim, Richard (2005), 'Henry McEwan of Glasgow: a forgotten astronomer? Part I: Moray Firth to Mount Florida (1864–1916)', *Journal of the British Astronomical Association*, 115, 13–24; McKim, Richard (2005), 'Henry McEwan of Glasgow: a forgotten astronomer? Part II: Cambuslang (1916–1955)', *Journal of the British Astronomical Association*, 115, 87–97.
3. ASG website. Available at: http://theasg.org.uk

Chapter 11

University Gardens

11.1 INTRODUCTION

Professor William Marshall Smart was appointed to the Regius Chair of Astronomy in 1937. Rather than making his home in the Horselethill Observatory, he shortly took up residence in one of the professors' houses in The Square at the University on Gilmorehill, living there until his retirement in 1959. He was a Scot, born in Doune, Perthshire, on 9 March 1889. His scholarly success at the McLaren High School, in Callander, brought him to Glasgow University. He took astronomy as a subsidiary subject under the tutelage of Professor Becker, his main subjects being mathematics, natural philosophy (physics) and chemistry, and graduated with first class Honours in 1911. He then became a scholar of Trinity College, Cambridge, taking pure and applied mathematics, which included astronomy. He graduated in 1914 with a triple first in the Mathematical Triposes and was awarded the Tyson Medal for astronomy. In 1916, he received the Rayleigh Prize.

As it turned out, both world wars affected the direction of his career. Most of his front-line research was undertaken at Cambridge, with an interruption during World War I. Soon after his return to Scotland to take the Chair at Glasgow, the themes of his astronomy teaching were affected by World War II, with the greater part directed to 'navigation'. At the end of the war, he took his turn as Dean of the Faculty of Science (1946–9). From 1949 to 1951, he served as the President of the Royal Astronomical Society. The last ten years of his professorial appointment allowed him to establish a small but regular school of undergraduates, to lecture around the country and to write several texts and popular books. His most famous text, *Spherical Astronomy*,[1] was reprinted many times in several editions; it was later revised and updated following his death. A portrait of Professor Smart is presented in Fig. 11.1.

11.2 THE CAMBRIDGE YEARS

Professor Smart's early research work was related to the 1906 discovery of the Trojan satellites associated with Jupiter. Previously, the triangular solution of the Lagrangian points related to binary motions had been purely

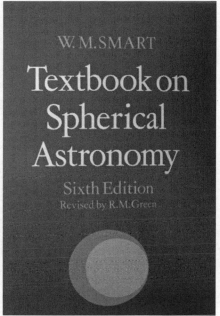

Figure 11.1 Professor William M. Smart with an image of the cover page of his famous and classic textbook entitled *Spherical Astronomy*. The depicted cover is that of a sixth edition with revision by Dr R. M. Green, who was on the staff of the Glasgow Department at the time of the production.

academic, but, with this observational discovery, such orbital issues had taken on practical significance. Smart was one of the first contributors[2] to explain the behaviour of the librations of the Trojans, publishing a paper in 1918. His other works within the field of celestial mechanics included a discussion of the possibility that the precession of the perihelion of Mercury might be caused by an undiscovered planet at a triangular point,[3] and an investigation into the evolution of binary star systems as a result of mass loss by radiation.[4] If World War I had not interrupted his research and changed its direction, he might have continued with such studies. His mathematical skills were obvious, as can be seen in a paper[5] of 1926 related to the determination of the perturbations caused by the gravitational influence of a planet on the orbit of another. His interest in this field remained latent, however, until he wrote a classical text[6] in 1953 on the principles of celestial mechanics.

In 1915, he left Cambridge for the Royal Naval College, Greenwich, and the following year was on active service as Instructor-Lieutenant, serving on HMS *Emperor of India*, a flagship of the division of battleships of the Grand Fleet. As a result of his naval experiences, which he seems to have remembered with affection, he acquired a lifelong interest in navigation.

A paper in 1919 entitled 'The Position Line in Navigation'[7] established his credentials in the field, and, with Commander F. N. Shearne, RN, in 1922 he published the *Admiralty Manual of Navigation* and in 1924, *Position Line Tables*. As it turned out, in the last[8] of his many papers was submitted in 1946 to the Royal Astronomical Society he returned to navigational matters, dealing with the departure formula for the spheroidal shape of the Earth. Between the first and last of his papers in this field, he wrote four books on navigation. During World War II, he gave intensive core courses on navigation to RAF cadets within Glasgow University.[9] Dr Tannahill also played a part here[10] as Flight-Lieutenant to the University Air Squadron. In 1941, Professor Smart was nominated[11] as Halley Lecturer, the chosen title of his talk being 'Sea and Air Navigation'.

Following his naval service, Smart returned to Cambridge as the John Couch Adams Astronomer at the University Observatory, then under the direction of Arthur S. Eddington. He later became Director of the Observatory. The post carried an obligation to pursue observational programmes, as far as practicable, the chief instrument being the Sheepshanks 12-inch coudé refractor. In 1921, he undertook photographic photometry[12] of Nova Cygni 1920, and later he published eight papers reporting on the light curves of several bright variable stars, the observations made with sodium/argon and potassium/argon photoelectric cells (see Appendix A). Although these studies have long since been superseded, the measurements were pioneering, particularly in the way the sources of error were investigated and the data reduced in respect of atmospheric extinction. His interest in general stellar variability may be noted from a paper[13] written in 1924. He was also part of the Cambridge Expedition to Norway to observe the solar eclipse of 1927; although the inclement weather ruled out any participation in the event, the account[14] of the thoroughness of the preparations to undertake observations makes interesting reading.

Around 1920, the concept of the Galaxy as a dynamical entity was only just beginning to emerge, with the realisation that the Sun was not at its centre. Professor Smart contributed to the debate by presenting a discourse[15] on the proper motions of galaxies and on the effects of internal rotations on the positions of stars within them. His interests in investigating the behaviour of proper motions of the stars in our galaxy provided the main theme of his research at Cambridge for over twenty years, with further reductional work continuing with his appointment to the Chair at Glasgow. This monumental study on stellar kinematics produced over thirty papers (see Appendix B), with the principal methods, and some of the results, brought together in his text entitled *Stellar Dynamics*.[16] It became a standard text on the subject and was reissued in 1968 as *Stellar Kinematics*.[17] The breadth of his interests shows the competence of the work that dealt with the techniques of observation, the principles of data reduction and the statistical interpretations of the data. It was his experimental astronomy work in stellar photometry and astrometry that led him to write a text[18] in 1958 on the analysis of experimental errors and their treatment.

Using photographic plates taken with the Cambridge Sheepshanks Telescope some twenty years previously, Smart employed the same instrument to take a second set, applying the 'back-to-back' method whereby the new plate would be exposed through the previous plate, with it in contact. The reliability of the method was assessed in a paper[19] of 1921, and values for the proper motions of stars in the Pleiades cluster were obtained in good agreement with studies made at other observatories. There then followed the major theme of his work, and that for which he is best remembered: the analysis of the systematic aspects of the observed motions of the stars, from both their proper motions and their radial velocities. Two initial proposals for the noted anisotropy of stellar motions were the 'two-stream' model of Eddington and the 'ellipsoidal' model of Schwarzschild. In retrospect, neither of the models had a strong physical basis, but Smart adhered to the two-stream model. He was, however, at pains to point out that, although the two models were not identical, there were, in certain circumstances, relationships between the parameters describing them. He studied several sky fields, some clearly illustrating the effects of star streaming and others providing values for the proper motions and parallaxes of star clusters. From the analyses, he provided a determination[20] of the solar apex, the point in the sky towards which the Sun is moving through its motion in the Galaxy. Occasional stars with large proper motions were identified,[21] and he was also able to identify occasional stars[22] as being members of a well-known cluster. Confirming the star Sirius as a member of the Ursa Major group, he determined[23] that the velocity of this cluster was $V = 19.1 \pm 0.2$ km s^{-1}.

From his photographic observations, he also contributed[24] to the measurement of some double stars, which prompted him to revamp[25] Kowalsky's method for determining orbits of such stars in a clear manner.

Professor Smart served as editor of the *Monthly Notices* of the Royal Astronomical Society in the early 1930s. During this period, he invited Father Georges Lemaître to republish an article of 1927 related to the general expansion of the Universe which, in its original form, had effectively calculated the expansion rate, or Hubble's constant, prior to the notion being accredited to Edwin Hubble. Having been translated from French to English, Lemaître's article published in the *Monthly Notices* omitted these key paragraphs, and it has been recently claimed that Smart might have been responsible for this censoring, so favouring Hubble as the discoverer of the expansion in preference to Lemaître. This notion has been scotched, however, by Livio[26] through the discovery of a letter written to Smart by Lemaître in which he reveals that it was the original author's idea to exclude the 'discussion on the radial velocities' of the galaxies because he thought it was of 'no actual interest'.

11.3 THE GLASGOW YEARS

Although Smart's research at the time of his appointment as the eighth holder of the Regius Chair at Glasgow lent a great deal to theoretical studies, or the analysis of data provided by others, he did have considerable experience with observational work and the development of instrumentation. However, he decided that the climate of Glasgow could not be considered as allowing meaningful research in observational matters, and it was confirmed that the Horselethill Observatory should be abandoned. A simple establishment, closer to the main University, was to provide the requirements for basic observational work for the support of undergraduate teaching. The Observatory at Horselethill, having stood for some 100 years, was demolished and the land sold for about £9,000, as reported in Chapter 9. Many of the old instruments appear to have been lost at this time, including the Ertel transit telescope.

The opening of the new Observatory in University Gardens was announced in the *Glasgow Herald* of 17 April 1939; an article[27] written by Professor Smart provided a comprehensive record of the history of astronomy within the University since 1760. It commented that the Observatory at Dowanhill had been abandoned, and that the new one was essentially for student teaching. The main instrument, a 7-inch telescope, was to be placed in a dome on the high ground. Before the dedication ceremony (see Fig. 11.2), Sir Arthur

Figure 11.2   The opening of the Observatory in University Gardens by Professor Sir Arthur S. Eddington at the top of the steps on 17 April 1939. Professor Smart is below, at the front and to the right. Also in the photograph are the Principal, Sir D. M. Stevenson, the architect, Mr John Stewart and, on the far right, Professor John Walton, Dean of the Faculty of Science.

Figure 11.3 The door of the Observatory at University Gardens incorporated the emblem of the 'snake, the star and the triangle', taken from the previous building at Horselethill, as shown in Figure 9.22.

Eddington gave a lecture entitled 'The Expansion of the Universe'. The opening was also reported[28] in the regular astronomical journals. From the outset of his tenure of the Chair, Professor Smart submitted Annual Reports of the activities of the Glasgow Department to the Royal Astronomical Society (see Appendix C), although these tended to be short during the years of World War II.

Details of the Observatory and of its opening were reported in his first two Reports.[29] It is noteworthy that the entrance door to the offices incorporated the stone 'snake, star and triangle' emblem taken from the Horselethill Observatory, as can be seen in Fig. 11.3. In addition to the telescope installed for student teaching, a small transit house was provided. The old Ertel workhorse instrument was not incorporated, and simply disappeared. The transit telescope installed was by Troughton & Simms, and in 1969 was transferred to the Acre Road Observatory. It was noted that the Director and Mr T. R. Tannahill were engaged in the studies of stellar proper motions using data from the Cape Photographic Proper Motions records. Professor Smart also investigated moving clusters, such as the Taurus Cluster. Of the thirty-nine papers related to stellar motions listed in Appendix B, the last fourteen were completed in Glasgow. Smart's last research paper, written in 1941, concerned his main lifelong study and dealt with the systematic corrections to the proper motions of the *General Catalogue*. From then on, most of his time was taken up by teaching, particularly during the war when

Figure 11.4   The Senior Honours Class of 1948–9
outside the Observatory in University Gardens.

he gave courses on navigation, and in writing books (see his bibliography in
Appendix D).

After the war, the Department attracted small but dedicated student
classes as suggested by the example depicted in Fig. 11.4. This photograph
of the Senior Honours class of 1948–9 identifies in the rear row from left to
right: James Good, Mavis M. S. Gibson, Neil McCorkindale, Archie E. Roy,
Alexander G. Robertson, Angela H. James, Anstruther P. MacFarlane; in the
front row from left to right: Anne Nicolson, Dr Russell Tannahill, Professor
W.M. Smart, Dr Peter A. Sweet, Agnes B. Thomson. On the recommendation of Professor Smart, both Mavis Gibson and Angela James became staff
members at the Greenwich Observatory at Herstmonceux, and married
colleagues there; in 1954, Mavis Gibson married Patrick Wayman, who
later became the Director of Dunsink Observatory in Ireland, and Angela
James married Keith Jarrett, a watchmaker in the workshop, in 1953. Neil
McCorkindale became headmaster at Ayr and Perth Academies, and later a
schools inspector; Anne Nicolson died at an early age, and her books were
donated to the departmental library. Agnes B. Thomson married a research
student in the Natural Philosophy Department, and they moved south.
Archie Roy was later appointed as Lecturer in the Astronomy Department,
and for a short time was Head of Department. Dr Tannahill shortly left the
Department but remained in educational activities elsewhere, maintaining an
interest in astronomy.

The beginning of the change of style of the operations within the Astronomy Department at the University can be clearly seen towards the end of Professor Smart's tenure. In previous times, research tended to be centred on a theme set by the Professor. It commenced this way in 1938, with the Assistant, Mr T. R. Tannahill, continuing to study stellar proper motions[30] using published catalogues, and with research students[31] following associated themes. After World War I there was a radical change as new staff were employed, bringing with them their own research themes. This was important for undergraduate teaching schemes, as it provided a broader and fuller spectrum of topics for the degree courses. The beginning of this can be seen in the appointment[32] in October 1947 of Mr P. A. Sweet, who remained in the Department until September 1952, when he returned south to take up a post as Assistant Director of the University of London Observatory. He was replaced[33] by Michael W. Ovenden, who was appointed Lecturer in the Department in 1952, his immediate interest being the behaviour of eclipsing binary stars. Through the 1950s, the Astronomy Department comprised the Professor, two Lecturers and occasional research students. After undertaking research on stellar structures, one such student, Dr Archie E. Roy, collaborated on several projects with Ovenden, including investigations[34] into commensurabilities associated with planetary and satellite orbits within the Solar System. Following the sudden resignation[35] of Dr Tannahill in September 1957, Dr Roy was appointed[36] Lecturer on 1 April 1958. Around this time, he was organising a 'Moonwatch Team' for recording artificial satellites at the Observatory (see Fig. 10.11).

As pointed out by Ovenden, in his obituary[37] for Professor Smart, the style of astronomical research changed radically after World War II, particularly with the advent of radio astronomy. Understanding of the physics of celestial sources grew at a pace, sometimes in speculative ways, very contrary to the approach of the likes of astronomers of the Smart school. Gone were the days of taking positional catalogues and performing meticulous calculations on their contents. Aspects of this kind of activity of course remain, but now with automated data collection of positional measurements and of brightness and spectral records over the whole of the electromagnetic domain being collated by electronic computers. The same obituary provides the details of Smart's life beyond his main research themes, covering the many positions he held and the public lectures he gave to communicate aspects of his knowledge to larger audiences beyond the confines of the professional scene.

11.4 NEPTUNE AGAIN

As related in Chapters 6 and 7, both Professors Nichol and Grant became involved in the controversies related to Adams, Leverrier, Airy and Challis, and the discovery of the planet Neptune. On the centenary of the tense

Figure 11.5   A photographic portrait of John Couch Adams that was on display at the University Gardens Observatory. His signature beneath may have come from correspondence with Professor Grant.

international dispute over priority of 'discovery', it seems ironic that the Professor at Glasgow should again take part in discussions of the intrigues. In an address to the Royal Astronomical Society, Professor Smart became involved in the continuing debate. His presentation took a fresh look at the episode and provided some new historical material that had come to light, so giving a new perspective to the events; a summary of his talk was published later.[38] The then Astronomer Royal, Sir Harold Spencer Jones, took exception to the portrayal of Sir George Airy's involvement in the affair, and wrote a letter[39] criticising Smart's interpretation. Smart replied,[40] making it clear how and why he had written such an account of the matter.

Glasgow's connection with John Couch Adams seems to have been particularly strong. One of the portraits (see Fig. 11.5) that hung within the Observatory at University Gardens, and possibly earlier at Horselethill, was of him, and its origin is likely to have been through correspondence connections with Professor Grant. As related in Chapter 7, Adams had supported Grant's application for the Glasgow Chair.

11.5 THE UNIVERSITY'S REFRACTOR

When the Observatory at University Gardens was abandoned to make way for the Queen Margaret Union Building, it was decided to dismantle the refractor telescope and to replace it with a more modern instrument. Its

Figure 11.6 At the end of its days, the refractor became a tracking platform for student experiments such as a simple stellar spectrometer.

pieces were preserved and later transported to the new Observatory at Acre Road. During its life at University Gardens, the refractor served a purpose of allowing students and the general public to view the regular celestial objects, but it also became a beast of burden, used as a platform for mounting student teaching projects. The whole arrangement became very cumbersome, as can be seen in Fig. 11.6. Many of the appended instruments, such as a small photographic spectrometer, resulted from ideas suggested by Dr Ovenden. A question arises, though, as to whether this telescope was the Corbett 7-inch Refractor, mentioned in Chapter 9 (see p. 282), or the 9-inch Ochtertyre Refractor.

At the time of its installation, and all through the early period of its use, as noted in the reports of the activities at University Gardens, the established refractor was referred to as a 7-inch. In the inventory[41] for 1939, it is noted that the Cameron Corbett 7-inch Refractor was housed in the dome. Further, a celebratory book[42] published on the occasion of the quincentenary of the University in 1951, from which the photograph of the telescope

Figure 11.7   The middle section of the telescope tube with the lens shutter cap displaying a label for the '8-inch' refractor.

has been reproduced in Fig. 7.11, refers to it as a 7-inch instrument, as was the case also of the announcement[43] of the opening of the University Observatory in 1939, and the Report[44] for 1938. In a 1952 article,[45] it was noted that a 7-inch refractor was available to visitors and school parties. An instrument of such a size had been donated to the University around 1900; it was initially housed under a primitive canvas covered dome but there is little evidence of its use. Later, it seems the University Gardens refractor was 'upgraded' to being an '8-inch'. An anonymous document within a Departmental file refers to the facilities available to the students and notes that: 'Among the instruments available to them are an eight-inch refractor with auxiliary photographic equipment, a transit telescope, sextants . . .'. Also, the notes of the technician, William Edgar, made around 1960 when the instrument was dismantled, refer to the refractor as being an 8-inch. A label (see Fig. 11.7) saying that it was of this size was attached to a section of a tube which found its way to the Observatory at Acre Road, formally opened in 1968. Around this time, a lens had been used in the construction of a primitive coelostat for student use to allow solar spectra to be recorded. As it turns out, this optic has a diameter of 9 inches (see Fig. 11.8), matching that of the Ochtertyre Telescope. It seems likely that this lens belonged to the instrument for many years frequently referred to as a 7-inch refractor, the latter diameter perhaps being taken erroneously from the main section of the telescope's tube.

Figure 11.8 The 9-inch Ochtertyre objective lens.

Appendix A

Papers on Photometry based on Observations made by W. M. Smart at Cambridge University

A1. Smart, W. M. (1929), 'Photoelectric Observations of two Low-temperature B-type Stars', *Monthly Notices of the Royal Astronomical Society* [hereafter *MNRAS*], 89, 545–8.

A2. Smart, W. M. (1933), 'The Atmospherical Extinction Factor in Photo-electric Photometry', *MNRAS*, 94, 115–29.

A3. Smart, W. M. (1934), 'A New Method of Determining the Atmospherical Extinction Factor in Photoelectric Photometry', *MNRAS*, 94, 839–46.

A4. Smart, W. M. and Green, H. E. (1934), 'Photoelectric observations of ζ Aurigae', *MNRAS*, 95, 31–3.

A5. Smart, W. M. (1935), 'Photo-electric Light-curves of δ Cephei and β Lyrae', *MNRAS*, 95, 644–50.

A6. Smart, W.M. (1936), 'On a Suspected Variation in the Light of β Canis Minoris', *MNRAS*, 96, 258–62.

A7. Smart, W. M. (1937), 'The Photoelectric Light-curve of Algol', *MNRAS*, 97, 396–406.

Appendix B

Papers on Measurements of Stellar Proper Motions based on Photographic Observations made by W. M. Smart at Cambridge University

B1. Smart, W. M. (1921), 'Proper Motions of Stars in the Pleiades', *MNRAS*, 81, 536–47.

B2. Smart, W. M. (1923), 'The Proper Motion of a Faint Star near η Cassiopeiae', *MNRAS*, 81, 501–2. See also Smart, W. M. (1923), 'A Faint Star with Large Proper Motion', *The Observatory*, 46, 380–1.

B3. Smart, W. M. (1923), 'The Systematic Motions of the Faint Stars in Four Photographic Regions', *MNRAS*, 84, 3–14.

B4. Smart, W. M. (1924), 'The Systematic Motions of the Faint Stars', Second Paper, *MNRAS*, 84, 123–33.

B5. Smart, W. M. (1924), 'The Systematic Motions of the Faint Stars', Third Paper, *MNRAS*, 84, 481–91.

B6. Smart, W. M. (1925), 'The Proper Motion of the Cluster NGC 2168 (M35)', *MNRAS*, 85, 257–67.

B7. Smart, W. M. (1925), 'The Systematic Motions of the Faint Stars', Fourth Paper, *MNRAS*, 85, 433–9.

B8. Smart, W. M. (1925), 'On the Effects of Long Exposures in Astronomical Photography', *MNRAS*, 85, 682–92.

B9. Smart, W. M. (1925), 'A Comparison of Schlesinger's Proper Motions of the Harvard A.G. Zone with proper Motions Photographically Determined at Cambridge', *MNRAS*, 86, 44–7.

B10. Smart, W. M. (1926), 'The Constants of the Star-Streams from the Photographic Proper Motions of 3029 Stars', *MNRAS*, 87, 122–38.

B11. Smart, W. M. (1927), 'The Errors of Photographic Proper Motions', *MNRAS*, 87, 446–58.

B12. Smart, W. M. (1927), 'The Constants of the Star-streams from the Groningen Proper Motions', *MNRAS*, 88, 144–54.

B13. Smart, W. M. (1928), 'On the Law of Stellar Distribution derived from Proper Motions', *MNRAS*, 88, 567–83.

B14. Smart, W. M. (1928), 'On the Frequency Distribution of Restricted Proper Motions', *MNRAS*, 89, 93–104.

B15. Smart, W. M. (1928), 'On Schwarzschild's Ellipsoidal Theory', *MNRAS*, 89, 105–15.

B16. Smart, W. M. and Green, H. E. (1928), 'The Analysis of Nechvile's Proper Motions', *MNRAS*, 89, 147–57.

B17. Smart, W. M. (1928), 'Capellas Begleiter', *Astronomische Nachrichten*, 232, 207–8.

B18. Smart, W. M. (1929), 'The Moving Foreground of the Large Magellanic Cloud', *MNRAS*, 90, 112–21.

B19. Smart, W. M. (1930), 'Star-streaming in Relation to Spectral Type', *MNRAS*, 90, 743–9.

B20. Smart, W. M. (1934), 'Some Theorems in the Statistical Treatment of Stellar Motions', *MNRAS*, 95, 116–31.

B21. Smart, W. M. (1935), 'Mean Stellar Parallaxes derived from Proper Motions', *MNRAS*, 96, 132–44.

B22. Smart, W. M. (1936), 'Radial Velocities and Star-Streaming', *MNRAS*, 96, 165–71.

B23. Smart, W. M. (1936), 'On the Determination of the Solar Motion', *MNRAS*, 96, 461–71.

B24. Smart, W. M. and Green, H. E. (1936), 'The Solar Motion and Galactic Rotation from Radial Velocities', *MNRAS*, 96, 471–80.

B25. Smart, W. M. (1936), 'On the K Term of the Radial Velocities of B-type Stars', *MNRAS*, 96, 568–74.

B26. Smart, W. M. and Tannahill, T. R. (1938), 'The Constants of the Star-Streams from the Photographic Proper Motions of 1775 Stars', *MNRAS*, 98, 563–70.

B27. Smart, W. M. and Chandrasekhar, S. (1938), 'A Method of Deriving the Constants of the Velocity Ellipsoid from the Observed Radial Speeds of the Stars', *MNRAS*, 98, 658–63.

B28. Smart, W. M. (1938), 'Determination of the Constants of the Velocity Ellipsoid from Radial Speeds', *MNRAS*, 99, 61–70.

B29. Smart, W. M. (1939), 'The Moving Cluster in Taurus', *MNRAS*, 99, 168–80.

B30. Smart, W. M. (1939), 'The Ursa Major Cluster', *MNRAS*, 99, 441–50.

B31. Smart, W. M. (1939), 'On the Relation Between the Stream and Ellipsoidal Constants', *MNRAS*, 99, 561–6.

B32. Smart, W. M. (1939), 'A New Member of the Ursa Major Cluster', *MNRAS*, 99, 700.

B33. Smart, W. M. and Tannahill, T. R. (1939), 'The Constants of the Star-Streams from the Cape Photographic Proper Motions of 18,323 Stars', *MNRAS*, 100, 30–44.

B34. Smart, W. M. (1939), 'The Scorpio-Centaurus Cluster (The Southern Stream)', *MNRAS*, 100, 60–85.

B35. Smart, W. M. (1939), 'The Ursa Major Cluster', Second Paper, *MNRAS*, 99, 710–22.

B36. Smart, W. M. (1940), 'The K Term and the Galactic Rotational Constants', *MNRAS*, 100, 370–7.

B37. Smart, W. M. and Ali, A. (1940), 'The Perseus Cluster', *MNRAS*, 100, 560–9.

B38. Smart, W. M. and Tannahill, T. R. (1940), 'Star-Streaming in Relation to Spectral Type from the Cape Photographic Proper Motions', *MNRAS*, 100, 688–92.

B39. Smart, W. M. (1941), 'The Systematic Corrections to the Proper Motions of the General Catalogue', *MNRAS*, 101, 37–42.

Appendix C

Annual Reports of the Glasgow University Observatory

C1. Smart, W. M. (1939), 'University Observatory, Glasgow', *MNRAS*, 99, 346.

C2. Smart, W. M. (1940), 'University Observatory, Glasgow', *MNRAS*, 100, 279.

C3. Smart, W. M. (1942), 'University Observatory, Glasgow', *MNRAS*, 102, 86–7.

C4. Smart, W. M. (1943), 'University Observatory, Glasgow', *MNRAS*, 103, 81.

C5. Smart, W. M. (1945), 'University Observatory, Glasgow', *MNRAS*, 105, 115.

C6. Smart, W. M. (1946), 'University Observatory, Glasgow', *MNRAS*, 106, 51.

C7. Smart, W. M. (1948), 'University Observatory, Glasgow', *MNRAS*, 108, 66.

C8. Smart, W. M. (1949), 'University Observatory, Glasgow', *MNRAS*, 109, 163.

C9. Smart, W. M. (1950), 'University Observatory, Glasgow', *MNRAS*, 110, 149.

C10. Smart, W. M. (1951), 'University Observatory, Glasgow', *MNRAS*, 111, 194–5.

C11. Smart, W. M. (1952), 'University Observatory, Glasgow', *MNRAS*, 112, 301.

C12. Smart, W. M. (1953), 'University Observatory, Glasgow', *MNRAS*, 113, 325.

C13. Smart, W. M. (1954), 'University Observatory, Glasgow', *MNRAS*, 114, 315–16.

C14. Smart, W. M. (1955), 'University Observatory, Glasgow', *MNRAS*, 115, 148–9.

C15. Smart, W. M. (1956), 'University Observatory, Glasgow', *MNRAS*, 116, 178–9.

C16. Smart, W. M. (1957), 'University Observatory, Glasgow', *MNRAS*, 117, 275.

C17. Smart, W. M. (1958), 'University Observatory, Glasgow', *MNRAS*, 118, 328–9.

C18. Smart, W. M. (1959), 'University Observatory, Glasgow', *MNRAS*, 119, 371.

Appendix D

Books Written by Professor W. M. Smart while at the Universities of Cambridge and Glasgow

D1. Smart, W. M. (1928), *The Sun, the Stars and the Universe*, London: Longmans, Green & Co.

D2. Smart, W. M. (1931), *Textbook on Spherical Astronomy*, Cambridge: Cambridge University Press. The fourth edition was published in 1960, the fifth in 1962 and the sixth, revised by R. M. Green, in 1977.

D3. Smart, W. M. (1938), *Stellar Dynamics*, Cambridge: Cambridge University Press.

D4. Smart, W. M. (1942), *Foundations in Astronomy*, London/New York: Longmans, Green & Co.

D5. Smart, W. M. (1950), *Some Famous Stars*, London: Longmans, Green.

D6. Smart, W. M. (1951), *The Origin of the Earth*, Cambridge: Cambridge University Press. The second edition was published in 1953; it was then published by Penguin Books in 1955, with a revised edition appearing in 1959.

D7. Smart, W. M. (1953), *Celestial Mechanics*, London: Longmans Green & Co.

D8. Smart, W. M. (1956), *Foundations of Analytical Geometry*, London: Longmans.

D9. Smart, W. M. (1958), *Combination of Observations*, Cambridge: Cambridge University Press.

D10. Smart, W. M. (1968), *Stellar Kinematics*, London : Longmans.

D11. Smart, W. M. (1969), *The Riddle of the Universe*, London: Longmans & Green.

References

1. Smart, W. M. (1931), *Spherical Astronomy*, Cambridge: Cambridge University Press. The sixth edition, published in 1977, was revised by R. M. Green.
2. Steg, L. and de Vries, J. P. (1966), 'Earth-Moon Libration Points: Theory, Existence and Applications', *Space Science Reviews*, 5, 210–33.
3. Smart, W. M. (1921), 'On the Motion of the Perihelion of Mercury', *Monthly Notices of the Royal Astronomical Society* [hereafter *MNRAS*], 82, 12–19.
4. Smart, W. M. (1925), 'The Cosmogonic Time-Scale and Binary Stars', *MNRAS*, 85, 423–33.
5. Smart, W. M. (1926), 'On the Fundamental Equations of Planetary Motion', *MNRAS*, 87, 76–93.
6. Smart, W. M. (1953), *Celestial Mechanics*, London: Longmans Green & Co.
7. Smart, Instructor-Lieut. W. M., M.A. D.Sc. R.N. (1919), 'The Position Line in Navigation', *MNRAS*, 79, 520–31.
8. Smart, W. M. (1946), 'On a Problem in Navigation', *MNRAS*, 106, 124–7.
9. Smart, W. M. (1942), 'University Observatory, Glasgow', *MNRAS*, 102, 86–7; Smart, W. M. (1943), 'University Observatory, Glasgow', *MNRAS*, 103, 81.

10. Smart, W. M. (1946), 'University Observatory, Glasgow', *MNRAS*, 106, 51.

11. Smart, W. M. (1942), 'University Observatory, Glasgow', *MNRAS*, 102, 86–7.

12. Smart, W. M. and Green, H. E. (1921), 'Nova Cygni III, 1920, Photographic magnitudes and effective wave-lengths of Nova Cygni (1920), from photographic plates taken at Cambridge University', *MNRAS*, 81, 179–81.

13. Smart, W. M. (1924), 'The Long-Period Variable Stars', *The Observatory*, 47, 267–76.

14. Newall, H. F., Carroll, J. A., Smart, W. M. and Butler, C. P. (1927), 'The Total Solar Eclipse of 1927 Jun 29: Report of an Expedition from the Solar Physics Observatory, Cambridge, to Aal (Hallingdal), Norway, *MNRAS*, 87, 676–84.

15. Smart, W. M. (1924), 'The Motions of Spiral Nebulae', *MNRAS*, 84, 333–53.

16. Smart, W. M. (1938), *Stellar Dynamics*, Cambridge: Cambridge University Press.

17. Smart, W. M. (1968), *Stellar Kinematics*, London: Longmans.

18. Smart, W. M. (1958), *Combination of Observations*, Cambridge: Cambridge University Press.

19. Smart, W. M. (1921), 'Proper Motions of Stars in the Pleiades', *MNRAS*, 81, 536–47.

20. Smart, W. M. and Green, H. E. (1936), 'The Solar Motion and Galactic Rotation from Radial Velocities', *MNRAS*, 96, 471–80.

21. Smart, W. M. (1923), 'The Proper Motion of a Faint Star near η Cassiopeiae', *MNRAS*, 81, 501–2. See also Smart, W. M. (1923), 'A Faint Star with Large Proper Motion', *The Observatory*, 46, 380–1.

22. Smart, W. M. (1939), 'A New Member of the Ursa Major Cluster', *MNRAS*, 99, 700.

23. Smart, W. M. (1939), 'The Ursa Major Cluster', *MNRAS*, 99, 441–50.

24. Smart, W. M. (1931), 'Photographic Observations of Double Stars', *MNRAS*, 92, 37–47; Smart, W. M. (1932), 'Photographic Observations of Double Stars (Second List)', *MNRAS*, 92, 37–47.

25. Smart, W. M. (1930), 'On the Derivation of the Elements of a Visual Binary Orbit by Kowalsky's Method', *MNRAS*, 90, 534–38.

26. Livio, Mario (2011), 'Lost in translation: Mystery of the missing text solved', *Nature*, 479, 171–3.

27. Smart, W. M. (1939), 'Astronomy in Glasgow – New Observatory to be Opened To-day by Sir Arthur Eddington – A Long Tradition', *Glasgow Herald*, 17 April 1939.

28. 'The New Observatory of the University of Glasgow' (1939), *Publications of the Astronomical Society of the Pacific* [hereafter *PASP*], 51, 184.

29. Smart, W. M. (1939), 'University Observatory, Glasgow', *MNRAS*, 99, 346; Smart, W. M. (1940), 'University Observatory, Glasgow', *MNRAS*, 100, 279.

30. Smart, W. M. (1939), 'University Observatory, Glasgow', *MNRAS*, 99, 346.

31. Smart, W. M. (1942), 'University Observatory, Glasgow', *MNRAS*, 102, 86–7.

32. Smart, W. M. (1948), 'University Observatory, Glasgow', *MNRAS*, 108, 66.

33. Smart, W. M. (1953), 'University Observatory, Glasgow', *MNRAS*, 113, 325; Smart, W. M. (1954), 'University Observatory, Glasgow', *MNRAS*, 114, 315–16.

34. Smart, W. M. (1955), 'University Observatory, Glasgow', *MNRAS*, 115, 148–9; Smart, W. M. (1956), 'University Observatory, Glasgow', *MNRAS*, 116, 178–9.
35. Smart, W. M. (1958), 'University Observatory, Glasgow', *MNRAS*, 118, 328–9.
36. Smart, W. M. (1959), 'University Observatory, Glasgow', *MNRAS*, 119, 371.
37. Ovenden, M. W. (1977), 'William Marshall Smart', *Quarterly Journal of the Royal Astronomical Society*, 18, 140–6.
38. Smart, W. M. (1946), 'John Couch Adams and the Discovery of Neptune', *Nature*, 158, 648–52. The same article was also reprinted in 1947 in *Popular Astronomy*, 55, 301–11.
39. Jones, Spencer H. (1946), 'G. B. Airy and the Discovery of Neptune', *Nature*, 158, 829–30.
40. Smart, W. M. (1946), 'G. B. Airy and the Discovery of Neptune', *Nature*, 158, 830.
41. Glasgow University Manuscript 117/27409. Catalogue of Apparatus in the Department of Astronomy – 1947 [Glasgow University Archives].
42. University of Glasgow (1951), *The University of Glasgow Through Five Centuries*, Glasgow: Glasgow University.
43. 'The New Observatory of the University of Glasgow' (1939), *PASP*, 51, 184.
44. Smart, W. M. (1939), 'University Observatory, Glasgow', *MNRAS*, 99, 346.
45. *Glasgow University Gazette*, 13 (February 1952), 6–7.

Chapter 12

To the New Millennium

12.1 INTRODUCTION

Peter Sweet returned to Glasgow following the retirement of Professor Smart in 1959. From the outset of taking the Chair, expansion of undergraduate teaching and research was very noticeable, this continuing to the present day. The expansion of astronomy and astrophysics was not just a local development, but was nationwide and universal. Not only did traditional universities grow in size, with an accompanying increase of activity in astronomy, but new universities became established, with peripheral studies of astronomy within their Departments of Physics, Mathematics or Chemistry; new astrophysics research groups also sprang up in existing universities. Locally, this was partly driven by the dawn of the space age, the opening of Jodrell Bank, the start of Sir Patrick Moore's *Sky at Night* TV series and the discovery and spectacular appearance of Comet Arend-Roland, all of which occurred in 1957. The number of astronomy papers published annually worldwide was nearly constant through the first half of the twentieth century, but in the second half it increased exponentially to over fifty times more. The period may also be described as the 'Age of the Acronym', as this chapter will bear out.

As part of this scene, Glasgow University saw a continual increase in student numbers, with intakes of over 100 for the first-year astronomy science course. This later fell for a while because of the introduction of competing courses such as statistics and computer science, but the numbers soon recovered. Although the Single Honours degree in astronomy was abandoned in the 1980s, there was a growth in the number of students undertaking joint degrees with physics and/or mathematics. Current graduations provide a tally of about twenty to twenty-five students per year, about ten times the levels of the 1960s. In addition, in the 1990s, elementary courses were provided with a less mathematical foundation, and these attracted several hundreds of students per session.

The period from 1960 onwards also saw an increase of staff to six, two of them short-term assistants, all involved with regular teaching and increasingly attracting grants to support postdoctoral research assistants (PDRAs) and research students. In 1970, the two assistantships were ceded to make a new permanent lectureship, the first appointee being J. C. Brown. A new teaching observatory was built in the late 1960s, and, in the mid-1980s, a dark site observing station was established at Cochno Farm, the Veterinary Field Station of the University, close to the Kilpatrick Hills.

It would be an impossible task to list all the people who have taken part in the activities from 1960 to the beginning of the new millennium. There have been so many researchers on short-term contracts and research students leaving after the award of a higher degree, but all have contributed to the general development and activity of astronomy in the University. A key member of staff was Professor Archie E. Roy who continued his research and writing for more than twenty years following his official 'retirement' in 1989. Sadly his passing on 27 December 2012 occurred during the last few days of the preparation of this book. Those who might be described as 'stable' or long-term staff, include, in addition to Professors Sweet and Roy, Professor John C. Brown and Drs Robin M. Green, David Clarke and John F. L. Simmons, all now retired. This team was appointed under Sweet's leadership partly to bring a very broad spectrum of specialist astronomy expertise to the teaching in the Department. In later years, the UK Research Council developed a policy of preferentially funding groups with critical mass of personnel to follow particular lines of study. Consequently, under Brown's leadership, after around 1990, replacement appointments were made to strengthen the research teams, particularly regarding research into solar and astrophysical plasmas; these included Professors Graham Woan and Martin Hendry, and Drs Declan A. Diver, Lyndsay Fletcher, Eduard Kontar and Nicolas Labrosse. Professors Hendry and Woan have since developed close research links with the Institute for Gravitational Research, within the Glasgow School of Physics and Astronomy, whose steady growth has more than doubled the total number of staff with astrophysical interests in the School. Research strengths in solar physics have also included the work of Dr Alec L. MacKinnon in the (former) Department of Adult and Continuing Education. Details of all these people can be found on the Glasgow University website under the School of Physics and Astronomy.

Again, it is an impossible task to describe in detail all the various pieces of research undertaken since 1960. The reader is referred to the Reports and Proceedings of the Observatory and of the Department/Group in publications of the Royal Astronomical Society, these covering an interval of some thirty-five years to the middle of the 1990s (see the Appendix to this chapter). References to the several hundred papers accruing can be obtained by clicking on the author's name in the NASA ADS website.[1] History, of course, is made every day, and it takes time for dust to settle before its essence can be moulded to form a coherent summary. Most of the characters on the astronomy stage in Glasgow over the last fifty years are still alive, and many are still strongly active in retirement. Part of that scene himself, the author finds it difficult to put the key stories into their proper perspective. So much appears to have happened that it is impossible to give the picture in full. It is hoped he will be forgiven if something that should be told has been missed. The recent and current times will be better written about fifty or a hundred years from now. The style of this last chapter is, therefore, a little different from that of the previous ones, but it takes us up to very recent times.

12.2 THE YEARS UNDER PROFESSOR SWEET

Professor Peter A. Sweet was the Ninth Regius Chair of Astronomy at Glasgow, and he held this office from 1959 until his retirement in 1982. Although he was not prolific in terms of the number of papers he published, their impact more than compensated for this. Within any quest, he was most thorough and meticulous, expressing his findings with brevity and absolute clarity. His two most famous and internationally renowned research contributions are referred to as 'Eddington-Sweet Circulation' and 'Sweet-Parker Reconnection'. For both these topics, he established a waymarked path which many researchers have followed, not only in Glasgow but around the world.

In his first fundamental work,[2] Sweet investigated the mixing of material within the Sun by circulation in meridian planes. A comparison of his results with those of Eddington revealed that the velocities involved were a factor of one million less than originally calculated, so considerably reducing the degree of mixing of solar material during its lifetime. By applying the theory to stars in general, he suggested that the evolutionary tracks of early type stars of a given mass would depend on individual equatorial rotation velocities. This was further investigated in a second paper,[3] in which it was shown that for a set of early type stars of a given mass, differences in rotational velocities could cause a spread in the luminosity by half a magnitude. As a result, stars of a given spectral type, but with a range of equatorial velocities, would exhibit a spread of location in colour-magnitude diagrams.

Sweet's proposed mechanism[4] for the production of solar flares was presented at an IAU Symposium in Stockholm and his idea was later taken up by Parker, who attended the same conference. The theory extended Dungey's formal discussion of a magnetic null point where opposing field lines can be squeezed together and reconnect in such a way as to produce a neutral or current sheet whose finite volume allows a finite (flare) energy release rate. Opposing field lines around a current-carrying thin plasma sheet drift together as the current dissipates, provided the plasma conductivity is not infinite. In practice the model fails to provide anything like observed flare rates because the plasma is highly conductive, and because the lane geometry severely restricts plasma flow rate. Nevertheless it was, and remains, the innovative canonical description of reconnection energy release on which most subsequent work has been built.

Following his retirement, Professor Sweet continued undertaking research and teaching for several years. He died on 16 January 2005. An obituary can readily be found on the internet as it appeared in *The Independent* on 9 February of that year.

At the time of taking the Chair, Professor Sweet was supported by two Lecturers, Ovenden and Roy, but the staff numbers were extended to include an Assistant Lecturer, Mr J. S. Griffith. The latter left the Department after a short time and was replaced[5] in August 1960 by Mr R. M. Green, who later became a Lecturer/Senior Lecturer.

The first Annual Report prepared by Professor Sweet[6] shows that, as part of an ordinary degree within the Faculty of Science, ninety-five astronomy students were registered in the first year and fourteen in the second; there were four Honours students. Collaborations between Drs Ovenden and Roy on commensurabilities within the Solar System continued.

Dr Ovenden spent some time at the Dominion Astronomical Observatory in 1961, his visit related to research into eclipsing binary stars. In 1964, he was promoted to Senior Lecturer at Glasgow, before departing permanently for Canada in 1966 as Professor of Astronomy in the Department of Geophysics and Astronomy at the University of British Columbia. While in Glasgow he continued the Departmental tradition of writing successful texts and produced three popular books.[7] In addition to his innovative work on the origins of the constellations,[8] he later invoked the principle of least interaction action, which states that a satellite system of *n* point masses will spend most time in a configuration for which the time-mean of the action associated with the mutual interactions is an overall minimum. Although Dr Roy's initial research had been under Professor Sweet, and was related to meridional circulation in stars, his associations with Ovenden carried him on to problems within celestial mechanics. He rapidly gained an international reputation for his work, particularly on commensurabilities within the Solar System, and many colleagues from abroad came to work with him in the Glasgow Department.

On Dr Ovenden's resignation, the newly appointed Lecturer, who had a bias towards practical astronomy, was Dr David Clarke, who took up the post in January 1967. He introduced the theme of polarimetry to the Department, which formed the basis for observational experiments related to stars, the Sun and the Earth's atmosphere, as well as providing a stimulus for a range of theoretical studies, led by Professor Brown, related to the geometries of stellar atmospheres and binary stars. Later, he became Director of the Observatory. Following his retirement, he produced a text[9] covering all aspects of stellar polarimetry.

Reports[10] of the early 1980s show the greatly enlarged activity of astronomy in Glasgow with various PDRAs employed, many overseas visitors, and contributions made by summer students. The strong factions within the research areas were celestial mechanics, solar physics, astrophysics, observational photometry and polarimetry, and polarimetric modelling. The publications list for the three years prior to 1 August 1986 shows that sixty-one papers were produced in that period.

12.3 CELESTIAL MECHANICS

Professor Roy's main research work centred on astrodynamics and celestial mechanics, although he had subsidiary interests in considering the behaviour of neural networks to explain how the brain develops and recalls memories, and in archaeo-astronomy. Further, he had a lifelong interest in things paranormal and was a Member and Past President of the Society for Psychical Research and Founding President of the Scottish Society for Psychical Research. Reference to such interests acts as a reminder of Professor Sinclair and his work on the witches of Glenluce, as described in Chapter 2. He was a prolific writer, author of twenty-seven books, many unrelated to astronomy, six of which are highly entertaining novels. An early book was a popular monograph[11] describing the stories behind some important astronomical discoveries. His classical text, entitled *The Foundations of Astrodynamics*, was published[12] in 1965. It was later republished[13] as *Orbital Motion*, which is now in its fourth edition. His international reputation was a passport to conferences around Europe, at which he served on the organising committees and edited the proceedings.[14]

In 1969, Professor Roy was one of the very few beneficiaries of the excellent odds offered by William Hill related to the date of the first man landing on the Moon. He had placed his bet several years prior to the Apollo missions, after travelling to the USA and visiting several of the teams involved with the development of the orbital mechanics associated with the great venture. The winning gamble gave him an amused satisfaction, and the honoured betting slip was pinned to the wall of his office.

One of the themes of his astrodynamical research was long-term commensurabilities within the Solar System. He was the international team leader for Project LONGSTOP[15] (the LONg-term Gravitational STudy of the Outer Planets), which began as an attempt to investigate, using numerical and analytical methods, the problem of the stability of the outer planets, from Jupiter to Pluto, over an interval of years, a time span approaching the known age of the system. The expertise gained from the study of the outer planets was later applied to the satellites of Uranus.

A second major contribution to astrodynamics was his work on the four-body problem,[16] referred to as the 'Caledonian Problem'. This project was conducted with Bonnie Steves who now continues the work at Glasgow Caledonian University, so bringing an element of astronomy to this younger educational establishment.

Continuing the tradition of publishing books, and based on the well-founded undergraduate courses within the Astronomy Department, two texts were produced in 1977 by Archie Roy and David Clarke: *Astronomy: Principles & Practice*[17] and *Astronomy: Structure of the Universe*.[18] In their original format, the two tomes had local and international success, and were reprinted and extended to three editions; the first book of the pair was revamped in a fourth edition in 2003.

Following Professor Sweet's retirement, Professor Roy acted as Head of the Department of Astronomy from 1982 until 1986. He retired in September 1989, and the occasion was marked by a day of special lectures entitled 'A Lad o' Pairts', with invited speakers covering his interests from archaeo-astronomy to celestial mechanics, and from the paranormal to science fiction. In the evening Heather Couper gave a public lecture which 800 people attended, filling three lecture theatres to capacity. To commemorate all his contributions to celestial mechanics, the asteroid no. 5806 was named in his honour, with the registration at the IAU's Minor Planet Center carrying the title Archieroy. It is an unusual Hungaria-type body about 5km in diameter, orbiting at 1.96AU with an eccentricity of 0.04 and inclination at 21°.

12.4 THE GARSCUBE OBSERVATORY

Soon after Professor Sweet's appointment, it was obvious that the Observatory at University Gardens, with its library and offices, was too small to accommodate the expansion of student teaching with increased staff numbers, PDRAs, research students, and so on. In addition, land within or close to the campus at Gilmorehill was at a premium.

In 1962, planning commenced to establish a new observatory at Acre Road, Garscube, about 3 miles north-east of the campus. The site at University Gardens was abandoned to make way for the Queen Margaret Building, which still remains for general student use, as seen in Fig. 12.1.

Figure 12.1 The current view of University Gardens from the top of the Boyd Orr Building. The Observatory site was on the near left, now occupied by the Queen Margaret Union Building.

Figure 12.2 The current site of 26 Ashton Road, the house now demolished, with the area used as a car park. The photograph was taken close to the point below the sculpture displayed in Figure 8.11.

Meanwhile, Departmental activities and offices were established in temporary accommodation at 26 Ashton Road, formerly the home of the Secretary of the University Court, Dr Hutcheson. The house was later demolished, together with the row of adjacent houses along the west side of Ashton Road, soon after the erection of the Mathematics Building, higher up in University Avenue. A car park currently occupies the former location of the house, as can be seen in Fig. 12.2. Undergraduate teaching and theoretical research moved to the top floor of the new Mathematics Building (see Fig. 12.3), while practical teaching and observational research developed later at Acre Road from 1970 onwards.

After a protracted building phase, the Garscube Observatory at Acre Road was opened on 17 March 1969. The official 'turning of the key' was performed by Professor Hermann Brück, Astronomer Royal for Scotland at that time. A photograph of the event taken in the Observatory Library is shown in Fig. 12.4. The occasion was reported not only in all the local papers (see Fig. 12.5) but also internationally.[19]

The main building principally comprised three large and three smaller (see Fig. 12.6) teaching laboratories, eight staff rooms, a library, a mechanical workshop, a photographic dark room and a tall tower at the west end for a coelostat to provide a solar image of about 20cm diameter for analysis by a large grating spectrometer that occupied the length of the basement. The solar observational scheme was abandoned as a result of insufficient tanking of the building, resulting in a substantial water flow seeping into the basement; the project was under-resourced in terms of both the finance and the

Figure 12.3 The top floor of the Mathematics Building, viewed from University Avenue, home to the Astronomy Department from c. 1970 to 1986.

Figure 12.4 The official party at the opening, on 17 March 1969, of the Garscube Observatory at Acre Road. From left to right: Professor Walter Stibbs of St Andrews; Professor Peter Sweet; Sir Charles Wilson, Principal of the University; Professor Hermann Brück, Astronomer Royal for Scotland; and Dr David Clarke.

GLASGOW: R.4 "s" Man 29/69 New observatory has key role to play

Professor Peter Sweet and the new Glasgow University Observatory.

A NEW chapter in astronomy in Glasgow opened this month with the commissioning of the new university observatory off the Maryhill Road on the north side of the city. This £100,000 installation, contained in an attractive crenellated building with two attendant domes, is Glasgow's fourth observatory since the found...

laboratories, and well-equipped workshop, the observatory will play an increasingly important part in the whole study of astronomy in Glasgow.

The head of the department is Professor Peter A. Sweet. "What we had, to have here was a kind of research instrument which we could demonstrate for teaching purposes. A

Moon landing may be seen in Glasgow

From Our Correspondent

Glasgow, March 16

A powerful 20in. reflecting telescope at the new observatory of Glasgow University will be able to track the American Apollo rocket ship from launching to touch down on the moon. The observatory, which cost more than £100,000, is being officially opened tomorrow on the Garscube estate, Maryhill Road, Glasgow, by Professor H. A. Bruck, Astronomer Royal of Scotland.

The telescope, which cost £16,000, can pick up areas as small as a quarter-of-a-mile wide on the moon. With super-sensitive photoelectric cell equipment, it could detect a point of light as small as a candle flame on the moon. Dr. David Clarke, lecturer in practical astronomy at the university, hopes to use the equipment to watch the first moon landing.

The observatory includes a solar tower and provision for a small radio telescope, research laboratories and a workshop. There is also a library, lecture room with small planetarium, and undergraduate laboratories. There is provision for "indoor observation in bad weather". This consists of plates with specially designed artificial stars and constellations together with instruments for observing and interpreting results.

Figure 12.5 Professor Sweet at the opening of the Observatory at Acre Road. Excitement was in the air prior to the first man landing on the Moon and newspaper articles tended to exaggerate the capabilities of the 20-inch telescope in terms of its use for observing this historic event.

STUDENTS MAY GET CLOSE-UP VIEW OF MOONSHIP LANDING

By a Staff Reporter

Student astronomers at Glasgow University hope to be among the few people on earth to watch the landing of the first manned spaceship on the moon.

A 20in. reflecting telescope installed at the university's new observatory will enable them to track the American Apollo rocketship from blast-off to the point of touch-down on the moon.

The observatory, with equipment, cost well over £100,000. It is to be opened to-day at Garscube Estate, Maryhill Road, by Professor H. A. Bruck, Astronomer Royal for Scotland. The telescope cost £16,000

and is powerful enough to pick up in detail areas as small as a quarter of a mile wide on the moon, 250,000 miles away.

But when a super-sensitive photo-electric cell equipment is fitted the telescope can detect the flame of a candle on the moon's surface.

It is this equipment which Dr David Clarke, lecturer in practical astronomy at the university, hopes to use to watch the first moon landing.

But he and Professor P. A. Sweet, of the department of astronomy, with their students, could be disappointed.

"The weather is so bad and there is so much cloud around Glasgow that you can use the telescope on only one night in seven," Professor Sweet said.

The observatory has a solar

tower and provision for a small radiotelescope and a workshop. There are library and lecture rooms with a small planetarium with undergraduate laboratories which can accommodate up to 30 students.

An innovation is equipment for indoor observation in bad weather. It includes plates with specially designed artificial stars and constellations, with instruments for observing them and interpreting results.

The two towers of the new observatory.

"GLASGOW HERALD" 17 March 1969

manpower required to establish it and then to maintain it in regular operation. The domed tower was later converted to carry a telescope on its upper level, with a Goto E5 Planetarium installed at ground level. The basement was eventually used for student experiments which required long optical paths in a dark environment.

At the time of the commissioning of the new Observatory, all talk was of the first Man Moon Landing scheduled for June of the same year. The local

Figure 12.6 A panoramic view of the Acre Road
Observatory soon after its opening in 1969.

media reported the official opening of the Observatory but, as frequently
happens, some casual remarks of the day provided slightly distorted portray-
als of what had been said. The reproduced articles reveal that students might
be able to see the Apollo lunar landings directly from Acre Road. An interesting comment made was that, with a photometer attached to the telescope,
it should be possible to detect a candle placed at a distance of a quarter of a
million miles, a figure just greater than the Moon's distance, not that such
an experiment was ever envisaged with respect to hunting for candles on the
surface of our nearest celestial neighbour.

A separate dome to the south of the main building housed a Grubb
Parsons 20-inch Ritchey-Chrétien telescope (see Figs 12.7 and 12.8). This
instrument was purchased following the collapse of the contract, with
some financial loss, with Cox, Hargreaves & Thompson, who were unable
to supply their 16-inch model. The new telescope arrived without a drive
and tracking system; this was designed and manufactured in house by Dr
Clarke, using a stepping motor controlled by a thermostated oscillator with
an output frequency selectable by relays under push-button control. The
declination movement used the original DC motor, with its control and
speed operated by relays. Although the telescope served well as a platform
for student observations and as a test bed for research photometric and
polarimetric instruments, it was eventually relocated in 1984 to the edge
of the Kilpatrick Hills at Cochno Farm, and reinstalled in a more spacious
dome. To the east of the telescope dome at Acre Road, a transit house was
established for teaching purposes. The old instrument (see Fig. 12.9) by
Troughton, with a diameter of 2 inches, was used by students until around
1990 when it was phased out of the scheme of observational practice.

In addition to being used regularly for undergraduate teaching, the
Observatory was, and remains, a base for many kinds of public outreach
activities. Visits from schools and adult associations were encouraged, and
the Astronomical Society of Glasgow has used the facilities on a monthly

Figure 12.7   The delivery of the 20-inch Grubb Parsons telescope in 1967.

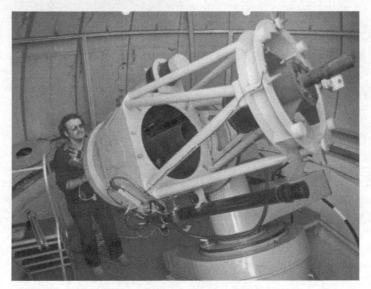

Figure 12.8   The Grubb Parsons 20-inch Ritchey-Chrétien telescope at the Acre Road Observatory. The observer is Dr Colin Aspin.

basis. The Visitors Book shows that about 1,000 people, adults and children, have signed in each year. There are two memorable events that deserve special mention.

Firstly, in July 1972, a summer school was organised, with about twenty teachers and other astronomy enthusiasts taking part. Included in the list of presenters were Dr Green, Mr Brown and Mr Black, and the list of participants included Mr White and Mr Blue; not all the spectrum was represented, but it was a most colourful occasion. The second event of note

Figure 12.9 The old Troughton telescope in the transit house. The technician is Mr Colin Hunter.

was the exhibition and week of public viewing arranged for the appearance of Comet Halley in November 1985. The event was oversubscribed, with entry tickets limited to 800 over the three viewing nights.

Observatory outreach activities by Glasgow astronomers laid the foundations for an ever-widening programme. Many of the exhibits and presentations for this work were supported by Mrs M. I. Morris. Her enthusiasm for the subject is unbounded. A feature of tea breaks in the Observatory workshop was the discussion of research projects, as well as talk related to personal matters and student life, all listened to with a sympathetic ear. Glasgow students, now based all over the world, remember her with affection.

12.4.1 Stellar Polarimetry

From the opening of the Observatory at Acre Road to the end of the millennium, optical observational astronomy was undertaken by Dr Clarke. During this period, several photometric and polarimetric instruments were developed and engineered by the technician, Mr William Edgar. Although they were used on the 20-inch telescope at home, providing data of sufficient import that papers ensued, their success came through use on the Yapp and 2.4m Isaac Newton telescopes at Herstmonceux, as well as on telescopes abroad. Several observing schemes were established, with Glasgow-made equipment shipped out to the Lowell Observatory at Flagstaff, Arizona, on several occasions and to the South African Astronomical Observatory at Sutherland. Figure 12.10 shows a Glasgow dual-channel polarimeter attached to the Morgan Telescope on Mars Hill at the Lowell Observatory in

Figure 12.10 Stellar polarimetric equipment from Glasgow attached to the Morgan Telescope at the Lowell Observatory, Flagstaff, Arizona, in December 1981. The research student was Hugo E. Schwarz.

December 1981. In addition to successful stellar polarimetric observations, including those of late-type stars,[20] the photometric effects of buoyancy waves in the Earth were discovered,[21] this work having repercussions on negating[22] conclusions relative to the then current notions that short period oscillations of the apparent solar diameter could be monitored by ground-based instruments.

Several early observational research programmes undertaken with the local 20-inch telescope achieved success with papers[23] reporting their results and conclusions. One instrument was designed to investigate the rapid spectral fluctuations that had been reported in emission line stars, particularly early-type stars. Instead of using a conventional slit entry, the specially constructed spectrometer allowed the whole of the seeing disc to pass through the optical system. Observations made with this equipment proved that the 'stellar flickering' reported by several groups of observers abroad were instrumental artefacts,[24] the degree of apparent variability depending on the steepness of gradients of the wings of the emission line profiles, and on the quality of the atmospheric 'seeing'.

Figure 12.11 The 20-inch Grubb Parsons reflector being placed in the new dome at the Cochno Observatory on 18 April 1984.

Following the move of the telescope to Cochno in 1984 (see Fig. 12.11), in addition to the stellar work carried out, polarimetric observations were made of the Moon[25] and of the Sun; for the latter, measurements were made at the solar limb[26] and of the total global light,[27] to see how the behaviour related to stars of similar type. Spectropolarimetry was also undertaken of the daytime sky to investigate the Ring effect.[28] Later, specially designed instruments were used to observe the blue sky[29] in order to reveal the temporal behaviour of rotational Raman scattering superimposed on Rayleigh scattering. A full record of references related to polarimetric observations made with Glasgow instruments and analyses of other data can be gleaned from the Reports listed in the Appendix below, and from the University and NASA websites.

Since the relocation of the 20-inch, various telescopes have been housed in the main dome at Garscube. These have included a 12-inch Newtonian reflector, a 10-inch Meade, a 12-inch Meade and, finally, a 16-inch Meade, all of these supporting student observational projects. With the arrival of Dr Graham Woan in 1996, the Observatory gained an additional facet through the introduction of instruments related to radio astronomy. A 3m dish radio telescope tuned to the 21cm hydrogen line (HI) is now sited on the laboratory roof and a 408MHz Yagi pulsar telescope is within the grounds. A photograph of the latter arrangement is shown in Fig. 12.12.

12.5 SOLAR, STELLAR AND LABORATORY PLASMA PHYSICS

The last ever formal Report[30] for astronomy at Glasgow University reveals the vibrant activities of the staff, visitors and research students. Paragraph

Figure 12.12   The Acre Road dome with a radio telescope antenna in the foreground.

headings were 'Celestial Mechanics', 'Cosmology & Relativity', 'Solar Physics', 'Astrophysics' and 'Photometry & Polarimetry'. With the exception of the first, these topics remain research themes today, but with the addition of plasma theory and radio astronomy. Emphases have changed, of course, as have the specific topics under the umbrella headings.

Under the topic of 'Polarization', the main aspects of study involved the establishment of the statistical behaviour of polarimetric measurements[31] according to the influence of experimental errors, the statistical behaviour of stellar polarimetric catalogues[32] and the modelling of circumstellar environments to explain the genesis of observed intrinsic polarization, its wavelength dependence and temporal behaviour. Much of the work in the last area was driven by Drs Brown and Simmons. Early in the establishment of these theoretical studies, two important papers appeared which now have canonical status. One[33] is related to the effect of the stellar atmosphere oblateness controlling any generated polarization. A second[34] provides the model whereby the inclinations, i, of binary stars undergoing mass transfer can be determined; the value of i is key to determining the masses involved and, for example, allowing discrimination to be made on whether a dark companion is a neutron star or a black hole. Brown and Simmons extended their studies of radiation transfer through their supervision of Colin McInnes's PhD study on light-sail spacecraft. This work launched Professor McInnes on a meteoric rise to world eminence in aerospace engineering, and he is now head of the Advanced Space Concepts team at Strathclyde University.

By far the strongest element in research from the mid-1970s onwards was theoretical solar physics, essentially established by Professor Sweet from 1959, though the Department had its origins in the solar work of Alexander Wilson from the 1760s (see Chapter 4). From 1971, one of Sweet's research students, John Campbell Brown, moved the hitherto theoretical solar work into applications associated with spacecraft data on solar activity, and progressively carried Glasgow's solar physics to an internationally renowned level in this field. Of notable importance were his papers[35] of the early 1970s; these set a benchmark in understanding the nature of solar flare spectra in the X-ray region. The first noted work has attracted by far the most citations (over 600) of any Glasgow work and the count is still rising. Amusingly, a plot of the citation rate versus year shows variations following the sunspot cycle, since the content of the paper is most useful when the Sun is active! Much of Brown's subsequent work in this area was carried out in international collaborations, including seventeen study leave periods of one month to one year, mainly in Holland and the USA, setting a working regime now followed by the young members of his built team. His main current research effort is in the physics of dramatic comet-sun impacts, bringing together his solar expertise and his early work on meteors as a summer student at Harvard in 1967.

With the launch of space projects such as SOHO, RHESSI, Hinode and SDO, the Glasgow astronomers have continued to be at the forefront of collating the data and physically interpreting solar events and their behaviour. This will no doubt continue as recently appointed young staff members (Drs Fletcher, Kontar and Labrosse) are very much involved in these areas, making substantial contributions to the theory and data interpretation associated with the NASA award-winning success of the RHESSI mission, and being closely involved in future mission plans such as ESA's Solar Orbiter. The standing of the solar team is reflected in recent awards by the Royal Astronomical Society: the Sir Harold Jefferies Lectureship to Dr Lyndsay Fletcher in 2011, and their highest accolade of a Gold Medal (Geophysics) to Professor John Brown in 2012. Dr Hugh S. Hudson of University of California, Berkeley, now spending six months of the year on this side of the Atlantic as an Honorary Member of the Glasgow Group, is a winner of the American Astronomical Society's top (Hale) prize in solar physics.

In a closely parallel strand, the appointment of plasma physicist Dr Declan Diver as a Lecturer in 1990 and his collaboration with Emeritus Professor E. W. Laing led to basic plasma theory work in solar physics, stellar physics and cosmology, as well as to fundamental laboratory work on cold plasmas currently in the near-market phase of commercial applications.

Going hand in hand with the interpretation of any accumulated astrophysical data are their inversions, or how well any outcome is established in reference to any other. This was a problem that came very much to the fore in the study of solar X-ray images and spectra. Yet again, the traditions of Glasgow astronomy were followed here through the publication[36] of a general approach to inversion strategies. Its appearance has offered a useful insight into how data may be handled and has stimulated strong debate on the meaningfulness of some 'determined' interpretations. This pioneering work by Craig and Brown has led to numerous Glasgow papers on such astronomical 'inverse problems' and to wide international recognition of their importance by, for example, drawing eminent mathematicians Piana and Massone into work on RHESSI data interpretation.

12.6 THE NEW PHYSICS AND ASTRONOMY DEPARTMENT

During the early 1980s there was a widespread feeling that astronomy should really be part of physics departments, and, with the departures of Regius Professors Sweet in 1982, Walter Stibbs (St Andrews) in 1989 and Malcolm Longair (Edinburgh) in 1990, there was a possibility that three Scottish astronomy departments might become minor players in these much bigger ponds. While recognising that the value of single astronomy degrees was weakening because of the lack of physics in them, and that there was scope for collaboration with nuclear physicists, particle physicists and gravitational wave experimenters, there was some hesitation on both sides regarding a merge with the Department of 'Natural Philosophy'. The arguments included the possible loss of the latter's historical name and a loss of astronomy's identity with no strong Chair post as its voice. (The recent sudden closure of the renowned Department of Astronomy in Utrecht, after 375 years, shows that such fears were not unfounded.) After several years of negotiations, the University still refused to refill the Regius Chair of Astronomy, but was persuaded to create and advertise a new Chair of Astrophysics with the remit of guiding the two Departments to much closer joint work in research and teaching. The advertisement attracted a very strong field, with John C. Brown winning the contest in 1984 (see Fig. 12.13). Details of the merger were thrashed out between Professors Brown, Roy, Hughes and Laing, and the Departments formally came together, under the name of 'Physics and Astronomy', on 1 August 1986. To mark the occasion, Professor Martin Rees (now Lord Rees and Astronomer Royal) gave a guest lecture entitled 'Relativistic Astrophysics and the Big Bang' on 21 November. Thus, after 225 years of independence, astronomy returned to being part of a larger entity with regard to studying the Universe and providing education on its physical aspects. After a little delay, the marriage was fully consummated by cohabitation, with the Astronomy & Astrophysics Group rehoused on a custom-built new floor (see Fig. 12.14) on top of the refurbished and renamed Kelvin (formerly 'Natural Philosophy') Building.

At the time of building the Garscube Observatory at Acre Road, one of the 'snake, triangle and star' emblems, originally over the front and rear doors of the Horselethill Observatory (see Fig. 9.22), was incorporated into the entrance façade. Some twenty years later, Professor Brown spotted the second carved block lying outside the Chemistry Building in thick moss and grass cover. Estates and Buildings undertook the renovation, and gave it a beautiful facelift, fitting it into the wall at the entrance of the Kelvin Building,

Figure 12.13 Professors John C. Brown, Peter A. Sweet and Archie E. Roy on the occasion of Professor Brown's appointment to the new Chair of Astrophysics in 1984.

Figure 12.14 A view from the top of the Boyd Orr Building shows the light-coloured Kelvin Building, with the top level added to accommodate the Astronomy Group.

as can be seen in the right-hand section of Fig. 12.15. The piece is at least 170 years old. As mentioned in Chapter 9, the origin of its design remains unknown, but the emblem of the Ouroboros is pertinent here as it provides the first image of the book, in the Preface, and the last in this final chapter.

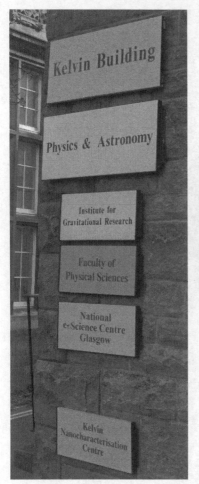

Figure 12.15   Labels at the entrance of the Kelvin Building announce the merger of the Natural Philosophy and Astronomy Departments. The emblem on the right was originally incorporated within the Observatories at Horselethill and University Gardens.

Professor J. C. Brown became Group Leader and coordinator of the astronomy and astrophysics research and astronomy teaching matters within the enlarged Department. Shortly afterwards, Professor E. W. Laing, originally Leader of the Plasma Physics Group, and Dr J. Cumming became members of the Astronomy and Astrophysics Group. This section is now under the leadership of Dr Declan Diver. In December 1996, Dr Graham Woan was appointed as Lecturer, bringing with him expertise in radio astronomy. Later, Dr Martin Hendry joined the staff, adding his enthusiasm for public outreach work to this already strong feature of the Glasgow scene; his chief research relates to the determination of cosmological distance scales. The interests of Professors Woan and Hendry have considerable overlap with the astrophysics associated with the world-eminent Institute for Gravitational Research, and very strong bridges are now in place between the two groups.

Three of the outcomes of the 'Observatory Wars' fuelled by the government's Particle Physics and Astronomy Research Council were the demise of the Royal Greenwich Observatory (RGO), the division of the Royal Observatory Edinburgh (ROE) into the UK Astronomy Technology Centre (UK ATC) and the University Institute for Astronomy, and the end of the Astronomer Royal for Scotland's connection to ROE, just as the Astronomer Royal post had been decoupled from RGO in 1972, with Sir Martin Ryle of Cambridge succeeding Sir Richard van der Riet Woolley.

Malcolm Longair had left the Scottish Astronomer Royal post and ROE, moving to the Cavendish Laboratory at the University of Cambridge in 1990. After a long interval, in February 1995 Professor John C. Brown was appointed the tenth Astronomer Royal for Scotland for his excellent research in and public promotion of astronomy. This was the first occasion of the prestigious position being released beyond Edinburgh's realm. From its establishment, it was automatically awarded to the Regius Chair holder at Edinburgh. It had been exclusive, not because of derogatory remarks made in the *Glasgow Herald* of 20 October 1863 (see Chapter 8, p. 246) regarding it being a 'mere title', but because of the status that Edinburgh had relative to Glasgow in the early part of the nineteenth century. When the telescope maker Short returned to Edinburgh in 1766, he built a small observatory on Calton Hill and it later grew in several stages. The site was eventually taken over by the city authorities and, following a loyal address to George IV during his visit to Scotland in 1822, it became the Royal Observatory. The first Astronomer Royal for Scotland, Regius Professor Thomas Henderson, was appointed in 1834. The office carries no pecuniary reward, but it provides an authority for expanding the 'cause' of astronomy within the community.

As holder, Professor Brown became even more active in this area, such as winning Scottish Higher Education Funding Council funding in 1997 for peripatetic inflatable planetaria in Glasgow, Edinburgh and St Andrews for use in local schools, and presenting countless public lectures, many of which have drawn on his skills as a magician. Much of his work as Astronomer Royal links astronomy to art, music and poetry, and he is involved in projects to save key historic sites such as Calton Hill Observatory and the remains of Brisbane's 1810 Observatory in Largs. In recognition of his status as Astronomer Royal, he was translated from the Chair of Astrophysics to the tenth Regius Chair of Astronomy within Glasgow University in 1996. He initiated the 2010 RAS National Astronomy Meeting in Glasgow in celebration of the 250 years of the Regius Chair. The meeting was highly successful, as had been the previous RAS Out-of-Town Glasgow meeting in

1990 masterminded by David Clarke. In 2007, Professor Brown moved to a half-time Chair appointment and retired in June 2010 to guarantee funds for the permanent appointment to a Lectureship of Dr Eduard Kontar, formerly a Science and Technology Facilities Council Advanced Fellow, in order to retain his outstanding solar expertise in Glasgow. In July 2010, the role of Group Leader was taken over by Professor Graham Woan.

12.7 THE GLASGOW SCIENCE CENTRE

After several decades of lobbying, particularly by Professors Roy and Brown and members of the Astronomical Society of Glasgow, a Public Science Centre at Pacific Quay on the south side of the Clyde was opened in June 2001, partly thanks to a large grant from the Millennium Commission. As well as housing over 250 interactive science-learning exhibits on three floors, the Centre holds a computer-controlled Zeiss Optical Planetarium with the best starry sky in the UK projected onto a 15m diameter dome. In addition to being programmable for special event talks, regular shows are presented to encourage the general public to learn about the night sky and its behaviour and to teach undergraduates about positional astronomy. The opening presentation was conducted by Professor Brown.

12.8 CONCLUSION

We are now well into the third millennium, and astronomical activity in Glasgow continues apace. Not only does the long-established University continue to lead in the academic field, there are also strong strands of research and teaching at the two other universities in the city centre, Glasgow Caledonian and Strathclyde. In addition, the Astronomical Society of Glasgow (as well as the Coates Observatory in Paisley, Renfrewshire, and the Airdrie Society) is well known for promoting the subject both in the city and beyond. Locally, the subject, in all its aspects, remains in good hands.

It is remarkable that in both education and commerce, Glasgow has such a history to relate. It is hard to imagine that the early teaching about the Universe in the College more than 550 years ago involved the concepts of an Earth-centred Solar System. Likewise, it is hard to grasp that just forty years ago, a whole Honours examination paper was devoted to positional astronomy, now relegated to one short first-year course as it has been overtaken by the explosion of knowledge about the physics, chemistry and biology of the cosmos, topics all researched to a high standard by members of the new Scottish Universities Physics Alliance.

Remarkable, too, are the stories of city people who came together, particularly to promote the subject, and established funds in 1810 and 1840 to build substantial observatories. The complete text above mentions eight observatories that have served some purpose, either for general education or for research. More importantly, however, the story provides an insight into some aspects of social history and some interesting characters and their lives. The future remains open!

Appendix

Proceedings of the Glasgow Observatory and Reports of the Department
The year in brackets immediately after the number refers to the period covered in the Proceedings or Report.

1. (1959) Sweet, P. A. (1960), 'University Observatory, Glasgow', *Quarterly Journal of the Royal Astronomical Society* [hereafter QJRAS], 1, 100–1.
2. (1960) Sweet, P. A. (1961), 'University Observatory, Glasgow', QJRAS, 2, 265–6.
3. (1961–2) Sweet, P. A. (1963), 'University Observatory, Glasgow', QJRAS, 4, 111–12.
4. (1963) Sweet, P. A. (1964), 'University Observatory, Glasgow', QJRAS, 5, 158–9.
5. (1963–4) Sweet, P. A. (1965), 'University Observatory, Glasgow', QJRAS, 6, 208–9.
6. (1969) Sweet, P. A. (1970), 'University Observatory, Glasgow', QJRAS, 11, 268–71.
7. (1969–72) Sweet, P. A. (1972), 'University Observatory, Glasgow', QJRAS, 13, 538–41.
8. (1972–5) Sweet, P. A. (1975), 'University Observatory, Glasgow', QJRAS, 16, 432–5.
9. (1975–7) Sweet, P. A. (1978), 'University Observatory, Glasgow', QJRAS, 19, 314–17.
10. (1977–8) Sweet, P. A. (1979), 'University Observatory, Glasgow', QJRAS, 20, 278–81.
11. (1978–9) Sweet, P. A. (1980), 'University Observatory, Glasgow', QJRAS, 21, 160–3.
12. (1979–80) Sweet, P. A. (1981), 'University Observatory, Glasgow', QJRAS, 22, 299–302.
13. (1981–2) Sweet, P. A. (1983), 'University Observatory, Glasgow', QJRAS, 24, 68–70.
14. (1982–3) Roy, A. E. (1984), 'Department of Astronomy, University of Glasgow', QJRAS, 25, 344–7.
15. (1983–6) Roy, A. E. (1987), 'Department of Astronomy, University of Glasgow', QJRAS, 28, 502–9.
16. (1986–91) Brown, J. C., Clarke, D. and Hough, J. (1992), 'Department of Physics and Astronomy, University of Glasgow', QJRAS, 33, 387–402.
17. (1991–2) Brown, J. C., Clarke, D., Hough, J. and Laing, E. W. (1994), 'Department of Physics and Astronomy, University of Glasgow', QJRAS, 34, 531–41.

18. (1993) Brown, J. C., Clarke, D., Hough, J. and Laing, E. W. (1994), 'Department of Physics and Astronomy, University of Glasgow', *QJRAS*, 35, 529–37.
19. (1994) Brown, J. C., Clarke, D., Hough, J. and Laing, E. W. (1995), 'Department of Physics and Astronomy, University of Glasgow', *QJRAS*, 36, 417–27.
20. (1995) Brown, J. C., Clarke, D. and Hough, J. (1996), 'Department of Physics and Astronomy, University of Glasgow', *QJRAS*, 37, 819–27.

References

1. Available at: http://adswww.harvard.edu
2. Sweet, P. A. (1950), 'The Importance of Rotation in Stellar Evolution', *Monthly Notices of the Royal Astronomical Society* [hereafter *MNRAS*], 110, 548–58.
3. Sweet, P. A. and Roy, A. E. (1953), 'The Structure of Rotating Stars', *MNRAS*, 113, 701–15.
4. Sweet, P. A. (1958), 'The Neutral Point Theory of Solar Flares', in Lehnert, Bo (ed.), *Proceedings from IAU Symposium No 6*, Cambridge: Cambridge University Press, pp. 123–34.
5. Sweet, P. A. (1961), 'University Observatory, Glasgow', *Quarterly Journal of the Royal Astronomical Society* [hereafter *QJRAS*], 2, 265–6.
6. Sweet, P. A. (1960), 'University Observatory, Glasgow', *QJRAS*, 1, 100–1.
7. Ovenden, M. W. (1957), *Looking at the Stars*, London: Phoenix House; Ovenden, M. W. (1960), *Artificial Satellites*, Harmondsworth: Penguin; Ovenden, M. W. (1962), *Life in the Universe*, New York: Doubleday.
8. Ovenden, Michael W. (1966), 'The Origin of the Constellations', *The Philosophical Journal*, 3, 1–18.
9. Clarke, D. (2010), *Stellar Polarimetry*, Weinheim: Wiley-VCH Verlag GmbH & Co. KGaA.
10. Sweet, P. A. (1983), 'University Observatory, Glasgow', *QJRAS*, 24, 68–70; Roy, A. E. (1984), 'Department of Astronomy, University of Glasgow', *QJRAS*, 25, 344–7; Roy, A. E. (1987), 'Department of Astronomy, University of Glasgow', *QJRAS*, 28, 502–9.
11. Roy, Archie E. (1963), *Great Moments in Astronomy*, London: Phoenix House Ltd.
12. Roy, Archie E. (1965), *The Foundations of Astrodynamics*, New York: Macmillan Press.
13. Roy, Archie E. (2005), *Orbital Motion*, 4th edition, London: Taylor & Francis.
14. Roy, A. E. (ed.) (1988), 'Long-Term Dynamical Behaviour of Natural and Artificial N-Body Systems', *Proceedings of the NATO Advanced Science Institutes (ASI) Series C: Mathematical and Physical Sciences, Cortina d'Ampezzo, Italy, 2–13 August 1987*, Dordrecht: Reidel; Steves, B. A. and Roy, A. E. (eds) (1999), *The Dynamics of Small Bodies in the Solar System: A Major Key to Solar Systems Studies*, NATO ASI Series: Series C, vol. 522, Dortrecht: Kluwer Academic Publishers.
15. Roy, A. E., Walker, I. W., MacDonald, A. J., Williams, I. P. and Fox, K. (1988), 'Project LONGSTOP', *Vistas in Astronomy*, 32, 95–116.

16. Steves, B. A. and Roy, A. E. (1998), 'Some special restricted four-body problems – I. Modelling the Caledonian Problem', *Planetary & Space Science*, 46, 1,465–74.

17. Roy, A. E. and Clarke, D. (1977), *Astronomy: Principles and Practice*, Bristol: Adam Hilger (2nd edition 1982; 3rd edition 1988; 4th edition 2003 (Institute of Physics)).

18. Roy, A. E. and Clarke, D. (1977), *Astronomy: Structure of the Universe*, Bristol: Adam Hilger (2nd edition 1982; 3rd edition 1989).

19. 'Scottish Astronomy – Seeing from Glasgow', *Nature*, 221 (22 March 1969), 1,092.

20. Clarke, D., Schwarz, H. E. and Stewart, B. G. (1985), 'Polarimetric measurements of late-type stars at Hβ, CaII K', *Astronomy & Astrophysics* [hereafter *A&A*], 145, 232–40.

21. Clarke, D. (1978), 'Brightness oscillations of daytime sky', *Nature*, 274, 670–1.

22. Clarke, D. (1978), 'Oscillations of the Blue Sky: A Source of Noise for the Measurement of Solar Oscillations', *Proceedings of the 2nd European Solar Meeting (Toulouse)*, CNRS, 282, 143–50.

23. Clarke, D., McLean, I. S. and Wyllie, T. H. A. (1975), 'Stellar Line Profiles by Tilt-scanned Narrow Band Interference Filters', *Astronomy & Astrophysics*, 43, 215–21; Clarke, D. and McLean, I. S. (1975), 'A Dual Narrow-Band Wavelength Scanning Polarimeter', *MNRAS*, 172, 545–56; Clarke, D. and McLean, I. S. (1974), 'Observations of the Linear Polarization in the Hβ Emission Feature of γ Cas', *MNRAS*, 167, 27P–30P.

24. Clarke, D. and Wyllie, T. H. A. (1977), 'On the Detection of Rapid Fluctuations in the Spectra of Be Stars', *The Observatory*, 97, 21–3.

25. Clarke, D. and Basurah, H. (1990), 'An Investigation of Lunar Luminescence by Spectropolarimetry', *Icarus*, 88, 396–406.

26. Clarke, D. and Ameijenda, V. (2000), 'Hα Polarimetry of the Solar Limb', *A&A*, 355, 1,138–45.

27. Clarke, D. and Fullerton, S. R. (1996), 'The Sun as a Polarimetric Variable Star', *A&A*, 310, 331–340.

28. Clarke, D. and Basurah, H. (1989), 'Polarization Measurements of the Ring Effect in the Daytime Sky', *Planetary & Space Science*, 37, 627–30.

29. Clarke, D. and Naghizadeh-Khouei, J. (2000), 'Polarimetric studies of rotational Raman scattering (the Ring effect) in the zenith sky', *Planetary and Space Science*, 48, 285–95.

30. Roy, A. E. (1987), 'Department of Astronomy, University of Glasgow', *QJRAS*, 28, 502–9.

31. Clarke, D., Stewart, B. G., Schwarz, H. E. and Brooks, A. (1983), 'The Statistical Behaviour of Normalized Stokes Parameters', *A&A*, 126, 260–4; Simmons, J. F. L. and Stewart, B. G. (1985), 'Point and Interval Estimation of the True Unbiased Degree of Linear Polarization in the Presence of Low Signal-to-Noise Ratios', *A&A*, 142, 100–6.

32. Clarke, D., Naghizadeh-Khouei, J., Simmons, J. F. L. and Stewart, B.G. (1993), 'A Statistical Assessment of Zero-Polarization Catalogues', *A&A*, 269, 617–26.

33. Brown, J. C., McLean, I. S. (1977), 'Polarisation by Thomson Scattering in Optically Thin Stellar Envelopes. I – Source Star at Centre of Axisymmetric Envelope', *A&A*, 57, 141–9.
34. Brown, J. C., McLean, I. S. and Emslie, A. G. (1978), 'Polarisation by Thomson Scattering in Optically Thin Stellar Envelopes. II - Binary and Multiple Star Envelopes and the Determination of Binary Inclinations', *A&A*, 68, 415–27.
35. Brown, J. C. (1971), 'The Deduction of Energy Spectra of Non-Thermal Electrons in Flares from the Observed Dynamic Spectra of Hard X-Ray Bursts', *Solar Physics*, 18, 489–502; Brown, J. C. (1972), 'The Decay Characteristics of Models of Solar Hard X-Ray Bursts', *Solar Physics*, 25, 158–77; Brown, J. C. (1972), 'The Directivity and Polarisation of Thick Target X-Ray Bremsstrahlung from Solar Flares', *Solar Physics*, 26, 441–59.
36. Craig, I. J. D. and Brown, J. C. (1986), *Inverse Problems in Astronomy, A guide to inversion strategies for remotely sensed data*. Bristol and Boston, MA: Adam Hilger.

Figures

There are instances where the author has been unable to trace or contact the copyright holders of quotations and images. If notified, the author and publisher will be pleased to rectify any errors or omissions at the earliest opportunity.

Chapter 6

Chapter 7

Colour Plates

Located between pages 182 and 183

1. Sighthill megalithic observatory. Courtesy of Linda Lunan, © Linda Lunan 2011. The images are also available in the Megalithic Portal; see www.megalithic.co.uk
2. The College grounds of the 1820s. Taken from David Smith's 1828 map. See Chapter 3, Reference 54. Reproduced with the permission of the University Librarian, University of Glasgow.
3. Watt's memorial statue. Author's photograph.
4. Portrait of Alexander Wilson. Courtesy of Hunterian Museum, University of Glasgow GLAHA 44340.
5. James Short telescopes used by Wilson. Telescope images courtesy of Hunterian Museum, University of Glasgow GLAHM 105684 and 105681; images of eyepiece rings from the library of Mrs M. I. Morris.
6. The High Possil meteorite. Image of the High Possil meteorite courtesy of Hunterian Museum, University of Glasgow GLAHM M172; image of the commemorative stone courtesy of John Faithfull.
7. Fulton's Orrery and poster. Upper image author's photograph, used courtesy of the Kelvingrove Art Gallery and Museum, City of Glasgow. Lower image by permission of University of Glasgow Library, Special Collections Eph. J/249.
8. 'Crimea' Simpson's colourwash picture of Horselethill Observatory. © Glasgow City Libraries. Licensor: www.scran.ac.uk
9. The main Bute Hall window. Author's photograph.
10. Inscriptions in the Bute Hall window. Author's photographs.
11. Professor Grant's book for the Transit of Venus. Taken from Grant (see Chapter 7, Reference 46). By permission of University of Glasgow Library, Special Collections.
12. The colourful Broomielaw. From the library of Mrs M. I. Morris.
13. Modern view of the Clyde. Author's photograph, 2007. Courtesy of the management of the BT building.
14. College tower clock mechanism and St George's Church. Author's photographs.
15. The Muirhead mean time clock. Author's photographs.
16. McGregor's business at 38 Clyde Place. Author's photographs; envelope from the library of Mrs M. I. Morris.

Index